MOBILE AND WIRELESS COMMUNICATIONS FOR IMT-ADVANCED AND BEYOND

MOBILE AND WIRELESS COMMUNICATIONS FOR IMT-ADVANCED AND BEYOND

Editors

Afif Osseiran

Ericsson Research, Sweden

Jose F. Monserrat

Universitat Politècnica de València, Spain

Werner Mohr

Nokia Siemens Networks, Germany

A John Wiley & Sons, Ltd., Publication

Library of Congress Cataloging-in-Publication Data

Mobile and wireless communications for IMT-advanced and beyond / editors, Afif Osseiran, Jose F. Monserrat, Werner Mohr.
    p. cm.
  Includes bibliographical references and index.
  ISBN 978-1-119-99321-6 (cloth)
1. Wireless communication systems. I. Osseiran, Afif. II. Monserrat, Jose F. III. Mohr, Werner, 1955-
  TK5103.2.M5694 2011
  621.382–dc22

                                                      2011012256

A catalogue record for this book is available from the British Library.

Print ISBN: 9781119993216
ePDF ISBN: 9781119976424
oBook ISBN: 9781119976431
ePub ISBN: 9781119977490
eMobi ISBN: 9781119977506

Set in 10/12pt Times by Thomson Digital, Noida, India

Cover image:
Designed by Shirine Osseiran

To those who Believe and Strive for a Just and Ethical World.
In memory of Mohamad Bouazizi

A. Osseiran

To my other half, Lorena, and the fruit of our love, Mireia.
Welcome to the world honey

J. F. Monserrat

To my parents

W. Mohr

Contents

About the Editors

Afif Osseiran

Dr. Osseiran received a B.Sc. in Electrical Engineering from Université de Rennes I, France, in 1995, and a DEA (B.Sc.E.E) degree in Electrical Engineering from Université de Rennes I and INSA Rennes in 1997, and a M.A.Sc. degree in Electrical and Communication Engineering from École Polytechnique de Montréal, Canada, in 1999. In 2006, he successfully defended his Ph.D thesis at the Royal Institute of Technology (KTH), Stockholm, Sweden. Since 1999 he has been with Ericsson, Sweden. In 2004 he joined as one of Ericsson's representatives the European project WINNER. During the years 2006 and 2007 he led in WINNER the spatial temporal processing (i.e. MIMO) task. From April 2008 to June 2010, he was the technical manager of the Eureka Celtic project WINNER+. His research interests include many aspects of wireless communications with a special emphasis on advanced antenna systems, on relaying, on radio resource management, network coding and cooperative communications. Dr. Osseiran is listed in *Who's Who in the World* and *Who's Who in Science and Engineering*. He has published more than 40 technical papers in international journals and conferences. In 2009, Dr. Osseiran coauthored *Radio Technologies and Concepts for IMT-Advanced* with John Wiley & Sons. Since 2006, he has been teaching advanced antennas at Master's level at the Royal Institute of Technology (KTH) in Stockholm.

Jose F. Monserrat

Dr. Monserrat received his MSc. degree with High Honors and Ph.D. degree in Telecommunications Engineering from the Polytechnic University of Valencia (UPV) in 2003 and 2007, respectively. He was the recipient of the First Regional Prize of Engineering Studies in 2003 for his outstanding student record, also receiving the Best Thesis Prize from the UPV in 2008. In 2009 he was awarded with the Best Young Researcher prize in Valencia. He is currently an associate professor in the Communications Department of the UPV. His research focuses on the application of complex computation techniques to Radio Resource Management (RRM) strategies and to the optimization of current and future mobile communications networks, such as LTE-Advanced and IEEE 802.16m. He has been involved in several European Projects, acting as task or work package leader in WINNER+, ICARUS, COMIC and PROSIMOS. In 2010 he also participated in an external evaluation group within ITU-R on the performance assessment of candidates for the future family of standards, IMT-Advanced.

Werner Mohr

Dr. Mohr graduated from the University of Hannover, Germany, with a Masters degree in Electrical Engineering in 1981 and a Ph.D. degree in 1987. He joined Siemens AG, Mobile Network Division, Munich, Germany, in 1991. He was involved in several EU-funded projects and ETSI standardization groups on UMTS and systems beyond 3G. In December 1996 he became project manager of the European ACTS FRAMES Project until the project finished in August 1999. This project developed the basic concepts for the UMTS radio interface. Since April 2007 he has been with Nokia Siemens Networks GmbH & Co. KG, Munich, Germany, where he is Head of Research Alliances. He was the coordinator of the WINNER Project in Framework Program 6 of the European Commission, Chairman of WWI (Wireless World Initiative) and the Eureka Celtic project WINNER+. The WINNER project laid the foundation for the radio interface for IMT-Advanced and provided the starting point for the 3GPP LTE standardization. He was also Vice Chair of the eMobility European Technology Platform in the period 2008–9 and is now eMobility (now called Net!Works) Chairperson for the period 2010–2011. He was Chair of the Wireless World Research Forum from its launch in August 2001 up to December 2003. He is a member of VDE (Association for Electrical, Electronic and Information Technologies, Germany) and Senior Member of IEEE. In 1990 he received the Award of the ITG (Information Technology Society) in VDE. He was a board member of ITG in VDE for the term 2006–8 and was re-elected for the 2009-11 term. He is coauthor of the books *Third Generation Mobile Communication Systems* and *Radio Technologies and Concepts for IMT-Advanced*.

Preface

Goal and Objective of the Book

This book was prompted by the desire to fill the gap between theoretical descriptions and a more pedagogical description of the main technological components of International Mobile Telecommunications Advanced (IMT-Advanced) such as Radio Resource Management (RRM), Carrier Aggregation (CA), improved MIMO support, and relaying. The book also covers the latest research innovations beyond the IMT-Advanced system, in particular, promising areas such as Coordinated Multipoint transmission or reception (CoMP), Network Coding (NC), Device-to-Device (D2D) and spectrum sharing.

Each chapter presents the basis of its topic in a simple way. A review of the latest research advances for that topic is then given. A special emphasis in each area is given to the state of the art of global standardization, in particular LTE-A. Finally, each chapter concludes by looking towards the future and discussing predictions. The reader is expected to have completed a basic undergraduate course in digital communications and mathematics (i.e. probability, calculus and linear algebra).

In order to help the reader extract the important information, the most relevant statements in every chapter are framed in a gray color, or emphasized in italic.

Structure of the Book

Chapter 1 provides an overview of cellular technology evolution and the latest market and technology trends. Chapter 2 presents innovative concepts for advanced RRM. Carrier Aggregation techniques are presented in Chapter 3. Chapter 4 explains spectrum sharing, especially within the context of femtocell, and game theory. Multiple-Input Multiple-Output (MIMO), and in particular Multi-User (MU)-MIMO are treated in Chapter 5. Chapter 6 thoroughly describes CoMP. Relaying for IMT-Advanced is addressed in Chapter 7. Chapters 8 and 9 address beyond IMT-Advanced areas, Network Coding and Device-to-Device communications, respectively. In Chapter 10 the end-to-end performance of LTE-Advanced (LTE-A) candidates is carefully presented. Chapter 11 discusses the future research trends within wireless communications. The appendices contain additional information related to Chapters 2, 3, 6, 8 and 10.

Background

Writing a book is a long journey involving substantial effort from a group of people. The core contributors of this book are from the "alumni" of the WINNER(+) project, which for more than six years provided an extraordinary forum for collaboration and contribution to global fora such as the Third Generation Partnership Project (3GPP) and the International Telecommunication Union (ITU). In 2004 the project paved the way toward International Mobile Telecommunications Advanced.

The WINNER project was a six-and-a-half year journey. It started with WINNER-I the first phase in 2004, continued with the second phase, WINNER-II (through years 2006–7), and was finalized in

WINNER+ (2008–mid-2010). These successful achievements were possible due to the close cooperation of project partners, respecting the interests of the different organizations. It was an excellent
experience to cooperate in such an environment, where problems were analyzed and discussed in a
trustful manner, blessed with consensus.

Finally, the editors would welcome any comments and suggestions for improvements or changes.
They can be reached at the following e-mail address: wire.comm.for.imta@gmail.com.

Acknowledgements

Only part of the material in this book has been extracted from or based on several of the public deliverables of the European Celtic project WINNER+. The completion of this project was supported by substantial additional material, which was originally not planned in the consortium. We would therefore like to thank all the colleagues involved in the project for their support and the good cooperation that made the book possible.

We are most grateful to the co-authors of this manuscript. They have shown incredible commitment and dedication during the writing process. Many were working in their free time, during evenings and week-ends. They have demonstrated an exemplary spirit of collaboration, always being available while dealing with professional and private constraints. During the period in which the book was being written we experienced the birth of at least three kids. We hope that the personal relationships forged during this project will continue and that there will be the potential for future collaboration.

We wish to thank those who reviewed the various chapters in this book. Most co-authors participated in that process. In particular, we are indebted to Mr. Petri Komulainen for his scrutiny and review of Chapter 6. We are thankful to Mr. Peter J. Larsson, as an external reviewer, for his review of Chapter 8.

We express our gratitude to Dr. Andrew Logothetis. His encouragement at the outset was significant in prompting us to start the work and to submit the book proposal.

Dr. Osseiran would like, in particular, to acknowledge Ericsson's generosity, through the persons of Dr. Magnus Frodigh, Dr. Claes Tidestav and Dr. Gunnar Bark, for giving ample resources to write the book. In addition, Dr. Göran Klang and Mr. Johan Lundsjö believed in the project and encouraged it from its inception.

We would like to thank Mr. Mark Hammond, Mrs. Sophia Travis and Mrs. Susan Barclay from John Wiley & Sons for their help to finalize this book. Mr. Hammond encouraged endlessly to initiate the proposal. Mrs. Travis has been always available, efficiently reacting to queries, and with a great sense of humor. We show our appreciations to the antonymous type editors, type setters, designers and proof readers. Finally, Mrs. Shirine Osseiran is greatly thanked for carefully designing the book front cover.

Afif Osseiran
Jose F. Monserrat
Werner Mohr

List of Abbreviations

1G First Generation
2G Second Generation
3G Third Generation
3GPP Third Generation Partnership Project
3GPP2 Third Generation Partnership Project 2
4G Fourth Generation
ABS Advanced Base Station
ACK Acknowledge
AF Amplify-and-Forward
AMBR Aggregated Maximum Bit Rate
AMC Adaptive Modulation and Coding
AMPS Advanced Mobile Phone System
AMS Advanced Mobile Station
AoA Angle of Arrival
AoD Angle of Departure
AP Access Point
APA Adaptive Power Allocation
APP Application
ARP Allocation and Retention Priority
ARQ Automatic Repeat-reQuest
AS Application Server
ASA Angle Spread Arrival
ASD Angle Spread Departure
AVC Advanced Video Coding
AWGN Additive White Gaussian Noise
BuB Busy Burst
BB Base Band
BC Broadcast Channel
BD Block Diagonalization
BE Best Effort
BER Bit Error Rate
BLAST Bell Labs Space-Time architecture
BLER BLock Error Rate
BPSK Binary Phase Shift Keying
BS Base Station
BSR Buffer Status Report
BWA Broadband Wireless Access

CA Carrier Aggregation
CAC Call Admission Control
CC Component Carrier
CCC Cognitive Component Carrier
CCI Co-Channel Interference
CDD Cyclic Delay Diversity
CDF Cumulative Density Function
CDM Code Division Multiplexing
CDMA Code Division Multiple Access
CEPT European Conference of Postal and Telecommunications Administrations
CESAR CEllular Slot Allocation and Reservation
CF Compress-and-Forward
CFI Control Format Indicator
CIR Carrier-to-Interference power Ratio
CJP Centralized Joint Processing
CLA Clustered Linear Array
CLO Cross-Layer Optimization
CoMP Coordinated MultiPoint transmission or reception
CP Cyclic Prefix
CPG Conference Preparatory Group
CPM Conference Preparatory Meeting
CPRI Common Public Radio Interface
CQI Channel Quality Indicator
CR Cognitive Radio
CRC Cyclic Redundancy Check
CRNTI Cell Radio Network Temporary Identifier
CRS Cognitive Radio System
CSG Closed Subscriber Group
CSI Channel State Information
CSIR CSI at the Receiver
CSI-RS Channel State Information Reference Signal
CSIT Channel State Information at the Transmitter

CSMA/CA Carrier Sense Multiple Access With Collision Avoidance
CU Central Unit
D2D Device-to-Device
dB decibel
DB Digital Broadcasting
DCI Downlink Control Indicator
DF Decode-and-Forward
DFT Discrete Fourier Transform
DJP Decentralized Joint Processing
DL Downlink
DLS Direct Link Setup
DM Demodulation
DmF Demodulate-and-Forward
DMO Direct Mode Operation
DM-RS DeModulation Reference Signal
DMT Diversity-Multiplexing-Tradeoff
DNC Diversity Network Codes
DOFDM Discontiguous OFDM
DPC Dirty Paper Coding
DRx Discontinuous Reception
DS Delay Spread
DSA Dynamic Sub-carrier Assignment
D.S.A Dynamic Spectrum Allocation
DSP Digital Signal Processor
DwPTS Downlink Pilot TimeSlot
DySA Dynamic Spectrum Access
EBF Eigen Beam Forming
ECC Electronic Communications Committee
EDCA Enhanced Distributed Channel Access
EDF Exponential Delay Fairness
EDGE Enhanced Data rates for GSM Evolution
EIRP Equivalent Isotropically Radiated Power
EM ElectroMagnetic
E-MBS Enhance Multicast Broadcast Service
eNB eNodeB
EPC Evolved Packet Core
EPS Evolved Packet System
ERO European Radio Communication Office
ETSI European Telecommunications Standards Institute
EU European Union

E-UTRA Evolved Universal Terrestrial Radio Access
E-UTRAN Evolved Universal Terrestrial Radio Access Network
FCC Federal Communications Commission
FDD Frequency Division Duplex
FDM Frequency Division Multiplexing
FDMA Frequency Division Multiple Access
FER Frame Error Rate
FFR Fractional Frequency Reuse
FFT Fast Fourier Transform
FIFO First Input First Output
FM Frequency Management
FSA Fixed Spectrum Assignment
FSS Frequency Selective Scheduling
FUE Femto-UE
GBR Guaranteed Bit Rate
GEK Global Encoding Kernel
GP Guard Period
GPRS General Packet Radio Service
GPS Global Positioning System
GSM Global System for Mobile Communications
GT Game Theory
GTP GPRS Tunneling Protocol
GW Gateway
HARQ Hybrid Automatic Repeat reQuest
H-BS Home Base Station
HeNB Home eNB
HetNet Heterogeneous Network
HHI Heinrich Hertz Institute
HII High Interference Indicator
HK Han-Kobayashi
HNB Home NB
HOL Head-of-line
HSA Hierarchical Spectrum Access
HSDPA High-Speed Downlink Packet Access
HSPA High-Speed Packet Access
HSUPA High Speed Uplink Packet Access
HUE Home User Equipment
HYGIENE HurrY-Guided-Irrelevant-Eminent-NEeds
ICI Inter-Cell Interference
ICIC Inter-Cell Interference Coordination
ICT Information and Communication Technologies

ID Identity
IEEE Institute of Electrical and Electronics Engineers
IEG Independent Evaluation Group
IF Intermediate Frequency
IFFT Inverse Fast Fourier Transform
IMS IP Multimedia Subsystem
IMT International Mobile Telecommunications
IMT-2000 International Mobile Telecommunications 2000
IMT-Advanced International Mobile Telecommunications Advanced
IP Internet Protocol
IRC Interference Rejection Combining
ISI Inter-Symbol Interference
ITU International Telecommunication Union
ITU-R International Telecommunication Union – Radiocommunication Sector
JD Joint Detection
JP Joint Processing
JQS Joint Queue Scheduler
JUS Joint User Scheduling
LA Link Adaptation
LAN Local Area Network
LDPC Low-Density Parity-Check
LMMSE Linear Minimum Mean Square Error
LoS Line of Sight
LRU Logical Resource Unit
LTE Long Term Evolution
LTE-A LTE-Advanced
LTE-Rel-8 LTE Release 8
LTE-Rel-10 LTE Release 10
M2M Machine-to-Machine
MAC Medium Access Control
M.A.C Multiple Access Channel
MARC Multiple Access Relay Channel
MBMS Multimedia Broadcast Multicast Service
MBR Maximum Bit Rate
MBSFN MBMS over Single Frequency Networks
MCI Maximum Carrier to Interference
MCS Modulation and Coding Scheme
MDNC Maximum Diversity Network Codes

MDS Maximum-Distance Separable
MET Multi-user Eigenmode Transmission
MI Mutual Information
MIMO Multiple-Input Multiple-Output
MISO Multiple-Input Single-Output
ML Maximum Likelihood
M-LWDF Modified-Largest Weighted Delay First
MME Mobile Management Entity
MMSE Minimum Mean Square Error
MOS Mean Opinion Score
MRC Maximum Ratio Combining
MSE Mean Square Error
MU Multi-User
MUE Macro UE
MVD Majority Vote Detection
NACK Negative Acknowledge
NAS Non-Access Stratum
NC Network Coding
NC-OFDMA Non-Contiguous OFDMA
N.E Nash Equilibrium
NGMN Next Generation Mobile Network
NLoS Non Line of Sight
NMT Nordic Mobile Telephone
NRT Non-Real-Time
OC Optimum Combining
OCA Opportunistic Carrier Aggregation
OFDM Orthogonal Frequency Division Multiplexing
OFDMA Orthogonal Frequency Division Multiple Access
OI Overload Indicator
OPEX OPerational EXpenditures
OSA Open Spectrum Access
OSI Open Systems Interconnection
OtoI Outdoor to Indoor
PAPC Per-Antenna Power Constraint
PAPR Peak-to-Average Power Ratio
PBCH Physical Broadcast CHannel
PC Power Control
PCFICH Physical Control Format Indicator CHannel
PDC Personal Digital Cellular
PDCCH Physical Downlink Control CHannel
PDN Packet Data Network
PDSCH Physical Downlink Shared CHannel
PF Proportional Fair

PHICH Physical Hybrid Automatic Repeat Request Indicator CHannel
PHR Power Headroom Report
PHY Physical
PJP Partial Joint Processing
PL Path Loss
PMI Precoding Matrix Indicator
P.M.I Preferred Matrix Index
PNC Physical Network Coding
PPF Predictive Proportional Fair
PRACH Physical Random Access CHannel
PRB Physical Resource Block
PRMA Packet Reservation Multiple Access
PRU Physical Resource Unit
PSE Peak Spectral Efficiency
PSK Phase-Shift Keying
PT A Project Team A
p-t-m point-to-multi-point
p-t-p point-to-point
PUCCH Physical Uplink Control CHannel
PUSCH Physical Uplink Shared CHannel
QAM Quadrature Amplitude Modulation
QCI QoS Class Identifier
QoE Quality of Experience
QoS Quality of Service
QPSK Quadrature Phase Shift Keying
RA Radiocommunication Assembly
RAN Radio Access Network
RB Resource Block
RCC Relay Coherent Combining
RCDD Relay Cyclic Delay Diversity
RCPC Rate Compatible Convolutional Codes
Rel-5 Release 5
Rel-6 Release 6
Rel-7 Release 7
Rel-8 Release 8
Rel-9 Release 9
Rel-10 Release 10
ReS Relay Selection
RF Radio Frequency
RI Rank Indicator
RIT Radio Interface Technology
RLC Radio Link Control
RMa Rural Macrocell
RMS Root-Mean-Square

RN Relay Node
RNTP Relative Narrowband Transmit Power
RoF Radio over Fiber
RoR Round Robin
RR Radio Regulations
RRC Radio Resource Control
RRH Remote Radio Heads
RRM Radio Resource Management
RRU Radio Remote Unit
RS Reference Signal
RSD Relay Selection Diversity
RT Real-Time
SAE System Architecture Evolution
SAO Spectrum Access Opportunities
SB Score Based
SC-FDMA Single Carrier – Frequency Division Multiple Access
SCH Synchronization CHannel
SCM Spatial Channel Model
SCME Spatial Channel Model Extended
SCTP Stream Control Transmission Protocol
SDMA Space Division Multiple Access
SDP SemiDefinite Programming
SDR Software Defined Radio
SE Spectrum Engineering
S.E Stackelberg Equilibrium
SpE Split-and-Extend
SF Shadow Fading
SFN Single Frequency Network
SFR Soft Frequency Reuse
S-GW Serving Gateway
SIC Successive Interference Cancellation
SIMO Single Input Multiple Output
SINR Signal to Interference plus Noise Ratio
SIP Session Initiation Protocol
SISO Single Input Single Output
SLNR Signal to Leakage and Noise Ratio
SMS Short Message Service
SNIR Signal-to-Noise-plus-Interference Ratio
SNR Signal to Noise Ratio
SOCP Second Order Cone Programming
SON Self-Organized Network
SR Source Relay
SRIT Set of Radio Interface Technologies
SRS Sounding Reference Signal

SRUS Separated Random User Scheduling
SSC Selection and Soft Combining
STBC Space Time Block Code
STTC Space-Time Trellis Codes
STTD Space Time Transmit Diversity
SU Single-User
SVC Scalable Video Coding
SVD Singular Value Decomposition
TACS Total Access Communications System
TASB Traffic-Aware Score Based
TC Turbo Code
TD-CDMA Time Division CDMA
TDD Time Division Duplex
TDLS Tunneled Direct Link Setup
TDM Time Division Multiplexing
TDMA Time Division Multiple Access
TETRA Terrestrial Trunked Radio
TLabs Deutsche Telekom Laboratories
TMO Trunked Mode Operation
TMSI Temporary Mobile Subscriber Identity
TP Throughput
TTI Transmission Time Interval
TUB Technical University of Berlin
TV Television
UCI Uplink Control Information
UE User Equipment
UEPS Urgency and Efficiency based Packet Scheduler
UG User Grouping

UL Uplink
ULA Uniform Linear Array
UMa Urban Macrocell
UMi Urban Microcell
UMTS Universal Mobile Telecommunication System
UpPTS Uplink Pilot Timeslot
UPS Utility Predictive Scheduler
URI Uniform Resource Indicator
URL Uniform Resource Locator
USB Universal Serial Bus
VAA Virtual Antenna Array
VLAN Virtual Local Area Network
VoIP Voice over IP
WARC World Administrative Radio Conference
WCDMA Wideband Code Division Multiple Access
WiMAX Worldwide Interoperability for Microwave Access
WINNER Wireless World Initiative New Radio
WINNER+ Wireless World Initiative New Radio +
WLAN Wireless Local Area Network
WP Working Party
WPAN Wireless Personal Area Networks
WRC World Radiocommunication Conference
ZF Zero Forcing

List of Contributors

Dr. Günther Auer
DOCOMO, Munich, Germany

Prof. Mats Bengtsson
Royal Institute of Technology (KTH),
Stockholm, Sweden

Dr. Mehdi Bennis
Centre for Wireless Communications,
University of Oulu, Finland

Prof. Slimane Ben Slimane
Royal Institute of Technology (KTH),
Stockholm, Sweden

Dr. Federico Boccardi
Bell Laboratories, Alcatel Lucent,
Vimercate, Italy

Mr. Mauro Boldi
Telecom Italia, Torino, Italy

Mr. Jorge Cabrejas
Universitat Politecnica de Valencia -
iTEAM, Valencia, Spain

Mrs. Valeria D'Amico
Telecom Italia, Torino, Italy

Dr. Klaus Doppler
Nokia Research Center, Helsinki,
Finland

Dr. Xavier Gelabert
Universitat Politecnica de Valencia -
iTEAM, Valencia, Spain

Mr. Alexandre Gouraud
Orange Labs, Paris, France

Dr. Eric Hardouin
Orange Labs, Paris, France

Mr. Pekka Jänis
Nokia Research Center, Helsinki, Finland

Dr. Volker Jungnickel
Fraunhofer Heinrich-Hertz-Institut,
Berlin, Germany

Mr. Petri Komulainen
Centre for Wireless Communications,
University of Oulu, Finland

Mr. David Martin-Sacristán
Universitat Politecnica de Valencia -
iTEAM, Valencia, Spain

Dr. Werner Mohr
Nokia Siemens Networks GmbH & Co. KG,
Munich, Germany

Dr. Jose F. Monserrat
Universitat Politecnica de Valencia -
iTEAM, Valencia, Spain

Mrs. Miia Mustonen
VTT Technical Research Centre, Oulu,
Finland

Mr. Magnus Olsson
Ericsson AB, Stockholm, Sweden

Dr. Afif Osseiran
Ericsson AB, Stockholm, Sweden

Dr. Cassio Ribeiro
Nokia Research Center, Helsinki, Finland

Dr. Peter Rost
NEC Laboratories Europe,
Heidelberg, Germany

Dr. Ahmed Saadani
Orange Labs, Paris, France

Mr. Krystian Safjan
Nokia Siemens Networks Sp. z.o.o,
Wroclaw, Poland

Dr. Hendrik Schöneich
Qualcomm, Nürnberg, Germany

Mr. Per Skillermark
Ericsson AB, Stockholm, Sweden

Mr. Pawel Sroka
Poznan University of Technology,
Poznan, Poland

Dr. Tommy Svensson
Chalmers University of Technology (CTH),
Gothenburg, Sweden

Mr. Lars Thiele
Fraunhofer Heinrich-Hertz-Institut,
Berlin, Germany

Dr. Antti Tölli
Centre for Wireless Communications,
University of Oulu, Finland

Mr. Jaakko Vihriälä
Nokia Siemens Networks Oy, Oulu,
Finland

Dr. Marc Werner
Qualcomm, Nürnberg, Germany

Dr. Ming Xiao
Royal Institute of Technology (KTH),
Stockholm, Sweden

1

Introduction

Afif Osseiran, Jose F. Monserrat and Werner Mohr

1.1 Market and Technology Trends

Social, economic and political factors determine the development of the mobile communications business. Consumer demand, the economic performance of operators and government policies are some of the aspects that affect technological advances, operators' capital investments and the regulatory environment. The mobile communications sector has been characterized by a worldwide rapid increase in the number of users. During the 1980s only a handful of people had a mobile phone. At the end of the 1980s, the number of cellular subscribers was merely around 5 million. With the introduction of the Second Generation (2G) cellular systems in 1991, the ambition was to popularize progressively the usage of mobile phones by making them affordable to a large part of the population. Progress in micro electronics then made it possible to produce cheaper mobile phones. The technology advanced and gradually increasing competition between mobile vendors made it necessary to reduce the cost of cellular infrastructures. The second part of the 1990s, witnessed an extraordinary surge in the number of mobile subscribers in the developed countries. In total, the number was close to half a billion. Progress continued worldwide at a frenetic pace. According to the International Telecommunication Union (ITU), in the last seven years the number of worldwide subscribers has grown from 1.7 billion to more than 5.3 billion (75.42% of the world population), which implies growing at a compound annual growth rate of 21%. Astonishingly, in 2002 and within only two decades, mobile subscribers surpassed fixed-telephone line subscribers (ITU n.d.). The evolution of the number of mobile and fixed line subscribers from the year 1996 to 2010 is shown in Figure 1.1.

Even though these numbers are quite significant, it is worth noting that the mobile communications sector has reached a saturation point in terms of the number of subscribers in a large number of markets, but new systems result in technology upgrades of networks and devices, which offer a significantly improved user experience and capabilities and provide new business opportunities. In the European Union, mobile penetration rate is over 110% of the total population, whereas in developed Asian countries has reached 80%, as in the United States and Eastern Europe where the growth of mobile services has been quite important in recent years. There is still room for mid-term growth of less-developed markets. Operators in saturated markets need to foster demand for new services to guarantee their revenues.

Mobile and Wireless Communications for IMT-Advanced and Beyond, First Edition.
Edited by Afif Osseiran, Jose F. Monserrat and Werner Mohr.
© 2011 Afif Osseiran, Jose F. Monserrat and Werner Mohr. Published 2011 by John Wiley & Sons, Ltd.

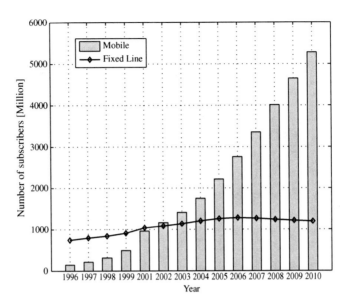

Figure 1.1   Evolution of mobile and fixed phone subscriptions from 1996 to 2010

That is why the mobile communications sector, today more than ever, seeks to put new telecommunication services on the market through mobile devices. Among these services social networks, location-based services, augmented reality, mobile TV, video on demand, interactive games and high quality music were applications added to mobile devices to ensure an upturn in usage of mobile services and, consequently, revenues. From 2007 there has been a quite significant increase in traffic demand. Apart from new services, several factors are fostering the mobile communications sector: the increase of Third Generation (3G) penetration as users rapidly migrate from 2G to 3G services; the increasing penetration of Universal Serial Bus (USB) modems and data cards, as well as smart phones and tablets, together with the increasing availability of easy-to-use data applications; the proliferation of flat-rate service bundles, which is also changing service mixes towards more usage-intensive services; and the increasing usage of 3G devices indoors, among other things. All these factors are making mobile data demand overload the capacity of 3G networks and will force next generation mobile systems to be designed according to take this trend into account. Rather than just requiring an increase in throughput, these developments will require improving the ubiquity of the Quality of Experience (QoE) Indicators, that is, allowing the mobile users to experience high QoE values in any geographical position, not only close to the Base Station (BS), while minimizing the radio resource and energy consumption.

In fact, current market forecasts predict mobile Internet penetration to double by 2015, which represents a real threat of congestion for current cellular networks. Indeed, recent analysis of the evolution of mobile broadband subscribers show that, starting in 2007, there has been a significant increase in their number and their traffic demands. In many countries where there are developed markets, mobile data consumption has increased from 2008 and 2010 and is growing exponentially as can be seen in Figure 1.2. Traditional asymmetric traffic – with more data in Downlink (DL) than in Uplink (UL) – is daily becoming more symmetrical. The increase in the usage of wireless systems is driving the industry to seek new methods to boost the capacity of cellular networks, that is, the number of users served or transmitted bits over the air interface.

The decision to adopt such methods, which may involve either improving a specific cellular standard or adopting a completely new technology through a standard change, has been seen as

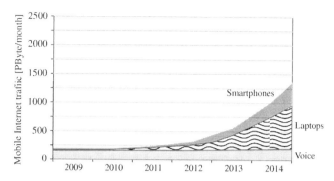

Figure 1.2 Exponential increase of mobile data traffic. UMTS Forum (UMTS-Forum 2010)

strategic and the decision is not based purely on technical or economic grounds. The influence of economic/political factors – through influential governmental and industrial players – is undeniable.

1.2 Technology Evolution

The first commercial analog mobile communication systems were deployed in the 1950s and 1960s, although with low penetration. The year 1981 witnessed the birth of the first commercial deployments of the First Generation (1G) mobile cellular standards such as Nordic Mobile Telephone (NMT), in Saudi Arabia and the Nordic countries, C-Netz in Germany, Portugal and South Africa, Total Access Communications System (TACS) in the United Kingdom and Analog Advanced Mobile Phone System (AMPS) in the Americas. The 1G standards are called the analog standards since they utilize analog technology. The beginning of the 1990s witnessed the introduction of 2G, characterized by the adoption of digital technology. This technology allowed considerable improvements in voice quality, capacity and growth potential towards advanced applications as well as the development of Short Message Service (SMS) messaging, a form of data transmission. The European Conference of Postal and Telecommunications Administrations (CEPT) decided in 1982 to develop a pan-European 2G mobile communication system. This was the starting point of the Global System for Mobile communications (GSM), the dominant 2G standard, which was deployed internationally from 1991. In the beginning, the main objective of GSM was the support of voice telephony and international roaming with a single-system across Europe. GSM is based on a hybrid Time Division Multiple Access (TDMA)/Frequency Division Multiple Access (FDMA) method, in contrast with 1G systems based only on FDMA (Hillebrand 2002). In parallel with GSM, other digital 2G systems were developed globally and competed with each other. These other main 2G standards include IS-136, also known as D-AMPS, IS-95A also known as CDMAOne – used mainly in the Americas – and finally Personal Digital Cellular (PDC) – used exclusively in Japan. In contrast to GSM, the IS-95 technologies are based on Code Division Multiple Access (CDMA) (Viterbi 1995).

The evolution of the 2G, called 2.5G, allowed the introduction of packet-switched services in addition to voice, the most significant circuit switched service. The main 2.5G standards, General Packet Radio Service (GPRS) and IS-95B, are basically an extension of GSM and IS-95A, respectively.

Shortly after the 2G became operational, industrial players were already preparing and discussing the next wireless generation standards. In January 1998, CDMA under two variants – Wideband Code Division Multiple Access (WCDMA) and Time Division CDMA (TD-CDMA) – was adopted by the European Telecommunications Standards Institute (ETSI) as a Universal Mobile Telecommunication System (UMTS) as the 3G mobile communication system, also called International Mobile Telecommunications 2000 (IMT-2000). As a member of the IMT-2000 family of standards, the Third Generation Partnership Project (3GPP) developed UMTS technology using both WCDMA and TD-CDMA modulation schemes (Holma and Toskala 2000) and is generally favored in Japan and countries using GSM. On the other hand CDMA2000, initially an outgrowth of the 2G CDMA standard IS-95, is mainly dominant in the Americas and Korea.

New specifications have been developed within the framework of 3GPP together known as 3G Evolution. For this evolution, two Radio Access Network (RAN) approaches and an evolution of the core network have been suggested. The first RAN approach is High Speed Packet Access (HSPA) – referred to as a 3.5G technology. HSPA comprises High Speed Downlink Packet Access (HSDPA), added in Release 5 (Rel-5), and High Speed Uplink Packet Access (HSUPA), added in Release 6 (Rel-6) of UMTS. Both enhance the packet data rate, respectively to 14.6 Mbps in DL and to 5.76 Mbps in UL. Again, HSPA is based on WCDMA and is completely backward compatible with UMTS. The philosophy behind this radio network approach is to add new features while still serving the old mobiles, and is further applied in HSPA Evolution, also known as HSPA+. This is a good solution for the mid-term future. The equivalent evolution in CDMA2000 are 1xEV-DO and 1xEV-DV. While CDMA 1xEV-DO started deployment in 2003, HSPA and CDMA 1xEV-DV entered into service in 2006.

The second UMTS evolution is called Long Term Evolution (LTE) (Dahlman et al. 2008; Sesia et al. 2011) and the evolved core network is known as Evolved Packet Core (EPC). The target of LTE is high performance and reduced cost for the radio access. LTE is a radio interface designed from scratch. Hence, in contrast to HSPA, LTE is not backward compatible with UMTS. However, the design is clearly influenced by earlier specification work done by 3GPP. At the end of 2007 first LTE specifications were approved. The LTE system has peak data rates of around 326 Mbps, increased spectral efficiency and significantly shorter latency than previous systems. LTE is based on Orthogonal Frequency Division Multiple Access (OFDMA) and advanced spatial processing Multiple-Input Multiple-Output (MIMO). The Next Generation Mobile Network (NGMN) initiative formulated requirements on further developments of mobile communications (NGMN Ltd n.d.). Such requirements are mainly related to a flat network architecture based on the Internet Protocol (IP) for cost reduction, higher spectral efficiency for better use of the available frequency spectrum, lower latency and higher peak data rates with flexible allocation of data rates to users. Additional requirements are a high cell average throughput and sufficiently high cell edge capacity in order to cover the expected increasing data traffic with growing user density. LTE was developed to meet these requirements, and this was an important step towards the next International Mobile Telecommunications Advanced (IMT-Advanced) standard (see section 1.3).

With regard to the cellular systems market, today, the GSM family (GSM, GPRS and Enhanced Data rates for GSM Evolution (EDGE)) is the dominant second-generation mobile communication standard with a global market share – at the end of 2010 – of more than 79% and 4.18 billion subscribers in more than 200 countries (GSA n.d.). On the other hand, the number of 3G subscribers including HSPA has risen to 619 million subscribers, which represents 11.7% of the market (GSA n.d.). The main competitor to GSM is IS-95 CDMA that serves the rest of the market. Currently, the main subscriber growth markets for the GSM system are emerging markets, such as China with about seven million new subscribers per month and India with about 16 million new subscribers per month. With respect to LTE, at the end of 2010 seven LTE networks were commercially launched, being still an incipient technology. The evolution of cellular mobile systems is shown in Figure 1.3.

With regard to the peak data rates of cellular systems, from the onset of the introduction of cellular systems and until the mid-1990s the data peaked at approximately around 10 kbps. The peak data rate was lifted to 160 kbps with the introduction of GPRS. Only few years later, the first UMTS systems supported peak data rates of 384 kbps. Nowadays, HSDPA supports peak data rates from 7.2 Mbps to about 14.6 Mbps (by using adaptive modulation and coding with higher-order modulation and multicode transmission (Holma and Toskala 2007). HSPA-Evolved specified by 3GPP

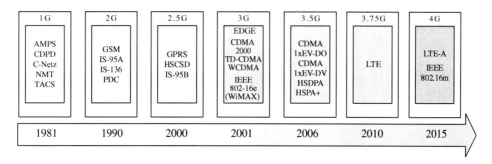

Figure 1.3 Evolution of wireless communication systems

Release 7 (Rel-7), the second phase of HSDPA, can achieve data rates of up to 42 Mbps (assuming 64-Quadrature Amplitude Modulation (QAM)). The coming technology, LTE, will see the peak data rate reaching 326 Mbps. Finally, in a few years, IMT-Advanced will theoretically push the peak rate to attain the huge throughput rate of 1.6 Gbps. The evolution of the peak data rate from years 1990 to 2015 is shown in Figure 1.4.

In parallel with these developments in the telecommunications industry, the wireless information and telecommunications sector provides different IP-based access systems for different application areas. Wireless Local Area Network (WLAN) systems, Institute of Electrical & Electronics Engineers (IEEE) standard 802.11, are used for local and short-range applications without mobility. WLAN systems are widely available globally. Wireless Personal Area Networks (WPAN) are standardized by IEEE 802.15 for very short ranges and high throughput. Broadband Wireless Access (BWA) systems, according to IEEE 802.16, target higher ranges including the support of user mobility (IEEE-ISTO n.d.; IEEE-SG n.d.). The BWA Worldwide Interoperability for Microwave Access (WiMAX) system is a member of the IMT-2000 family (WiMAX n.d.). As with UMTS and LTE, an evolution process

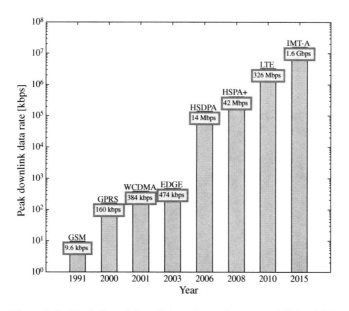

Figure 1.4 Evolution of downlink peak rate from years 1990 to 2015

has taken place within IEEE. In fact, the IEEE wireless system will evolve from the current WiMAX toward the new IEEE 802.16m standard, which is an IMT-Advanced technology.

1.3   Development of IMT-Advanced and Beyond

The radio spectrum is a scarce resource that has considerable economic and social importance. In general, governments of every country decide on the spectrum allocation. On the other hand, global coordination of spectrum usage is in the responsibility of ITU, which, through spectrum regulation, aims to facilitate spectrum harmonization for global roaming to reduce equipment cost by means of global economies of scale. Since 1992, and in the framework of the International Telecommunication Union - Radiocommunication Sector (ITU-R), the World Administrative Radiocommunications Conference (WARC 1992) has reached quite significant agreements at a global level to designate specific frequency bands to International Mobile Telecommunications (IMT) standards. The objective of this initiative is to specify a set of requirements in terms of transmission capacity and Quality of Service (QoS), in such a way that if a certain technology fulfills all these requirements then the technology is included by ITU in the IMT-2000 standards. This inclusion is an official endorsement of the technologies that might motivate the concerned players (e.g. operators, telecommunications providers, etc.) to take the technologies into account and to consider investing in them. Furthermore it allows these standards to make use of the frequency bands designated for IMT. With the aim of coordinating the global use of spectrum, every three to four years ITU-R holds the World Radiocommunication Conference (WRC), where ITU radio regulations that govern spectrum distribution are adopted.

World Radiocommunication Conference (WRC)-07 identified additional frequency spectrum for mobile and wireless communications. The first step towards this new spectrum allocation was performing an in-depth study of the mobile market forecast and the development of spectrum requirements for the increasing service demand. Reports predicted the total spectrum bandwidth requirements for mobile communication systems in the year 2020 to be 1280 MHz and 1720 MHz for low and high user-demand scenarios, respectively. Bearing in mind that the spectrum bandwidth designed by ITU as IMT was much lower than this forecast (693 MHz in Region 1 (Europe, Middle East and Africa, and Russia), 723 MHz in Region 2 (Americas) and 749 MHz in Region 3 (Asia and Oceania)), and given that the time elapsed between the adoption of the radio regulations and the definitive allocation of a frequency band to operators takes from 5 to 10 years, the WRC-07 that took place in Geneva ended with the identification of new frequency bands for IMT technologies.

Figure 1.5 depicts the current state of the frequency bands reserved for IMT. Despite not fully corresponding to what was targeted, the new spectrum allocated for mobile communications will allow operators to satisfy the initial needs with the deployment of technologies towards IMT-Advanced. Furthermore, the increasing demand for mobile services has been progressively recognized with additional spectrum, a trend that is expected to be maintained in future WRCs.

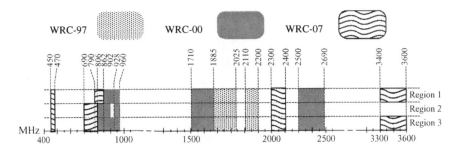

Figure 1.5   Mobile frequency bands allocated for IMT technologies

With this strong endorsement, the race towards IMT-Advanced successfully reached its end in the autumn of 2010. Anticipating the invitation from ITU, in March 2008 3GPP initiated a study item on LTE-Advanced (LTE-A) (also called LTE-Release 10 (Rel-10)). At IEEE, the IMT-Advanced candidate – IEEE 802.16m – was finalized in September 2009. These two technologies, LTE-A (3GPP 2010) and IEEE 802.16m (IEEE 2010) were submitted to ITU as IMT-Advanced technology candidates. Based on the evaluation results submitted to ITU-R in June 2010, ITU-R announced in October 2010 that both LTE-A and IEEE 802.16m proposals successfully met all of the criteria for the first release of IMT-Advanced.

Compared to its predecessor, the IMT-Advanced technologies rely on several new features such as Carrier Aggregation (CA), improved MIMO support, relaying and improved support for heterogeneous deployments. These features are described and analyzed in this book. The most promising ideas within those features for IMT-Advanced and beyond are explained and illustrated.

Advanced Radio Resource Management (RRM) While, from technological point of view, physical layer improvements are already close to their upper limit and only advanced antenna systems seem to be able to improve system performance, there is still a high potential to maximize efficiency in radio resource and interference management. Medium Access Control (MAC) aspects are attracting huge attention. Chapter 2 presents some innovative concepts for advanced RRM that have been identified by the research community for potential inclusion in IMT-Advanced and beyond.

Spectrum and Carrier Aggregation (CA) IMT-Advanced requirements establish a minimum support of 1 Gbps and 100 Mbps peak rates for low-mobility and high-mobility users, respectively. In order to fulfill these challenging requirements, wider channel bandwidth than legacy 3G systems have to be supported. However, as shown in Figure 1.5, the available spectrum resources are spread out over different frequency bands and with different bandwidths. Hence, CA, the concept of aggregation of continuous or discontinuous spectrum will be necessary in order to achieve wider effective carrier bandwidth. In addition, spectrum sharing, due to the scarcity of wide contiguous frequency bands, is crucial in order to optimize spectrum usage. These two concepts, CA and spectrum sharing, will be treated in Chapters 3 and 4, respectively.

MIMO In MIMO communications, multiple antenna elements are employed both in the transmitter and the receiver, in order to obtain increased data rates or improved reliability compared to single-antenna transmission. In order to address the challenge to offer very high data peak rates, IMT-Advanced has moved the emphasis from simple transmit diversity modes to spatial multiplexing and beamforming. In fact, LTE-A and IEEE 802.16m have evolved in the same direction: up to eight transmit antennas at the BS and up to four transmit antennas at the User Equipment (UE) are supported. Further, the standards are progressing toward the full adoption of Multi-User (MU)-MIMO transmissions, that is, the BS can spatially multiplex data streams intended for different UEs. Chapter 5 describes the latest advances in MIMO techniques.

Relaying With the growth of data traffic and the emergence of new services, there is a need to enhance coverage and/or capacity in specific locations. In addition, fast roll-outs are sometimes required to extend a network, implying a very dynamic and fast change of operator's infrastructure. Relaying techniques are emerging as an attractive solution to fill these needs. In fact, they are by construction characterized by ease of deployment (due to in-band backhauling) and reduced deployment cost compared to a regular BS. Chapter 7 overviews relaying techniques in general and describes them within the IMT-Advanced framework in particular.

Although IMT-Advanced systems will offer high data rates, even higher bit-rate demands and hungry applications such as high-definition TV are foreseen. Hence, novel techniques are required to meet the expected performance in terms of higher cell edge throughput and increased average cell capacity. Some of the candidate techniques include Coordinated Multipoint transmission or reception (CoMP), Network Coding (NC) and Device-to-Device (D2D) communications. These techniques are seen as necessary to further extend IMT-Advanced capabilities. Chapters 6, 7 and 9 will overview and describe them.

Coordinated Multipoint transmission or reception (CoMP)   Future cellular networks will need to provide high data-rate services for a large number of users, which requires a high spectral efficiency over the entire cell area. Hence it is important to ensure that the radio interface is robust to interference, especially the presence of Inter-Cell Interference (ICI), which degrades the performance of users located in the cell-edge areas. Recently, CoMP has attracted interest in its attempt to curb the ICI by relying on tight interference coordination. CoMP refers to a system where the transmission and/or reception at multiple, geographically separated antenna sites is dynamically coordinated in order to improve system performance. Chapter 6 treats thoroughly CoMP systems.

Network Coding (NC)   In a classical network, data streams originating from a source and intended for a desired destination are routed through intermediate nodes before reaching their final destination. By contrast, NC manipulates those data streams at an intermediate node by combining them before forwarding the resulting data to the destination. Network Coding is well suited to broadcast and cooperative wireless networks. Chapter 8 deals with NC in general and with its application to wireless communication in particular.

Device-to-Device (D2D) Communications   One aspect of the design of IMT-Advanced systems that has not received sufficient attention so far is the emergence of high data-rate local services. Such local services can provide the high data rates needed to consume rich multimedia services through mobile computers such as tablets, laptops, netbooks and smart phones. Cellular operators may offer such cheap access to spectrum with controlled interference enabled D2D communication as underlay to the cellular network. The licensed spectrum may be used as the only resource for communication or it may be complemented by license-exempt spectrum. The D2D concept is presented and studied in Chapter 9.

IMT-Advanced Performance Evaluation   The basic framework for IMT-Advanced performance evaluation has been well defined by ITU-R (ITU-R 2009). These guidelines have been published by ITU-R to perform the assessment of IMT-Advanced technology candidates by external evaluation groups. In this framework system parameters, channel models and deployment scenarios are defined. Moreover, the IMT-Advanced technology proponents have presented their own self-evaluation based on these scenarios. External evaluation groups like the Wireless World Initiative New Radio + (WINNER+) European Eureka Celtic project have performed an extensive performance evaluation of IMT-Advanced. In Chapter 10 the end-to-end performance of the LTE-A candidate is carefully presented, including the entire calibration process followed in WINNER+.

Future Directions   The technologies above offer further potential to significantly improve system performance. Therefore, research is needed in areas such as radio resource allocation, heterogeneous networks, MIMO and CoMP, relaying, network coding, device-to-device communications, green and energy efficient communications. Chapter 11 discusses these research trends.

References

3GPP 2010 Further Advancements for E-UTRA Physical Layer Aspects. Technical Specification Group Services and System Aspects 36.814, 3rd Generation Partnership Project (3GPP), http://www.3gpp.org/ftp/Specs/html-info/.

Dahlman E, Parkvall S, Sköld J and Beming P 2008 *3G Evolution: HSPA and LTE for Mobile Broadband* 2nd edn. Academic Press, New York.

GSA n.d. The Global Mobile Suppliers Association, http://www.gsacom.com.

Hillebrand F 2002 *GSM and UMTS: The Creation of Global Mobile Communication*. John Wiley & Sons, Ltd, Chichester.

Holma H and Toskala A 2000 *WCDMA for UMTS- Radio Access for Third Generation Mobile Communications*. John Wiley & Sons, Ltd, Chichester.

Holma H and Toskala A 2007 *HSDPA/HSUPA for UMTS*. John Wiley & Sons, Ltd, Chichester.

IEEE 2010 IEEE 802.16m System Description Document (SDD). Technical Specification 09/0034r3, Broadband Wireless Access Working Group, http://ieeexplore.ieee.org.

IEEE-ISTO n.d. The IEEE Industry Standards and Technology Organization (IEEE-ISTO) http://www.ieee-isto.org.

IEEE-SG n.d. IEEE Standards Groups, http://grouper.ieee.org/groups/802/.

ITU n.d. Mobile Cellular: Subscribers per 100 People, International Telecommunication Union, http://www.itu.int/ITU-D/ict/statistics/.

ITU-R 2009 Guidelines for Evaluation of Radio Interface Technologies for IMT-Advanced. Report ITU-R M.2135-1, International Telecommunications Union Radio (ITU-R), http://www.itu.int/publ/R-REP/en.

NGMN Ltd n.d. The Next Generation Mobile Networks, www.ngmn.org.

Sesia S, Baker M and Toufik I 2011 *LTE - The UMTS Long Term Evolution: From Theory to Practice* 2nd edn. John Wiley & Sons, Ltd, Chichester

UMTS-Forum 2010 Recognising the Promise of Mobile Broadband. White paper, UMTS-Forum, http://www.umts-forum.org.

Viterbi AJ 1995 *CDMA: Principles of Spread Spectrum Communication*. Addison Wesley Longman Publishing Co., Inc., Redwood City, CA.

WiMAX n.d. Worldwide Interoperability for Microwave Access (WiMax) Forum, www.wimaxforum.org/home/.

2

Radio Resource Management

Jose F. Monserrat, Gunther Auer, David Martin-Sacristan and Pawel Sroka

2.1 Overview of Radio Resource Management

The concept of Quality of Service (QoS), which entails service differentiation and the possibility of adapting resource allocation to the specific service requirements, has opened the door to an important increase in the provision of advanced services to the mass market. In this framework, Radio Resource Management (RRM) has become a key element of current and future wireless and wired communication networks to provide the negotiated QoS to the end users. In the end, RRM algorithms have a direct impact on the performance experienced by each individual user and, furthermore, on the overall network performance. In consequence, they have to make the most of the available resources for the benefit of users and operators. When specifically addressing Orthogonal Frequency Division Multiple Access (OFDMA) systems, the importance of RRM algorithms becomes crucial since the system performance to a large extent depends upon these mechanisms. Besides, the complexity of the system, encompassing time, frequency and space dimensions, and the increasing need of reducing the OPerational EXpenditures (OPEX) of the network, encourages all the agents involved in the wireless market to optimize the operation of the RRM algorithms.

In future wireless systems, like 3GPP Long Term Evolution (LTE) or the evolution of Worldwide Interoperability for Microwave Access (WiMAX) IEEE 802.16m, RRM comprises various techniques that can be grouped into three categories: resource allocation, load control and mobility control. Within the OFDMA-based paradigm, resource allocation includes frequency, time and/or space allocation, scheduling and resource reservation. The scheduling mechanisms are especially designed to favor those users with best channel conditions, which entails an increase in the system throughput while augmenting the number of users. This effect is also known as multi-user diversity. At the same time, the allocated power must be controlled to reduce interference as much as possible. Secondly, load control manages the access to the network and plays a major role in the avoidance of system congestion. Finally, mobility control schemes manage the subsequent User Equipment (UE) mobility, attempting to guarantee the continuity of the service. They also play a major role in the avoidance of system congestion.

In next-generation wireless systems, the Base Station (BS) is in charge of the dynamic resource allocation, both in Uplink (UL) and Downlink (DL). Concerning the protocol stack, cooperation among Physical (PHY) layer, Medium Access Control (MAC) and Application (APP) layers is required to

Mobile and Wireless Communications for IMT-Advanced and Beyond, First Edition.
Edited by Afif Osseiran, Jose F. Monserrat and Werner Mohr.
© 2011 Afif Osseiran, Jose F. Monserrat and Werner Mohr. Published 2011 by John Wiley & Sons, Ltd.

guarantee QoS in a highly unstable environment, such as that associated with wireless mobility. Most important RRM functionalities and their mapping to different Open Systems Interconnection (OSI) layers are (Holma and Toskala 2009):

- **PHY**. PHY comprises several processes, the most relevant being: (1) the channel state process, which aims at processing the measurement reports and the pilot signals in DL and UL respectively; (2) the physical signalling process, responsible for updating the physical control channel that provides information about the specific frequency-time chunks allocated to each UE together with the modulation, coding scheme and precoding matrix used in the transmission; finally, (3) the power control process, which coordinates transmitted power in both links.
- **MAC**. The MAC layer (1) manages the Hybrid Automatic Repeat-reQuest (HARQ) process; (2) carries out the dynamic resource allocation in space (with multiple antennas), frequency and time; (3) is also the entity that performs link adaptation by choosing the proper modulation and coding scheme for each UE; (4) manages the UE's QoS; finally, (5) it executes Call Admission Control (CAC) from the radio interface point of view.
- **APP**. In order to maintain good end-to-end performance the APP layer must be flexible enough to follow the variations on the lower layers state seamlessly, most of all, on the radio interface. For example, video streaming servers can modify the video codec used in the transmission when the system reduces the amount of resources allocated to a certain UE.

All these algorithms and processes are complemented with strategies for Inter-Cell Interference Coordination (ICIC), essential to fulfill QoS requirements at the cell edge. ICIC mechanisms are highly related to the power and resource allocation and they must be studied together.

Recently, in the standardization process of International Mobile Telecommunications Advanced (IMT-Advanced) candidates, several new concepts have been proposed to produce a highly flexible and efficient high-speed communications system. These concepts include improved Multiple-Input Multiple-Output (MIMO) communications, Coordinated Multipoint transmission or reception, support for contiguous or non-contiguous spectrum aggregation or efficient Multimedia Broadcast Multicast Services (MBMSs). These techniques allow high system spectral efficiency and robustness to interference to be achieved, simultaneously providing the negotiated QoS to the UEs.

This chapter analyzes all those ideas together with some innovative concepts in the area of Advanced RRM algorithms introduced within the framework of next-generation wireless systems. Detailed innovative techniques constitute the concepts described in this chapter. State-of-the-art reviews are provided for each identified technique, as well as an evaluation of their performance, their benefits and the impact of their usage in signalling and architecture. Section 2.2 poses the constraints of the RRM mechanisms reviewing the main characteristics of the new IMT-Advanced technologies. Section 2.3 addresses techniques for dynamic resource allocation concentrating on scheduling with QoS constraints and the inclusion of relays. This section also encompasses a detailed analysis of cross-layer aspects of QoS management with cross-layer optimization involving the application layer. Section 2.4 deals with interference coordination in next-generation mobile networks. Section 2.5 presents some techniques for efficient MBMS. Finally, Section 2.6 concludes the chapter, summarizes the main contributions and highlights the future research trends in the topic.

2.2   Resource Allocation in IMT-Advanced Technologies

The development of efficient and effective RRM algorithms is a key differentiation element among vendors and operators. That is why the definition of RRM algorithms is out of the scope of the specifications of any IMT-Advanced Radio Interface Technology (RIT). In fact, the characteristics of the IMT-Advanced RITs propitiate the development of different algorithms. However, some specific characteristics have been included in the standards to enforce the use of concrete families of algorithms.

The first objective of this section is to highlight the characteristics of the IMT-Advanced standards that allow the use of RRM algorithms. With this foundation, the second objective is to indicate the common algorithms currently used in these standards. The focus of this section is specifically on Third Generation Partnership Project (3GPP) LTE and Institute of Electrical and Electronics Engineers (IEEE) 802.16m standard, hereinafter called WiMAX.

2.2.1   Main IMT-Advanced Characteristics

This subsection highlights the main characteristics of IMT-Advanced technologies related to RRM. LTE-Advanced (LTE-A) and WiMAX technology proposals submitted and approved by International Telecommunication Union (ITU) are (ITU-R 2009a) and (ITU-R 2009b), respectively. In these documents, the reader can find the main characteristics commented on below and also additional information. Furthermore, the reader can find references in these documents to the standards and other documents of the technology proponents, that is, the 3GPP and IEEE.

QoS Differentiation

IMT-Advanced RITs must support a wide range of services, each one with different QoS requirements. That is why IMT-Advanced RITs allows QoS differentiation among packet flows. Different QoS profiles are defined, then these profiles are used to perform RRM operations.

In LTE, each LTE Evolved Packet System (EPS) bearer has an associated QoS profile that includes the following parameters:

- **Allocation and Retention Priority (ARP)**. A number that identifies the priority of the bearer. This number is used to set priorities in the call admission control and also to make call dropping decisions in case of congestion.
- **QoS Class Identifier (QCI)**. An index to a table of additional QoS parameters, namely priority, allowable delay and packet error rate. These parameters are used to configure the Radio Link Control (RLC) and MAC behavior. The QCI also identifies if the bearer is a Guaranteed Bit Rate (GBR)-bearer or a non-GBR-bearer. Several profiles have been included in the standard but additional profiles can be defined.
- **GBR and Maximum Bit Rate (MBR)**. For GBR-bearers these two parameters indicate both the guaranteed and maximum bit rate of the bearer.
- **Aggregated Maximum Bit Rate (AMBR)**. For non-GBR-bearers this parameter establishes the maximum bit rate for the set of flows mapped on the bearer.

In WiMAX the QoS attributes are: the traffic priority, maximum sustained traffic rate, minimum reserved traffic rate and maximum latency. A service flow is a MAC transport service associated with a set of QoS parameters. Service flow classes with predefined QoS profiles can be defined by vendors and operators.

Orthogonal Frequency Division Multiplexing (OFDM) Modulation

Both LTE and WiMAX are based on OFDM. Different sets of OFDM parameters are defined in each system to support a variety of propagation environments. These sets are different in both standards. In the UL, LTE uses Single Carrier – Frequency Division Multiple Access (SC-FDMA).

OFDMA Multiplexing

OFDMA is the main multiplexing method in IMT-Advanced, where several UEs can share time-frequency radio resources.

In LTE a physical unit is defined, called Physical Resource Block (PRB). PRBs consist of a number of contiguous subcarriers in the frequency domain during a slot of 0.5 ms. The minimum allocable resource unit is a pair or PRBs contiguous in time during a subframe (1 ms) but not necessarily occupying the same set of subcarriers.

In WiMAX, the Physical Resource Unit (PRU) is defined as the basic physical unit for resource allocation and comprises a number of contiguous subcarriers during a subframe.

MIMO Capabilities

IMT-Advanced RITs present Single-User (SU)-MIMO and Multi-User (MU)-MIMO capabilities as a means to enhance cell and UE performance. Cell spectral efficiency and cell-edge UE spectral efficiency can be improved using MIMO. Moreover, UE data rate and UE coverage can be enhanced with MIMO.

The use of MIMO implies more degrees of freedom in RRM. Multiple UEs can be scheduled simultaneously on the same resources by employing MU-MIMO. On the other hand, SU-MIMO permits multiple streams to be transmitted on the same resources to a single user.

LTE Release 8 (Rel-8) MIMO schemes feature are: transmit diversity (based on space-frequency block coding), open-loop spatial multiplexing, closed-loop spatial multiplexing and MU-MIMO (the last three being codebook-based precoding schemes) and beamforming (that is noncodebook-based). In LTE Release 10 (Rel-10) the codebook-based MIMO modes have been substituted by noncodebook-based ones that are more powerful. Similar MIMO schemes are found in WiMAX: transmit diversity (based on space-frequency block coding or vertical encoding with a single stream), spatial multiplexing (open loop and closed loop) and MU-MIMO. It should be noted that Chapter 5 addresses MIMO techniques in detail.

Adaptive Modulation and Coding

Different modulations (Binary Phase Shift Keying (BPSK), Quadrature Phase Shift Keying (QPSK), 16-Quadrature Amplitude Modulation (QAM) and 64-QAM) and coding schemes are available to fit upper layer packet data units into the different PHY channels and to make the system capable of being adapted to varying channel conditions.

Reference Signals and Channel Quality Indicators

In IMT-Advanced RITs different reference signals are available in DL and UL in order to estimate the quality of the link between the BS and the UE. In Frequency Division Duplex (FDD), in order to estimate the DL channel quality, the BS needs some feedback from the UE that measures the link quality using the DL reference signals. However, the UL quality is directly measured by the BS using the UL reference signals. In the Time Division Duplex (TDD) mode, DL channel quality may be estimated by the BS from UL signals due to channel reciprocity. However, interference observed at the BS is different from that observed at the UE, which may result in suboptimum performance.

In LTE there are three indicators sent from the UE as DL quality feedback: the Rank Indicator (RI), which indicates the rank of the channel, the Precoding Matrix Indicator (PMI), which provides information about the best precoder that can be used in the transmission, and the Channel Quality Indicator (CQI), which indicates the best combination of modulation and coding rate for that UE. All these reports can be measured and reported considering either wideband transmission over the system bandwidth or narrowband transmissions.

In WiMAX there are also indicators defined to perform the link adaptation. The Advanced Mobile Station (AMS) may feedback, for example, a CQI, a Preferred Matrix Index (P.M.I) used for codebook-based precoding and a long-term Channel State Information (CSI) including an estimate of the transmit spatial correlation matrix.

Other Uplink Signaling

In LTE, Buffer Status Reports (BSRs) and Power Headroom Reports (PHRs) are transmitted by the UE to the BS. BSRs indicate the amount of data buffered at the UE. PHRs provide information about the power spectral density used by the UE.

Communication among Base Stations

LTE allows communication among the BSs through the so-called X2 interface. Three different indicators may be sent from one BS to another:

- **High Interference Indicator (HII)**. An UL indicator that reports per PRB the occurrence of high interference sensitivity as seen from the sending cell. That is to say, whether the sending cell plans to schedule cell-edge UEs in each PRB.
- **Overload Indicator (OI)**. An UL indicator that provides information about the interference level (high, medium, or low) experienced by the sending cell per PRB.
- **Relative Narrowband Transmit Power (RNTP)**. A DL indicator that reports per PRB if the transmitted power exceeds a threshold. The power threshold is also exchanged among BSs.

Note that the standard does not specify either how often these indicators are sent nor the behavior of receiving BSs.

WiMAX, (IEEE 802.16 2010a) specifies that the Advanced Base Stations (ABSs) can exchange information among themselves or with another control element in the backhaul network. The following information may be exchanged among ABSs:

- Interference measurement results (as the recommended P.M.I subset to be restricted or to be applied in neighbor cells).
- Precoder and resource allocation applied in each cell.
- Soft decision information (for macrodiversity combining).
- Channel state information.

2.2.2   Scheduling

The scheduling is the process that performs the allocation of resources to the UEs. Scheduling is highly related to Link Adaptation (LA), whose goal is to adapt the characteristics of the transmissions to the link quality. Interaction between these two processes is very close. In fact, the scheduler cannot make optimum decisions without considering the information provided by the LA. The LA must provide the scheduler with the rate achievable by a UE given a specific resource allocation indicated by the scheduler. Furthermore, the LA should indicate the format of the transmission (modulation, coding, MIMO scheme, etc.) to reach that rate.

Based on all those characteristics, the inputs needed by the scheduler are:

- **QoS demanded per flow**. The scheduler needs to know the QoS demanded per flow in order to ensure that this demand is satisfied.
- **State of RLC, MAC and upper layers**. The state of RLC and MAC buffers should be taken into account to make optimum decisions. For example, the scheduler can give higher priority to a UE whose RLC has packets with tighter delay constraints.
- **State of HARQ entities**. The scheduler must decide per UE if a new transmission or a retransmission is handled. It is therefore necessary to know the state of the HARQ entities of that UE. In some scheduling implementations higher priority can be given to retransmissions.
- **Link adaptation information**. This information is needed to select the optimum format for each transmission.

Scheduling is usually performed every subframe. In IMT-Advanced technologies the high number of available degrees of freedom (time, frequency, space, users, and service flows) implies that finding optimal resource allocations is a very complex task. Therefore, low complexity schedulers are preferred in practice to perform the scheduling at a fast pace. For example, scheduling is commonly divided in two processes: time-domain scheduling and frequency-domain scheduling. In the first step a low number of UEs is selected to be scheduled based on some criterion; then the frequency resources are allocated based on another criterion.

Both in WiMAX and in LTE, the scheduling algorithms are implementation specific. Nevertheless, both standards have been designed to support efficient scheduling. Sub-band and full-band channel quality indicators are included in the standard together with sounding reference signals (in UL) to perform link adaptation and frequency domain scheduling. In LTE, BSR and PHR also facilitate the UL scheduling.

The concepts discussed above are further developed in Section 2.3.

2.2.3   Interference Management

Various interference management mechanisms have been proposed for IMT-Advanced technologies in order to enhance the performance in interference-limited scenarios. These mechanisms are especially interesting for the UEs experiencing high interference levels and for those UEs capable of using MIMO.

The following interference management schemes have been proposed for WiMAX in (IEEE 802.16 2010a):

- **Fractional Frequency Reuse (FFR)**. It allows the configuration of several frequency partitions (portions of the system bandwidth), each one with a different frequency reuse factor and power level. For example, for a reuse-3 scenario, the FFR concept in WiMAX allows the definition of three different power-frequency patterns with four frequency partitions. One of these partitions is

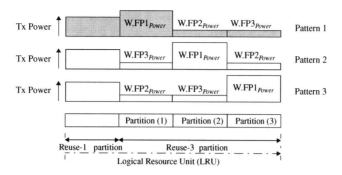

Figure 2.1 Fractional frequency reuse for reuse-3 scenario (IEEE 802.16 2010b). IEEE Std 802.16m. Reproduced by permission of © 2010 IEEE

used with reuse 1 while the others are used with reuse three. Clusters of three BSs are defined and each one selects one of the different power-frequency patterns, as can be seen in Figure 2.1.

- **Schemes using advanced antenna technologies**. Two different classes of these schemes are considered in WiMAX. The first class is known as single-cell antenna processing with multi-ABS coordination. This class refers to the interference mitigation techniques based on the MIMO schemes in which only one ABS transmits data to one AMS. This mode relies on inter-ABS coordination and interference measurement. In this class, coordination does not require data forwarding. When precoding is applied in interfering cells, coordination of the precoders can reduce the interference experienced in those cells. For example, an AMS can estimate, based on its measurements of the channel from interfering ABSs, which is the most or least interfering precoder that these ABSs can use. Then, this information can be reported to the serving ABS being subsequently shared with the coordinated ABSs. With this information each ABS can choose its precoders taking into account the interference produced over the other ABSs. This process is called DL P.M.I coordination. UL P.M.I coordination can be achieved too. In this case, the ABS measures the channel from interfering AMSs using sounding signals. Based on these measurements, the less interfering precoders can be calculated and reported to other ABSs. The second class of interference mitigation scheme using advanced antenna technology is called multi-ABS joint antenna processing. In this class, joint MIMO transmission or reception across multiple ABSs is possible. Collaborative MIMO or CoMP is one of the main techniques within this class. According to this technique, a number of ABSs perform a joint transmission to a group of AMSs. In Closed-Loop Macro Diversity several ABSs perform a joint transmission to a single AMS.
- **Power control and scheduling**. Several techniques are included under this common title, such as, for example, subchannels scheduling, dynamic transmit power control, dynamic antenna patterns adjustment and dynamic Modulation and Coding Scheme (MCS). MCS has a significant effect on UL transmit power and hence on UL interference. Therefore, dynamic selection of MCS can be used to manage UL interference. The same is also true for dynamic transmit power control. Subchannel scheduling can be used by different ABSs to schedule high mutual interference AMSs in different subchannels. With this aim, ABSs need to exchange some interference and channel measurements and also scheduling information.
- **Cell/sector specific interleaving**. This is used to randomize the transmitted signal to perform interference suppression at the receiver.

In LTE Rel-8 no mechanism is explicitly included in the specifications. Nevertheless, some algorithms are easily implemented based on the characteristics of the standard. For example, FFR and similar schemes can be implemented due to the great flexibility of the resource allocation. Static and dynamic mechanisms have been proposed, that is, mechanisms with fixed frequency partitions and power allocations and mechanisms whose parameters are adaptive. The information exchanged among BSs using the X2 interface is used to perform this kind of interference mitigation in both directions of the communication:

- **In uplink**. If a received HII indicates high interference sensitivity in a PRB, the receiving eNodeB (eNB) will try to not use this PRB for a cell edge UE. If a received OI indicates a high interference level, the receiving eNB can reduce its transmitting power (by adjusting its open loop power control) to keep the interference under a desired value.
- **In downlink**. If the RNTP indicates that a eNB plans to transmit in a PRB with high power, this information can be used by another eNB to select a UE with a good channel, that is, less prone to suffer from interference.

In LTE Rel-10 the Coordinated Multipoint transmission or reception (CoMP) concept was presented as one step forward in the interference management roadmap. One of the CoMP types, coordinated scheduling/beamforming, is an interference mitigation scheme where the coordination is performed basically at the RRM level. Hence, the scheduler tries to minimize the interference among BSs. More information about CoMP can be found in Chapter 6.

Similar interference management mechanisms have been included both in LTE and WiMAX. The plethora of available mechanisms ranges from the simplest ones, which try to mitigate the interference, to the more complex, which exploit this interference to obtain a benefit. By moving from the simplest to the more complex it is possible to achieve greatest performance improvements at the expense of increased complexity.

More information about interference coordination mechanisms can be found in Section 2.4.

2.2.4 Carrier Aggregation

One of the IMT-Advanced minimum requirements in (ITU-R 2008b) is related to bandwidth. First, it is stated that IMT-Advanced RITs must have the ability to operate with different bandwidth allocations. Besides, it is required the support of a scalable bandwidth up to 40 MHz. Finally, proponents are encouraged to support operation in wider bandwidths, for example, up to 100 MHz.

Both IMT-Advanced RITs support Carrier Aggregation (CA). Moreover, in both systems link adaptation feedback is required from every carrier used. Assignment of carriers may be based on system load, peak data rate or QoS demand. In conclusion, CA increases the availability of resources to perform RRM. Nevertheless, each standard implies different implementation and hence different constraints on CA use.

Detailed information about CA can be found in Chapter 3.

2.2.5 MBMS Transmission

International Telecommunication Union – Radiocommunication Sector (ITU-R) does not explicitly require the capability of IMT-Advanced to support multicast or broadcast transmission. Nevertheless, in the technology description template presented in (ITU-R 2008a) there is a section about unicast, multicast and broadcast capability. Specifically, it is required to describe how each proposal supports this kind of transmissions.

In LTE, multicast and broadcast services are supported. When multicast/broadcast services span multiple contiguous cells a MBMS over Single Frequency Networks (MBSFN) can be applied in order to improve the spectral efficiency. Mixed carriers with multicast/broadcast and unicast services are supported. Coordination of the allocation of radio resources used by the BSs in MBSFN areas is required.

MBMS transmission is called in WiMAX Enhance Multicast Broadcast Service (E-MBS). E-MBS supports two types of access: single-ABS access and multi-ABS access. Single-ABS access is implemented over multicast and broadcast transport connections within one ABS, while multi-ABS access is implemented by transmitting data from service flow(s) over multiple ABSs, using the concept of E-MBS zone. This transmission is supported either in the nonmacro diversity mode or in macro diversity mode as a wide-area multi-cell MBSFN. E-MBS service may be delivered via either a dedicated carrier or a mixed unicast-broadcast carrier.

The enhanced MBMS transmission found in IMT-Advanced technologies is explored in Section 2.5 with special focus on LTE.

2.3 Dynamic Resource Allocation

The wireless shared channel in cellular networks is a medium over which many UEs compete for resources. In such a scenario, spectral efficiency and fairness are crucial aspects for resource allocation. From a cellular operator's perspective, it is very important to use the channel efficiently because the available frequency spectrum is scarce and revenue must be maximized. From the UEs' point of view, it is more important to have a fair resource allocation so that they can meet their QoS requirements and maximize their satisfaction. Providing QoS, in particular meeting the data rate and packet delay constraints of real-time data UEs, is one of the requirements in emerging next-generation wireless networks. Efficient data scheduling is one of the ways to address this issue. However, the time-varying nature of the wireless environment, coupled with different channel conditions for different UEs, poses significant challenges to accomplishing these goals. Thus, the objectives of fair and efficient resource allocation cannot be achieved simultaneously and a tradeoff must be achieved. With the introduction of some innovative techniques, such as OFDMA, MU-MIMO, relaying and spectrum sharing, the resource allocation problem became even more complicated and multidimensional. Consequently, novel sophisticated algorithms have to be formulated to provide support for the new functionalities of the next-generation wireless networks. This section summarizes recent advances in the area of scheduling with QoS support in a heterogeneous traffic scenario using utility theory. Moreover, analysis of the new challenges in resource allocation related to the introduction of relays is presented. Detailed information about resource allocation techniques with MU-MIMO support can be found in Section 2.4 and Section 5.1.3. Finally, the investigation on the spectrum allocation and spectrum sharing techniques is carried out in Chapter 4.

2.3.1 Resource Allocation and Packet Scheduling Using Utility Theory

Next-generation wireless networks, such as LTE-A and WiMAX are based on the OFDMA modulation, which offers many degrees of flexibility for resource management compared to traditional single-carrier networks. Taking advantage of the knowledge of the CSI at the BS, OFDMA-based systems can employ the following adaptive resource allocation techniques:

- Adaptive Modulation and Coding (AMC) .
- Dynamic Sub-carrier Assignment (DSA), in which the BS dynamically assigns subcarriers according to CSI and QoS requirements.
- Adaptive Power Allocation (APA), in which the BS allocates different power levels to UEs to improve the performance of OFDMA-based networks, which is called multiuser water-filling (Cheng and Verdu 1993).

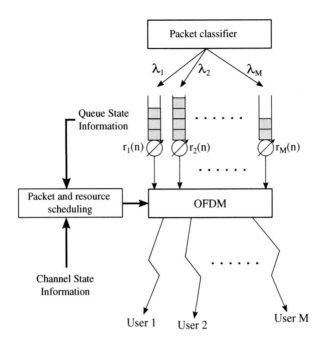

Figure 2.2 Functional structure of the considered scheduler

The major issue is how to effectively assign subcarriers (or groups of them called PRBs) and to perform the multiuser power allocation. This allocation should exploit knowledge of the CSI and the traffic characteristics to improve spectral efficiency and guarantee diverse QoS. Both LTE and WiMAX are expected to support heterogeneous traffic, and traffic volume and the QoS requirements of each UE shall be taken into account in the scheduling process, as shown in Figure 2.2. Therefore, a proper scheduler design is crucial, to provide the support for various QoS traffic classes.

In a real system, DSA requires much less feedback information than APA, because APA requires the exact Signal-to-Interference-plus-Noise Ratio (SINR) value of each subcarrier while DSA only needs to know the achievable data rate at each subcarrier. APA is therefore often neglected in the scheduler design with the assumption of equal power allocation among all subcarriers (Song and Li 2005).

Proportional Fair (PF) scheduling has been well studied in the literature in the past. Although, it provides an excellent tradeoff between maximization of the long-term throughput of UEs and fairness of resource allocation, it is less efficient for delay-sensitive applications. Therefore, Modified-Largest Weighted Delay First (M-LWDF) has been proposed for Code Division Multiple Access (CDMA) networks in Andrews et al. (2001), where both channel and queueing states are taken into account to provide a statistical delay guarantee. However, when heterogeneous traffic is to be supported, simple QoS provisioning based on measurements of delay of the Head-of-line (HOL) packet may not suffice to differentiate the delay-sensitive services. Thus, utility theory has been employed by many researchers to deal with support of multiple QoS classes (Lei et al. 2007; Miao and Himayat 2008; Ryu et al. 2005; Song and Li 2005). The basic idea of utility pricing is to map the resources (bandwidth, power, etc.) or performance criteria (data rate, delay, etc.) into the corresponding utility or price values and optimize the established utility-pricing system. Different applications (or traffic classes) may correspond to different utility function curves or even different parameters. Specifically, the QoS requirement of Real-Time (RT) services can be represented using:

$$Pr\{W_i > T_i\} \le \delta_i \tag{2.1}$$

where W_i is HOL delay of UE i, and T_i and δ_i are the delay threshold and the maximum probability of exceeding it, respectively. Therefore, the related utility function is, with respect to delay $U_i(W_i)$ (Song and Li 2005).

On the other hand, when considering Best Effort (BE) applications, the QoS requirement is given as follows:

$$R_i > r_i \tag{2.2}$$

where the average throughput R_i provided to UE i is not less than some predefined value r_i, and the utility function is, with respect to throughput $U_i(R_i)$ (Song and Li 2005).

As resource management in the next generation wireless systems has the BS as a central controller, the optimization objective in resource allocation can be specified as the one that maximizes the aggregate utility in the system, as given by:

$$\max \sum_i U_i \tag{2.3}$$

Developing the resource allocation algorithms is very challenging because the utility functions are usually nonlinear and the DSA problem can be regarded as a nonlinear combinatorial optimization. It has been outlined in Song et al. (2004) that if the utility functions are concave with respect to the instantaneous data rates, the utility-based optimization can be solved effectively using gradient algorithms. Therefore, in case the subcarrier assignment is independent for different subcarriers, the DSA can be performed using Weighted Max Algorithm as:

$$m(k, n) = \arg \max_i \left(\frac{U_i'}{\lambda_i} \cdot c_i[k, n] \right) \tag{2.4}$$

where $m(k, n)$ represents that subcarrier k is assigned to UE m at time n and $c_i[k, n]$ is the achievable data rate for subcarrier k at time n for user i. U_i' is the so-called marginal utility function (the derivative of utility function) of UE i, and λ_i is the average arrival bit rate of UE i (Song et al. 2004).

A similar approach was proposed in Ryu et al. (2005), where the Urgency and Efficiency-based Packet Scheduler (UEPS) can support heterogeneous traffic. Here the idea of soft deadline was introduced, which is a time constraint that yields diminishing values to the system when the deadline has passed. Therefore, the authors proposed an optimization criterion, given in Equation (2.5), based on the marginal utility function U_i' and the efficiency factor $\frac{C_i(t)}{\overline{C_i(t)}}$, where $C_i(t)$ and $\overline{C_i}(t)$ represent current and average channel state, respectively:

$$m(k, n) = \arg \max_i \left(|U_i'| \frac{C_i(t)}{\overline{C_i(t)}} \right) \tag{2.5}$$

Another algorithm that tries to maximize the aggregate utility of the system is the Traffic-Aware Score Based (TASB) scheduler proposed in the framework of the Wireless World Initiative New Radio + (WINNER+) project (WINNER+ 2009a). It extends the idea of Score Based (SB) scheduler proposed in (Bonald 2004) including the utility factor of scheduled packets in the optimization metric as follows:

$$m(k, n) = \arg \min_i \left(\frac{s_i(n)}{1 + \alpha \cdot \sum_{j=1}^{L} \beta_{i,j} \cdot U_{i,j}'(t)} \right) \tag{2.6}$$

where $\beta_{i,j}$ is the priority class factor of packet j, $U'_{i,j}(n)$ is the marginal utility function of packet j at slot n, L is the total number of packets from UE i queued for scheduling and α is a constant defining the impact of packet urgency on the resource allocation process. $s_i(n)$ is the score of UE i at slot n that represents the user's throughput statistics, defined as:

$$s_i(n) = 1 + \sum_{l=1}^{W-1} 1_{r_i(n) < r_i(n-l)} + \sum_{l=1}^{W-1} 1_{r_i(n) = r_i(n-l)} \cdot X_l \qquad (2.7)$$

where W is the length of observation window in samples, $r_i(n)$ is the rate of UE i at slot n and X_l is a random variable on the discrete set $\{0,1\}$, with $Pr(X_l = 0) = Pr(X_l = 1) = 0.5$.

The well-known sigmoidal and parabolic utility functions are considered for RT services whereas a linear function is proposed for Non-Real-Time (NRT) services. One can notice that the utility factor in Equation (2.6) depends not only on the utility of the HOL packet but all (or specifically L) packets in UE i queues. Such approach, according to (WINNER+ 2009a), lead to a slight improvement of QoS satisfaction in a heterogeneous traffic scenario.

An interesting approach using the utility functions has been proposed in WINNER+ project (WINNER+ 2009a,c), extending the idea of Predictive Proportional Fair (PPF) scheduler (Ekman et al. 2008). The Utility Predictive Scheduler (UPS) employs the rate-based utility function to determine the potential rate increase when allocating a PRB to a UE. Additional information in the UPS-based utility function can be found in Appendix A.1.1.

Utility-based resource allocation is a tool that provides full support for QoS provisioning in a heterogeneous traffic scenario. Thanks to the elastic approach of utility functions one can gain a flexible scheduler capable of dealing with the requirements of wireless networks.

2.3.2   Resource Allocation with Relays

In order to extend the cell coverage and maintain the Signal to Interference plus Noise Ratio (SINR) with low upfront investment, LTE-A and WiMAX networks are designed with the assistance of relaying techniques, where the traffic signal can be relayed through one or more Relay Nodes (RNs). This poses significant challenges for the resource allocation algorithms, namely:

- CSI updates are received periodically every D frames and UE scheduling and resource allocation needs to be jointly optimized for UL and DL. In case of rapid changes of channel characteristics, the CSI may become outdated and, therefore, the resource allocation may become inefficient. Moreover, CSI has to be collected for all hops (BS to RN and RN to UE) thus increasing the signaling.
- The optimal transmission path between BS and UE has to be established, with the BS selecting the transmission path and signaling it to the UE.
- The resource allocation pattern has to be optimized in advance for a given number of slots, depending on the number of hops, and the information on the allocation has to be transmitted to the UEs, as shown in Figure 2.3 for the two-hop TDD transmission case. For each frame (0–7) the BS signals the resource allocation pattern to UE either directly (single-hop transmission – user 1) or via the RN (two-hop transmission – user 2).

The last issue becomes even more challenging when considering the half-duplex TDD operation of LTE and WiMAX. The BS must ensure that the RN is in a phase where it is acting as a UE when it is scheduled by the BS to receive the resource map. As shown in Figure 2.3, there is a gap of one frame between the DL and UL phases of both duplex groups and a gap of three frames between

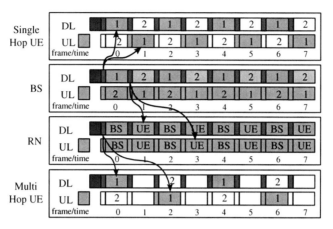

Figure 2.3   Signaling procedure in half-duplex transmission with relay node

consecutive DL and UL phases, respectively. This leads to an increased delay and a lower maximum data rate for UEs owing to the gaps mentioned (WINNER+ 2009a). When investigating the resource map information flow, shown in Figure 2.3 with arrows, one can observe that, for the relayed UEs, the resource allocation map is delayed two frames for the DL and four frames for the UL. The multihop UEs suffer from the reduced capabilities, therefore, compared with the single-hop UEs. Moreover, the performance of the multihop UEs is further reduced when half-duplex transmission mode is used compared with the full-duplex (WINNER+ 2009a). Figure 2.4 shows the relation between offered and carried throughput in the case of UL multihop transmission.

In the scenario considered here, one full-duplex and one half-duplex single-hop UE are present in the system, as well as one full-duplex and one half-duplex two-hop UE, which are connected to the BS via a RN. A simple Round Robin scheduler is used to allocate the resources. Figure 2.4 can be divided into three sections. To the left, none of the UEs has reached its maximum data rate and the throughput of all UEs increases. Then the throughput still continues to rise and decreases for single-hop half-duplex UE until all UEs reach their saturation. The saturation value for the single-hop full-duplex UE is around three times as high as the one for the single-hop half-duplex UE, because the half-duplex UE gets half of all resources every second frame and no resources at all other frames. In the low traffic cases the single-hop full-duplex UE does not request all of the 50% resources that are entitled to it in the frames it shares with the single-hop half-duplex UE and the half-duplex UE gets

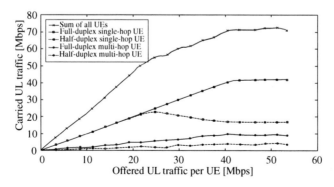

Figure 2.4   UL throughput in multi-hop scenario. Reproduced by permission of © 2008 IEEE

more than half of the available resources. But as soon as the full-duplex UE needs more resources, it accesses them up to the saturation point, so that the resources reserved for the half-duplex UE decrease down to 50% every second frame, which equals 25% in total.

It is therefore necessary to design a resource-allocation scheme able to deal with all of these issues. Such a methodology has been proposed in Chang et al. (2008) as a centralized QoS Guaranteed Throughput Enhancement (GTE) scheduling scheme. The proposed approach consists of three steps:

1. A transmission time-based path selection algorithm, which is based on the total transmission time that a BS-UE path experiences. The total transmission time depends on the allocated rates as shown in Equation (2.8):

$$\frac{1}{t_{m,i}} = \frac{1}{R'_{m,i}} + \frac{1}{R''_{m,i}}, \quad i \neq 0 \tag{2.8}$$

where $R'_{m,i}$ is the allocated data rate for UE m from BS to RN i and $R''_{m,i}$ is the allocated data rate for UE m from RN i to UE m. Finally, the path with minimal transmission time is selected as the transmission path for UE m.
2. A service order-based resource-allocation algorithm, which aims at maintaining the QoS requirement. A service-order parameter is defined that depends on the priority and urgency of the UE's packets.
3. Transmission concurrency decision algorithm, which seeks for more than one RN that can transmit concurrently in order to improve the spectral efficiency.

The performance analysis given in Chang et al. (2008) shows that the QoS GTE scheduling scheme provides a mechanism for simultaneous QoS satisfaction and spectral efficiency improvement thanks to the use of RNs.

An interesting approach to resource allocation with QoS support in relay-enhanced networks was proposed in WINNER+ (2009b). The HurrY-Guided-Irrelevant-Eminent-NEeds (HYGIENE) scheduling algorithm prioritizes urgent UEs and then relayed UEs. HYGIENE brings significant improvement for the delay-sensitive RT services, such as the Voice over IP (VoIP). More details about HYGIENE and related performance can be found in Appendix A.1.2.

The analysis provided in this section concerning relays show that a significant gain can be accomplished by using proper techniques. However, a detailed and careful analysis of the signaling scheme is necessary to further illustrate the usefulness of relays.

2.3.3   Multiuser Resource Allocation Maximizing the UE QoS

With high envisaged data rates of IMT-Advanced systems, multi-media broadband applications can be offered to UEs. Multi-media applications are characterized by a multitude of data-rate and QoS requirements. On the other hand, owing to the nature of the mobile radio channel, the attainable spectral efficiency for the UE changes over time due to UE mobility. Consequently, providing QoS with stringent rate and delay requirements over a time-varying shared wireless channel is infeasible.

The APP layer outputs encoded applications, for example, a video stream. For the Scalable Video Coding (SVC) extension (ITU-T 2007; Schwarz and Marpe 2007) of the advanced video coding (AVC) standard H.264 the stream may be received with a variable information bit-rate. Other kinds of video streams may be encoded or transcoded (Ahmad et al. 2005) with the desired data rate. In general, any application may be delivered with variable information bit rate, allowing UE perceived quality to be traded with data rate. The high level of flexibility and adaptability offered by emerging system architectures provides an opportunity for dynamic allocation of resources across UEs and applications, so as to increase the network resource usage and to enhance the UE satisfaction. We have to allocate resources to a set of users simultaneously. The algorithm is cross-layer because PHY-layer modulation and coding is adjusted based on the APP-layer performance.

A more realistic measure of UE perceived quality than objective QoS parameters, such as data rate and delay, is identified by the Quality of Experience (QoE) paradigm. The ITU defines the Mean Opinion Score (MOS) (ITU-T 1996) as utility metric to quantify UE experience. In Khan et al. (2006) a framework is established that allows the MOS for multiple applications, such as voice, video streaming and file download, to be formulated mathematically.

Application-Driven Cross Layer Optimization

A wireless OFDMA downlink shared by K UEs is considered here. An application server is transferring multi-media applications via a core network and BS to UEs. There are K applications, associated with K different UEs. Figure 2.5 shows a block diagram of the Cross-Layer Optimization (CLO) framework and illustrates the signal flow of the exchanged control information between optimizer and layers. The exchanged parameters between the CLO unit and the system layers are cast in vectors of dimension K, where the k-th entry is the data rate, share of resources and MOS of UE k, denoted by $\{\mathbf{R}\}_k = R_k$, $\{\boldsymbol{\alpha}\}_k = \alpha_k$, and $\{\mathbf{MOS}\}_k = MOS_k$, respectively.

A major challenge in cross-layer design is the abstraction of parameters exchanged as control information (Saul and Auer 2009). In order to limit the amount of control information a layer model is introduced at the optimizer that emulates the relevant characteristics of the layer. The parameters of the layer model are determined at the corresponding layer and only these parameters are sent as control information to the optimizer. The optimizer then tunes the model so as to identify the operating modes that maximize the chosen utility, which are then fed back to the system layers.

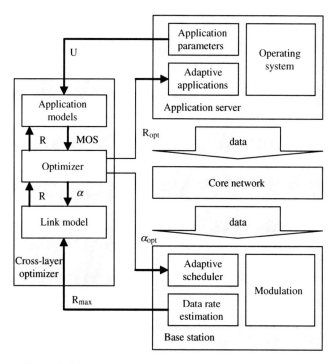

Figure 2.5 Block diagram of the considered CLO framework

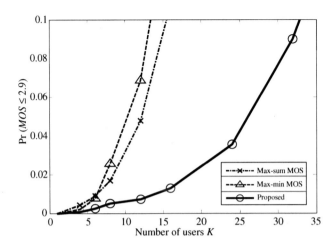

Figure 2.6 Probability that a minimum data rate of 1 Mbps, corresponding to a perceived quality of MOS = 2.9, is not achieved (Saul 2008). Reproduced by permission of © 2008 IEEE

The PHY, MAC, APP models as well the CLO optimization algorithm are described in detail in Appendix A.2.

2.3.4 Optimization Problems and Performance

The share of resources for each UE, α_k, and the associated application data rates, R_k, are determined by solving an optimization problem between link and APP layer, as illustrated in Figure 2.5. One commonly used utility function is to maximize the sum of the MOS of all UEs, thereby maximizing the average perceived service quality. Alternatively, the max-min approach distributes resources so that all UEs experience the same utility degradation.

Provision of max-min fairness in terms of UE perceived quality implies that one single UE who cannot achieve a good perceived quality forces all other UEs to share this poor experience. From an operator's point of view it might be desirable to provide premium services with a higher quality only to some UEs, which contradicts the idea of 'equal loss'. To mitigate these shortcomings, a resource allocation scheme is presented in (Saul 2008). The scheme ensures that the MOS of a premium service is superior to ordinary service quality.

Figure 2.6 plots the outage probability that MOS = 2.9 (corresponding to 1 Mbps data rate) cannot be achieved against the system loads, that is, number of UEs, K. For outage probabilities of 1% the combination of max-min MOS with admission control can serve more than twice the number of users with the guaranteed service quality than the conventional max-min MOS and the max-sum MOS techniques. The simulation assumptions as well as additional results regarding the MOS can be found in Appendix A.2.4.

2.4 Interference Coordination in Mobile Networks

One fundamental issue when enhancing the bandwidth efficiency of wireless networks is how to deal with *interference*. Unlike *ad hoc* networks, resource allocation in cellular networks is typically carried out in a centralized manner, so that the BS is in full control of assigning slots to different UEs. By doing so, interference within a cell can be completely avoided. On the other hand, Inter-Cell

Interference (ICI) is inevitable in cellular networks, as the intended signal for one receiver interferes with other simultaneously active receivers that use the same radio resources.

In this section the focus is on decentralized interference management; for centralized approaches the interested reader is referred to the CoMP Chapter 6. Self-organizing RRM aims at supporting future wireless networks that are largely uncoordinated, random and hierarchical in nature, for which ICI constitutes a major limiting factor for system performance (Elayoubi et al. 2008; Necker 2008).

A fundamental challenge in interference management allows the amount of available radio resources, B, to be shared among neighboring cells to accomplish certain goals, which may include maximization of the sum-rate, subject to QoS constraints. Given a set of active receivers $\mathcal{R}$, and treating interference as white Gaussian noise, the sum-rate per unit area A is upper bounded by the Shannon capacity:

$$C_{sys} = \frac{1}{A} \sum_{y \in \mathcal{R}} C_y \quad [\text{bit/s/m}^2] \tag{2.9}$$

$$= \frac{1}{A} \sum_{y \in \mathcal{R}} B_y \log_2\left(1 + \frac{S_y}{I_y + N}\right)$$

where S_y and I_y denote the received signal power of the intended signal and aggregate interference, respectively. Furthermore, N accounts for Additive White Gaussian Noise (AWGN). The sum-rate C_{sys} depends on several factors: the number of active links in the network $|\mathcal{R}|$, the amount of resources B_y allocated to UE $y \in \mathcal{R}$, and the achieved SINR $S_y/(I_y + N)$.

Besides maximizing the sum-rate C_{sys}, the fair distribution of resources to all UEs is equally important. In particular, UEs that are located close to the cell boundary experience relatively low channel gains towards their desired BS, *and* are exposed to high ICI. To this end, allowing user y to access all resources $B_y = B$, may result in destructive interference to users located in adjacent cells, expressed by a diminishing SINR. Since cell-centre users experience relatively high SINR, they contribute overproportionally to the sum-rate; therefore maximizing the sum-rate and guaranteeing an equal share of user rate are contradicting goals.

In the following sections interference management schemes are discussed that are based on the power control, resource partitioning, and adaptive antenna arrays. These approaches may be combined in an almost arbitrary manner to optimize the tradeoff between maximizing the sum rate, while maintaining a certain UE rate.

2.4.1 Power Control

Power control is essential for power constraint wireless nodes, such as wireless *ad hoc* and sensor networks (Bambos 1998), as well as on the uplink of cellular networks (Himayat et al. 2010). By reducing the transmit power the interference to competing links decreases, so that more links can coexist with acceptable SINR levels. On the other hand, when maximizing the sum rate of a wireless network is the primarily concern, nodes should transmit with maximum power (Behzad and Rubin 2006). In fact, for a two–link wireless network, the sum rate is maximized by either one of three cases (Sinanović et al. 2008): either both links transmit with maximum power or one link transmits with maximum power while the other link is silent. The extension to an N-link network is a binary selection between maximum and zero transmit power. While this approach does *not* maximize the sum rate for an N-link network, the problem of resource allocation is grossly simplified while the sum-rate degradation is within acceptable limits (Gjendemsj et al. 2008).

When maximizing the sum rate is not the primary concern, power control is an effective way to enhance fairness by letting UEs with good channel conditions transmit with less power. Then, cell-edge UEs may observe less interference, which results in an SINR boost. In any case, power control alone is unable to guarantee a certain minimum rate to all UEs (Bambos 1998). To understand this, consider two UEs, y and z, with relatively poor SINR, competing for resources. While reducing the transmit power for UE y boosts the SINR for UE z, its own SINR $S_y/(I_y+N)$ is further reduced because the received signal power S_y decreases accordingly. Power control therefore needs to be complemented with resource partitioning.

2.4.2 Resource Partitioning

Resource partitioning is a common means for interference management, where the available bandwidth B is subdivided into S sub-bands or slots. A simple resource-partitioning scheme is Soft Frequency Reuse (SFR), which is implemented in today's cellular networks, such as 3GPP LTE (Himayat et al. 2010). For SFR each BS may transmit with maximum power in one subband, while it must transmit with reduced power on the remaining $S-1$ subbands, as illustrated in Figure 2.7. Typical values for SFR are $S = 3$ and a transmit power reduction of 6 dB.

Another conventional resource partitioning is FFR, which typically involves a sub-band commonly used by all cells (i.e. with a frequency reuse of 1) that is exempt from any transmit power restriction, while the allocation of the remaining sub-bands is coordinated among the neighboring cells, so to create, for each cell, one sub-band with a lower intercell interference level (Halpern 1983). As illustrated in Figure 2.7, FFR means that UEs are not allowed to transmit on the cell-edge sub-band assigned to adjacent cells. Studies suggest that the effective spatial reuse of FFR is about 50% (Sternad et al. 2003). Compared to SFR, FFR achieves a higher cell-edge UE throughput, while the bandwidth utilization as well as the sum rate are somewhat lower.

Both interference mitigation by SFR and FFR have severe limitations. Since the resource partitioning is static, SFR and FFR are unable to adapt to changes in the traffic load or the distribution of UEs within the cellular network. To this end, at times the majority of UEs are close to the center of the cell, so that the reservation of exclusive frequency bands mean a waste of resources. On the other hand, when UEs are mainly located near the cell border, cell-edge sub-bands tend to be overcrowded, so that cell-edge UEs partly need to resort to transmitting in unprotected parts of the bands, so that ICI again compromises user throughput.

Given a central controller, inter-cellular resource allocation may be enhanced by invoking graph-coloring algorithms (Necker 2008). In this case, resource partitioning is performed dynamically in

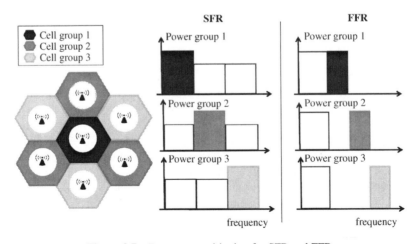

Figure 2.7 Resource partitioning for SFR and FFR

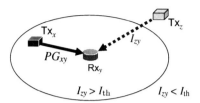

Figure 2.8 The receiver of the intended link between transmitter Tx$_x$ and receiver Rx$_y$ is protected from interference by an exclusion region. Solid and dashed arrows denote intended and interfering links

the way that a pre-defined interference level is not exceeded or that a certain SINR observed at any active receiver is maintained.

Decentralized Interference Avoidance using Busy Bursts

In order to overcome the aforementioned problems of FFR and SFR, a decentralized and dynamic resource partitioning scheme is presented in the following that allows for interference aware resource allocation. This dynamic resource partitioning concept does not impose *a priori* constraints on the number of sub-bands that may be assigned to one BS; rather rules on the allocation of sub-bands are established that ensure that the imposed interference to victim receivers in adjacent cells stays within acceptable limits. Moreover, the number of subbands may be as large as the number of Resource Blocks (RBs).

To protect active receivers from excessive interference, an exclusion range around active receivers is introduced in Hasan and Andrews (2007): a transmitter Tx$_z$ is granted access to a given RB, if the potential interference I_{zy} induced at active receivers $y \in \mathcal{R}$, is below a predefined maximum interference threshold I_{th} (see Figure 2.8):

$$I_{zy} = P_z G_{zy} \leq I_{th} \tag{2.10}$$

where P_z is the transmit power of Tx$_z$ and G_{zy} accounts for the channel gain between Tx$_z$ and Rx$_y$.

One fundamental challenge for distributed resource allocation approaches is the ability to cope with the *hidden node problem*: potential transmitters are unaware about the interference their transmission would cause to active *victim* receivers in adjacent cells.

Receiver feedback is a common means for decentralized interference mitigation: transmitters that sense a strong feedback signal are denied access to a given resource unit. Therefore, receivers are no longer *hidden* from potential interferers. The feedback signal establishes an exclusion region around an active receiver according to Equation (2.10). Applications include medium access control based on RTS/CTS (request to send, clear to send) handshaking (Karn 1990) and busy-tone protocols, where out-of-band busy tones (Haas and Deng 2002; Tobagi and Kleinrock 1975) and in-band busy bursts (Ghimire et al. 2009a; Omiyi et al. 2007; Zhao et al. 2006) are distinguished. In-band receiver feedback utilizing time-multiplexed busy bursts (Ghimire et al. 2009a; Omiyi et al. 2007) possesses the appealing property that channel reciprocity can be exploited to impose an exclusion range: the strength of the observed feedback signal $I_{yz}^{bb} = G_{yz} P^{bb}$ is equivalent to the potential interference this transmitter causes to an active receiver (transmitter of the busy burst) of a competing link. Provided channel reciprocity holds, that is, $G_{zy} = G_{yz}$, the interference $I_{zy} = G_{zy} P_z$ that Tx$_z$ imposes on Rx$_y$, is equivalent to Rx$_y$'s feedback measured at Tx$_z$, so that $I_{yz}^{bb} / P^{bb} = I_{zy} / P_z$. A feedback signal from an active receiver $y \in \mathcal{R}$, is detected as strong if the following condition holds:

$$\max_{y \in \mathcal{R}} I_{yz}^{bb} \cdot \frac{P_z}{P^{bb}} > I_{th} \tag{2.11}$$

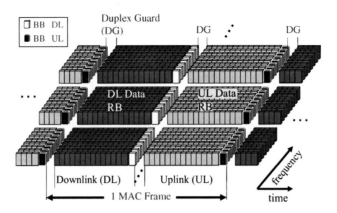

Figure 2.9 Frame structure for OFDMA-TDD with BuB signaling (Ghimire et al. 2009a). Reproduced by permission of © 2009 IEEE

Hence, an exclusion range around active receiver Rx_y as defined in Equation (2.10), is established by denying potential transmitters of competing links $z \neq x$ access to a given RB if Equation (2.11) is met.

This effectively enables receivers to adjust the size of the exclusion region by tuning the maximum tolerable interference level I_{th}. By tuning the size of the exclusion region around active receivers, the tradeoff between maximizing the sum rate and maintaining fairness can be optimized. Besides interference mitigation, receiver feedback is also mandatory when maximizing spectral efficiency by means of link adaptation.

The Busy Burst (BuB) protocol may be embedded within a time-frequency slotted OFDMA-TDD air interface with successive Downlink (DL) and Uplink (UL) subframes, as illustrated in Figure 2.9. The available bandwidth B is divided into N_{RB} mutually orthogonal resource blocks; associated to each RB is a BuB slot that reservation indicator for that RB.

The busy burst protocol applied to OFDMA-TDD is summarized as follows (Ghimire et al. 2009a):

- Given successful reception of a RB and that transmitter Tx_x has more data to transmit, receiver Rx_y emits a busy burst in a time-multiplexed BuB slot.
- If transmitter Tx_z, $z \neq x$ senses a strong busy burst according to Equation (2.11), the RB is occupied, so that Tx_z needs to reschedule its transmission to another RB.

Initial Access in Contention

The principle of imposing an exclusion range around victim receivers, is closely related to reservation protocols designed for random access channels, commonly used with Local Area Networks (LANs), such as the wireless LAN protocol IEEE 802.11 (Karn 1990). When several UEs in a random access channel simultaneously attempt to access a given time-frequency slot, collisions due to ICI are encountered. This is known as the *contention* problem. Reservation protocols, such as reservation ALOHA (R-ALOHA) (Crowther et al. 1973) and Packet Reservation Multiple Access (PRMA) (Goodman et al. 1989), divide the available resources to idle and reserved slots. For R-ALOHA idle slots are allocated in contention and reserved slots are protected from ICI as follows (Crowther et al. 1973):

Contention Before transmission the channel is scanned. In case of outage the packet is retransmitted with probability q in a subsequent unreserved slot. This is repeated until the receiver acknowledges successful reception. In case of collision the packet is retransmitted in subsequent idle slots until the receiver acknowledges successful reception.

Reservation Upon successful reception the receiver broadcasts an acknowledgment. This acknowledgment reserves the slot, so that all other UEs refrain from using that slot in future transmissions. The UE with the reservation thus has uncontested use of the slot. R-ALOHA therefore limits the occurrence of collisions to the contention phase.

In wireless networks, slot reservation translates to an exclusion region around an active receiver (Hasan and Andrews 2007). A competing communication link is denied access to a reserved slot if its transmitter is located within the exclusion region; otherwise the slot may be accessed concurrently by both links. An efficient realization of R-ALOHA to OFDMA-TDD systems is provided by the busy signal concept (Ghimire et al. 2009a), discussed in section 2.4.2.

A well known means to reduce the probability of simultaneous access of unreserved chunks in contention is to impose a p–persistent chunk-allocation policy, where idle slots may be accessed with probability $p_{acc} \in (0, 1]$ (Kleinrock and Tobagi 1975).

Contention-Free Intercellular Slot Reservation

A distributed reservation protocol tailored for cellular wireless networks is presented that facilitates contention-free intercellular slot allocation and reservation. While a busy burst-enabled reservation protocol ensures uncontested use of reserved slots, collisions in contention are encountered, caused by simultaneously accessed idle slots from adjacent cells. The Cellular Slot Allocation and Reservation (CESAR) protocol (Auer et al. 2009) combines dynamic slot reservation with intercell coordination by resource partitioning. CESAR grants access to unreserved slots based on two conditions: the slot is sensed idle, that is, the interference caused to previously reserved slots in neighboring cells is sufficiently low, and a predefined resource-partitioning pattern issues adequate access to a given cell. While resource partitioning ensures that at most one cell may access an idle slot at a time, cyclically shifting the partitioning pattern allows each cell successively to contend for all slots. Hence, CESAR imposes no restrictions on the amount of resources one cell may allocate, and therefore overcomes the limitations of classical intercell resource partitioning based on static frequency reuse planning. CESAR and a busy burst-enabled reservation protocol (Ghimire et al. 2009a) perfectly complement each other; the former mitigates collisions due to simultaneous access of idle slots, while the latter controls the spatial reuse of reserved slots. CESAR therefore facilitates interference aware dynamic slot allocation in cellular networks.

For intercell coordination by resource partitioning, cells are organized into R pre-defined cell groups, such that adjacent cells are in different cell groups G, $1 \leq G \leq R$, similar to conventional frequency reuse, illustrated on the left-hand side of Figure 2.7. The essential principle of CESAR is to successively access unreserved slots by virtue of resource partitioning patterns, provided the interference induced to already reserved slots in adjacent cells is sufficiently low. Associated with cell group G is one of R resource partitioning patterns, which are constructed according to the following rules:

1. All R patterns are mutually orthogonal.
2. All patterns point to each slot once every R frames.

The first rule avoids collisions due to simultaneously accessed slots, provided that cells within the same group G experience low interference. The second rule ensures that all N_s slots may be assigned

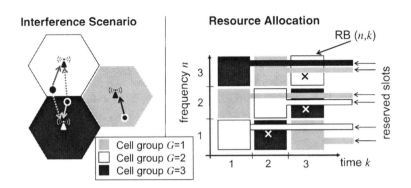

Figure 2.10 CESAR principle: RBs are accessed by virtue of $R = 3$ cyclically shifted resource partitioning patterns. The transmitter in cell 1 is separated from receivers in cells 2 and 3. Cell 1 may access all RBs within $R = 3$ frames. Nodes in cells 2 and 3 mutually interfere, hence resources reserved by cells 2 may not be accessed by cell 3 (marked by ×), and vice versa (Auer et al. 2009). Reproduced by permission of © 2009 IEEE

to all R cell groups within R frames. Considering OFDMA, a slot resembles a RB at frequency index n of subframe k. These rules are satisfied by the cyclically shifted pattern:

$$g_{n,k} = (n + k) \mod R, \quad 1 \leq n \leq N_s \tag{2.12}$$

which associates RB n of frame k to cell group $G = g_{n,k}$.

CESAR accomplishes two objectives: UE z retains all previously reserved RBs, identified by the reservation indicator $\varrho_{n,k}^z = 1$; in addition, UE z is granted contention-free access to idle RBs that satisfies Equation (2.12). UE z may access RB (n, k) if the following condition is met

$$\left(g_{n,k} = G \quad \text{and} \quad \frac{P_z}{P^{bb}} \cdot \sum_{y \in \mathcal{R}} I_{yz}^{bb} > I_{th} \right) \quad \text{or} \quad \varrho_{n,k}^z = 1 \tag{2.13}$$

The working mechanism of CESAR is explained in Figure 2.10. An unreserved RB (n, k) for UE z is sensed to be idle, if all active out-of-cell receivers are outside the hearable region of UE z, which resembles the threshold test according to Equation (2.11). Otherwise RB (n, k) is sensed to be busy. Then UE z is denied access for RB (n, k), regardless the outcome of Equation (2.12). Unlike conventional resource partitioning in cellular networks based on static frequency planning, such as SFR or FFR, the partitioning pattern exclusively controls the contention free allocation of idle RBs – reserved RBs are governed by a distributed slot reservation protocol.

CESAR is compared with a p-persistent variant of the busy signal concept, referred to as p-persistent RB allocation and reservation (p-PSAR). While CESAR controls access of idle RBs by the resource partitioning pattern, p-PSAR transmits on idle RBs with access probability $p_{acc} \in (0, 1]$. All other assumptions for CESAR and p-PSAR are identical, to allow for a fair comparison.

The probability of outage over time in the uplink is plotted in Figure 2.11. Outage occurs if the achieved SINR of RB (n, k) is below the target, which is set to 10 dB. Initially at $k = 1$ all N_s RBs are idle, and cells attempt to access RBs dependent on the chosen resource-allocation policy. While CESAR exhibits diminishing outage, p-PSAR initially suffers from a significant collision probability, especially when p is high, due to the random allocation of RBs that are sensed idle. The residual outage for CESAR is due to interference from distant cells that belong to the same cell group G.

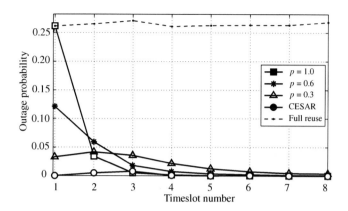

Figure 2.11 Probability of outage over time k for RBs that fail to achieve their SINR target of 10 dB (Auer et al. 2009). Reproduced by permission of © 2009 IEEE

2.4.3 MIMO Busy Burst for Interference Avoidance

By extending the BuB protocol to MIMO systems, a decentralized coordinated beamforming concept is established in Ghimire et al. (2009b). Unlike other CoMP approaches, exchange of control information between cells is completely avoided. For omnidirectional transmissions the observed BuB power from entities in neighboring cells. I_{yz}^{bb} in Equation (2.11), is solely determined by the channel gain. For beamforming approaches I_{yz}^{bb} is typically dependent on the selected spatial precoder $\mathbf{v}_i$ of spatial stream i and the spatial processing at the receiver end $\mathbf{u}_i$. This implies that the exclusion range around an active receiver depends on the selected spatial precoder of the interfering source $\mathbf{v}_i$. Hence the BuB concept is extended to a multicell MIMO system by ensuring that the effective channel (the channel including spatial processing at transmitter and receiver) is reciprocal. This is accomplished by (Ghimire et al. 2009b):

- a transmitter in an adjacent cell must scan the BuB slot using the same antenna weights $\mathbf{v}_i$, which it would apply in the subsequent data slot for transmitting data,
- a victim receiver emits the busy burst using the spatial precoder $\mathbf{u}_i$ that is used for spatial receive processing of the data signal.

Provided that the potential interferer Tx_z intends to use spatial precoder $\mathbf{v}_i$, the received BuB power emitted from victim receiver Rx_y amounts to:

$$I_{yz}^{bb,(i)} = P^{bb} \cdot \left| \mathbf{v}_i^T \mathbf{H}_{zy}^T \mathbf{u}_i \right|^2 = P^{bb} G_{zy}^{(i)} \tag{2.14}$$

where $G_{zy}^{(i)}$ and $\mathbf{H}_{zy}$ represent the effective channel gain and the unweighted MIMO channel matrix between Tx_z and Rx_y. The threshold test is then applied to the BuB power received as in Equation (2.11), in order to determine whether Tx_z may commence transmission using precoder $\mathbf{v}_i$.

As a BS typically serves multiple UEs, there are several spatial precoders $\mathbf{v}_i$ available at the BS to choose from. The *interference awareness* property of the BuB protocol enables the BS to determine the most appropriate UE whose associated spatial precoder $\mathbf{v}_i$ causes the least ICI to the active link. If all available spatial precoders fail to maintain a certain ICI, then the chunk remains idle. In this way, interference management by resource partitioning is combined with multiuser MIMO-OFDM.

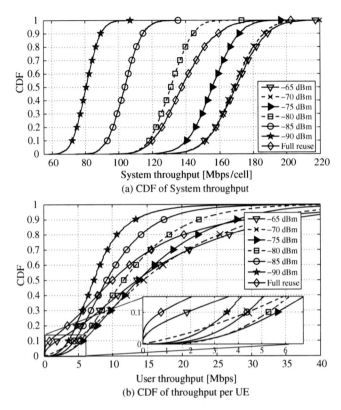

Figure 2.12 CDF of the system throughput and UE throughput achieved with BB-enabled beam selection against conventional beam selection with full-frequency reuse

Results

The performance of the proposed BuB-enabled beam-selection algorithm is compared against the state-of-the-art approach, where beams are selected based on the channel gain of the served UEs, that is, the interference caused to adjacent cells is ignored for beam selection. The performance metrics of interest are user throughput and sum rate. The simulation assumptions can be found in Appendix A.3.

Figure 2.12 shows the performance of BuB-enabled beam selection. Results for conventional beam selection are included for comparison. It can be observed that BuB-enabled beam selection outperforms the conventional beam selection approach, both in terms of system throughput and user throughput. Furthermore, Figure 2.12 reveals some important properties: using a low threshold (e.g. $I_{th} = -90$ dBm), resources are spatially reused in a given sector only if no UE in an adjacent cell would suffer. Consequently, at this interference threshold, the system exhibits a high level of fairness, demonstrated by the steepness of the user throughput curve in Figure 2.12(b). However, the system becomes overcautious and restricts the spatial reuse of RBs, which results in a compromised system throughput in Figure 2.12(a). On increasing the threshold I_{th}, both user and system throughput increase relative to $I_{th} = -90$ dBm, due to an increase in spatial reuse; until at $I_{th} = -75$ dBm the cell-edge users throughput, measured as the lower 10%-ile of the user throughput, attains its maximum. At this threshold, BuB-enabled beam selection achieves a median system throughput of 155.7 Mbps/cell together with 5.68 Mbps at the cell edge.

Compared to conventional beam selection, this is a 13% increase in median system throughput together with a 7.3-fold increase in the lower 10%-ile of user throughput. By increasing the

Table 2.1 Spatial reuse and achieved bit/symbol per served RB. The spatial reuse is defined as the ratio between the number of served RBs divided by the total number of RBs

Interference threshold I_{th}	Spatial reuse [%]	Bit/symbol per served RB
−90 dBm	15	6.38
−85 dBm	21	5.95
−80 dBm	30	5.17
−75 dBm	47	4.09
−70 dBm	67	3.02
−65 dBm	84	2.43
Full Reuse	100	2.41

threshold further to $I_{th} = -70$, the users closer to the cell center benefit from an increased spatial reuse, whereas cell-edge users increasingly suffer from ICI. At this point, median system throughput of 168.6 Mbps/cell and a cell-edge user throughput of 4.39 Mbps is achieved, which is a 25% increase in median system throughput and a 5.8-fold increase in cell-edge user throughput compared to conventional beam selection.

How the chosen interference threshold I_{th} affects the spatial reuse (i.e. the ratio between the number of served RBs divided by the total number of RBs) and the achieved spectral efficiency in bit/symbol of served RBs is examined in Table 2.1. As I_{th} increases the spatial reuse increases, but the interference protection diminishes. Reduced interference protection is exemplified by the decreasing spectral efficiency per served RB, as the interference threshold I_{th} increases.

2.5   Efficient MBMS Transmission

Traditionally, cellular systems used to focus on unicast services. However, nowadays multicast and broadcast transmission is one of the most promising enhancements in the downlink of next-generation wireless systems. This way, all IMT-Advanced technologies comprise in their portfolio the improvement of the technologies related to Multimedia Broadcast Multicast Service (MBMS). In general, three emission modes can be considered: broadcast, multicast and unicast. The broadcast mode consists of a unidirectional point-to-multi-point (p-t-m) transmission of MBMS data from a single source to all users in an MBMS area. The multicast mode allows for p-t-m transmission of the same service in the same area from the same source but to a multicast group. Finally, the unicast mode facilitates bidirectional point-to-point (p-t-p) transmission to one dedicated user. Multicast and broadcast are more appropriate transport technologies to cope with high numbers of users simultaneously consuming the same service. However, they cannot support a high number of personalized services and the transmission has to be designed for the worst-case user.

In MBMS, multimedia content is delivered either as a streaming service or as a file download service to the end user. In streaming services, a continuous data flow of audio and/or video is delivered to the end user's handset. In the specific case of file downloading, a finite amount of data is delivered and stored into the UEs as a file. Applications that fall within this category are: video clips, digital newspapers, software download, etc. Common to all these services is the requirement of an error-free transmission of the files, as even a single bit error can corrupt the whole file and make it useless for the receiver. As it cannot be guaranteed that each and every user will be able to recover the file after the MBMS p-t-m transmission, given that some users might have experienced reception conditions that were too bad for this, a post-delivery repair phase can be performed to complete the file download. The repair phase employs by default p-t-p transmissions, although it is also possible to employ a p-t-m transmission in case too many users fail to receive the file. **This combination of p-t-m for the first attempt and p-t-p for the repair phase is called hybrid delivery** and is potentially the most efficient configuration, as in a realistic scenario there will always be some users that experience significantly worse reception conditions than the majority and it may be more efficient to serve them through p-t-p connections.

This section explores the benefits of MBMS features of IMT-Advanced technologies. Temporal or frequency multiplexing of unicast and MBMS services is used to allow an eventual delivery of the same file to a large number of users. Given this assumption, this section specifically assesses the joint performance of unicast and MBMS services in LTE.

2.5.1   MBMS Transmission

In order to improve the efficiency of MBMS transmission, IMT-Advanced technologies include new features inherited from other successful broadcast networks, such as terrestrial digital television (TV). The main new feature included in the standard is a higher flexibility in the management and deployment of Single Frequency Networks (SFNs), referred to as MBSFN. In MBSFN a set of synchronized BSs, that constitute the SFN area, transmit in the same resource block, that is, on the same subcarriers and time slots. This combined transmission significantly improves the received SINR, as compared with the non-SFN operation. As the SFN area can be defined by just one cell, this option is preferred for the file delivery services.

The MBSFN operation entails the following benefits:

- An increase in the received signal level, especially at the cell border within a MBSFN cluster.
- A reduction in the interference level, again, especially at the cell border within a MBSFN cluster, since the signals received from neighboring cells do not appear as interference but as constructive signals.
- An additional diversity gain against signal fading since data is received from different paths.

However, in order to be constructively combined, signals from the MBSFN cluster of cells must reach the users within the Cyclic Prefix (CP) and before the start of the next OFDM symbol. Otherwise, the signal suffers from intersymbol interference. In order to allow big cluster sizes, the CP is usually extended for the MBSFN operation.

Apart from the extended CP, the pilot signal pattern is also modified in the MBSFN operation. The pilots are closer in the frequency domain, which reduces system capacity. However, the increased frequency selectivity of the channel response, caused by the large deviation in propagation delays between a mobile user and the serving sites, essentially requires more pilots in the frequency domain.

Concerning the use of carriers, MBMS services can be provided on a specific radio frequency carrier dedicated to MBMS or on a carrier shared with other non-MBMS services, that is, unicast services. The former case, with dedicated carriers, is only for downlink transmissions without unicast channels, whereas the latter case is known as mixed carriers. In both cases, SFN operation mode was feasible for MBMS transmissions.

On a mixed carrier, the way of multiplexing MBMS and unicast services is different in LTE and IEEE 802.16m. In LTE both services are multiplexed in time, allocating different subframes to each mode. The subframes assigned to MBMS within a radio frame are defined with a bitmap. Among the ten subframes included in a frame, only six subframes can be allocated to MBMS: #1, #2, #3, #6, #7 and #8. In IEEE 802.16m MBMS services are allocated together with unicast services in

Table 2.2 System level simulation parameters

Parameter	Value
Intersite distance	500 m
Cell layout	IMT-Advanced, Urban macrocellular
User speed	30 km/h
Frequency reuse	Universal (1)
Scheduling	Proportional Fair
MBMS traffic	File delivery: 2 MBytes
Unicast traffic	Full Buffer
Handover margin	1 dB
Transmitted power	46 dBm
Antenna tilt	12°
Interference modeling	Explicit
Noise spectral density	−174 dBm/Hz

the same time slots, just occupying different frequency sub-bands. Again, a zone-allocation bitmap identifies the use of resources. Except for the control information, the remaining spectrum may be fully employed by MBMS services.

2.5.2 Performance Assessment

This section presents some results related to the LTE standard, although conclusions are also valid for IEEE 802.16m given the significant similarities between both technologies. In Release 9 (Rel-9) of the LTE standard, only two alternatives for MBMS inclusion are considered. The former allows Time Division Multiplexing (TDM) of unicast transmission and multicast transmission, whereas the latter consist in separating both transmission modes in different carriers, that is, applying Frequency Division Multiplexing (FDM). As explained before, in case of mixed carriers, the system can choose among reserving from one to six subframes at most for the MBMS transmission. More details on service operation can be found in Appendix A.4.1.

The simulations parameters and assumptions follow ITU-R guidelines (ITU-R 2009c) and 3GPP specifications as summarized in Table 2.2.

The macrocellular scenario was chosen because this is the most likely deployment for MBMS provision. 57 sectorized cells transmitting in the same band provide service to a set of uniformly distributed users. The serving cell is, by default, the closest one, unless its received signal power is exceeded by more than 1 dB by any other cell. The antenna pattern is modeled according to (ITU-R 2009c), taking into account azimuth, elevation and tilt.

In the case of TDM, a mixed carrier of 10 MHz is considered with unicast and MBMS users jointly allocated. As shown in the left part of Figure 2.13, the higher the number of subframes allocated to MBMS users, the lower the average quality perceived by unicast counterparts. On the other hand, MBMS users increase their QoS with a higher number of sub-frames. This additional QoS can be measured with the time required for file delivery. In this case, three sizes for that file have been considered: 2, 4 and 8 MB. Figure 2.13 represents, on the right, the time required to download the file versus the number of subframes allocated to MBMS users. From Figure 2.13, and given a certain number of unicast users and a desired QoS for these users, the maximum number of subframes to be allocated to MBMS can easily be derived. In case of flexibility, Figure 2.13 demonstrates that most benefit for MBMS occurs for three subframes, whereas for a higher number of subframes the benefit is not so high. Further when FDM is applied (assuming the same number of unicast and multicast users), the performance degrades compared to using a unicast system (see Figure A.6 in

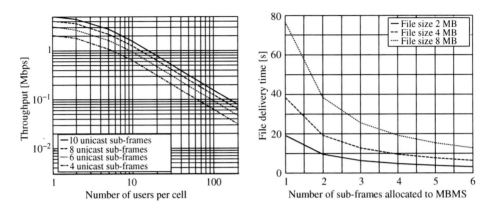

Figure 2.13   Unicast (left) and MBMS (right) performance with TDM

Appendix A). Therefore, TDM is always preferred for unicast and MBMS users given its higher flexibility in the resource utilization

Finally the transmission of the same file to static MBMS users is studied. In this situation hybrid transmission is especially interesting because some users could hardly be reached by MBMS services. Indeed, the most robust MCS is the only one that is able to deliver the file to all users in the cell in this particular case. Thus, it is necessary to calculate the time required to serve 95% of users in the next three operation modes:

1. Transmit the file to MBMS users using p-t-p connections.
2. Transmit the file using only one MBMS transmission with the most robust MCS.
3. Transmit the file using the hybrid delivery.

In this case, simulation considers a set of 30 unicast users plus a variable number of MBMS static users. Figure 2.14 shows the service time required for the three operation modes, when one

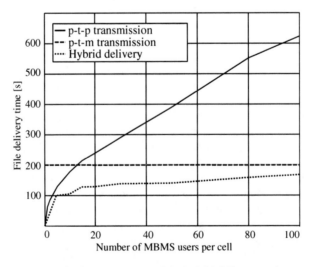

Figure 2.14   Assessment of the hybrid delivery mode

subframe is allocated to the MBMS transmission. It is observed that with few MBMS users in the cell, p-t-p transmission is the best option. However, starting from a certain number of users (13 in this particular case), it is more efficient to use MBMS to serve the file. The figure shows that the best option to minimize the service time is the hybrid delivery, obtaining time gain for every number of MBMS users. The gain increases until the crossing point of the two reference curves using unicast and MBMS transmission separately. This is the point with maximum gain, yielding 40.3% of time reduction. After this point, the gain decreases with the number of users per cell. The mean gain achieved with the hybrid delivery is 27.4%.

2.6 Future Directions of RRM Techniques

Future wireless systems will need to provide highly flexible and reliable data-rate services to guarantee QoS to the large number of end users. The principal techniques to be included in development of future radio systems include several aspects of Advanced RRM, which have a direct impact on each user's performance, as well as the overall performance of the network. In the development process of IMT-Advanced systems, several concepts have been proposed to produce a highly flexible and efficient high-speed communications system. These concepts include MIMO communication, CoMP transmission, support for contiguous or noncontiguous spectrum aggregation, efficient MBMS and dynamic multidimensional resource allocation. These techniques allow high system spectral efficiency and robustness to interference to be achieved, simultaneously providing the negotiated QoS to the users.

Regarding dynamic resource allocation, we showed that an efficient and flexible scheduling and spectrum allocation process improves the spectral efficiency. Moreover, the QoS support allows heterogeneous services to be provided in the network, such as VoIP and streaming video. Several techniques were proposed showing substantial gains compared to baseline schemes. These techniques include a scheduling approach applicable to a mixed service classes scenario and a framework for cross-layer design. In particular, a powerful scheduler was presented that can flexibly support a mix of realtime and nonrealtime traffic and a QoS-aware scheduler based on a utility function that reflects the user-perceived quality measured by the MOS. All the above approaches for QoS control inherently assume that QoS demands are known to the scheduler.

Concerning interference coordination, this chapter focused on the use of the Busy Burst (BuB) mechanism. Decentralized interference avoidance using BuBs yields a system performance gain, which corresponds to around 17% increase in median system throughput and over sevenfold increase in cell-edge user throughput. This comes at the cost of introducing a low latency feedback channel and intercellular synchronization. Moreover, the BuB concept is tailored for the TDD mode. On the other hand, as a distributed scheme, it does not require any extension in backhaul infrastructure or changes to the current systems architecture, which is rather remarkable as compared with other CoMP techniques.

This chapter has offered a clear insight into the main features of the MBMS transmission mode. IMT-Advanced systems specifications allow optimum transmission of multicast and broadcast information, defining a set of possible network configurations to adjust the system to either streaming or file-delivery services. For file-delivery services, some results were presented regarding the distribution of resources among unicast and multicast users. Once unicast and multicast users are put together, the inclusion of MBMS in the same carrier affects the performance of unicast users. As expected, this degradation is proportional to the number of subframes allocated to MBMS. Actually, the reduction in the mean user throughput matches exactly the percentage of stolen subframes, which simplifies the decision making. A complementary analysis for file delivery using hybrid configurations was also offered. Unicast transmission could be employed for error repair of the MBMS transmission. Hybrid delivery has proven to be the most efficient configuration, although this will further damage unicast performance.

This chapter does not represent the end point for Advanced RRM related research. In fact, this chapter only covers a subset of topics whereas others chapters deal with other interrelated issues like selection of users in a multiuser scenario (see Chapter 5), coordination among base stations (see Chapter 6) and peer-to-peer communications (see Chapter 9). Moreover, the frequency-time-space-transmitter selection is a quite a difficult optimization problem that has only been tackled tangentially in this chapter. There are many QoS management procedures to be designed, concerning the interworking of layers and real-time flow management.

References

Ahmad I, Wei X, Sun Y and Zhang YQ 2005 Video transcoding: An overview of various techniques and research issues. *IEEE Trans. Multimedia* **7**(5), 793–804.

Andrews M *et al.* 2001 Providing quality of service over a shared wireless link. *IEEE Commun. Mag.* **39**(2), 150–154.

Auer G, Videv S, Ghimire B and Haas H 2009 Contention free inter-cellular slot reservation. *IEEE Commun. Lett.* **13**(5), 318–320.

Bambos N 1998 Toward power-sensitive network architectures in wireless communications: concepts, issues, and design aspects. *IEEE Personal Commun. Mag.* **5**(3), 50–59.

Behzad A and Rubin I 2006 High transmission power increases the capacity of ad hoc wireless networks. *IEEE Trans. Wireless Commun.* **5**(1), 156–165.

Bonald T 2004 A score-based opportunistic scheduler for fading radio channels *Proc. European Wireless Conference 2004*.

Chang C, Yen C, Ren F and Chuang C 2008 QoS GTE: A centralized QoS guaranteed throughput enhancement scheduling scheme for relay-assisted WiMAX Networks *Proc. ICC 2008 – IEEE Int. Conf. Commun.*, pp. 3863–3867.

Cheng RS and Verdu S 1993 Gaussian multiaccess channels with ISI: Capacity region and multiuser water-filling. *IEEE Trans. Inform. Theory* **39**(3), 773–785.

Crowther W, Rettberg R, Walden D, Ornstein S and Heart F 1973 A system for broadcast communications: reservation – ALOHA *Proc. 6th Hawaii Int. Conf. Sys. Sci.*, pp. 371–374.

Elayoubi SE, Haddada OB and Fourestie B 2008 Performance evaluation of frequency planning schemes in OFDMA-based networks. *IEEE Trans. Wireless Commun.* **7**, 1623–1633.

Ekman T, Gesbert D and Bang H 2008 Channel predictive proportional fair scheduling. *IEEE J. WCOM* **7**(2), 482–487.

Ghimire B, Auer G and Haas H 2009a Busy Bursts for Trading-off Throughput and Fairness in Cellular OFDMA-TDD *EURASIP J. Wireless Commun. Networking*, pp. 1–14.

Ghimire B, Auer G and Haas H 2009b OFDMA-TDD Networks with Busy Burst Enabled Grid-of-Beam Selection *Proc. ICC 2009 – IEEE Int. Conf. Commun.* pp. 1–6.

Gjendemsj A, Gesbert D, Oien GE and Kiani SG 2008 Binary power control for sum rate maximization over multiple interfering links. *IEEE Trans. Wireless Commun.* **7**(8), 3164–3173.

Goodman DJ, Valenzuela RA, Gayliard KT and Ramamurthi B 1989 Packet reservation multiple access for local Wireless Communications. *IEEE Trans. Commun.* **37**(8), 885–890.

Haas ZJ and Deng J 2002 Dual Busy Tone Multiple Access (DBTMA) – a multiple access control scheme for ad hoc networks. *IEEE Trans. Commun.* **50**(6), 975–985.

Halpern S 1983 Reuse partitioning in cellular systems, *Proc. 1983 Vehicular Technology Conference* pp. 322–327.

Hasan A and Andrews JG 2007 The guard zone in wireless ad hoc networks. *IEEE Trans. Wireless Commun.* **6**(3), 897–906.

Himayat N, Talwar S, Rao A and Soni R 2010 Interference management for 4G cellular standards. *IEEE Commun. Mag.* **48**(8), 86–92.

Holma H and Toskala A 2009 *LTE for UMTS: OFDMA and SC-FDMA Based Radio Access* first edn. John Wiley & Sons, Ltd., Chichester.

IEEE 802.16 2010a System Description Document (SDD). Document IEEE 802.16m-09/0034r3, IEEE 802.16 Broadband Wireless Access Working Group, Task Group m, http://www.ieee802.org/16/tgm/core.html.

IEEE 802.16 2010b System Description Document (SDD). Document IEEE 802.16m-09/0034r3, IEEE 802.16 Broadband Wireless Access Working Group, Task Group m.

ITU-R 2008a Requirements Evaluation Criteria and Submission Templates for the development of IMT-Advanced. Report ITU-R M.2133, International Telecommunications Union Radio (ITU-R), http://www.itu.int/publ/R-REP/en.

ITU-R 2008b Requirements related to Technical Performance for IMT-Advanced Radio Interface(s). Report ITU-R M.2134, International Telecommunications Union Radio (ITU-R), http://www.itu.int/publ/R-REP/en.

ITU-R 2009a Acknowledgement of Candidate Submission from 3GPP Proponent (3GPP Organization Partners of ARIB, ATIS, CCSA, ETSI, TTA and TTC) under Step 3 of The IMT-Advanced Process (3GPP Technology). Document IMT-ADV/8, International Telecommunications Union Radiocommunication (ITU-R) Study Groups Working Party 5D, http://www.itu.int/md/R07-IMT.ADV-C-0008/en.

ITU-R 2009b Acknowledgement of Candidate Submission from IEEE under Step 3 of The IMT-Advanced Process (IEEE Technology). Document IMT-ADV/4-E, International Telecommunications Union Radiocommunication (ITU-R) Study Groups Working Party 5D, http://www.itu.int/md/R07-IMT.ADV-C-0004/en.

ITU-R 2009c Guidelines for Evaluation of Radio Interface Technologies for IMT-Advanced. Report ITU-R M.2135-1, International Telecommunications Union Radio (ITU-R), http://www.itu.int/publ/R-REP/en.

ITU-T 1996 *Methods for Subjective Determination of Transmission Quality* International Telecommunications Union Geneva, Switzerland. ITU-T Recommendation P.800.

ITU-T 2007 Advanced Video Coding for Generic Audiovisual Services. Recommendation ITU-T H.264, International Telecommunications Union Telecommunication Standardization Sector (ITU-T).

Karn P 1990 MACA: A New Channel Access Protocol for Packet Radio *Proc. ARRL/CRRL Amateur Radio 9th Computer Networking Conference*, pp. 134–140.

Khan S, Duhovnikov S, Steinbach E, Sgroi M and Kellerer W 2006 Application-driven cross-layer optimization for mobile multimedia communication using a common application layer quality metric *Proc. of the 2006 International Conference on Wireless Communications and Mobile Computing*, pp. 213–218 IWCMC '06.

Kleinrock L and Tobagi FA 1975 Packet switching in radio channels: Part I – carrier sense multiple-access modes and their throughput-delay characteristics. *IEEE Trans. Commun.* **23**(12), 1400–1416.

Lei H *et al.* 2007 A packet scheduling algorithm using utility function for mixed services in the downlink of ofdma systems *Proc. VTC 2007 Fall - IEEE 66th Vehicular Technology Conf.*, pp. 1664–1668.

Miao GW and Himayat N 2008 Low complexity utility based resource allocation for 802.16 ofdma systems *Proc. WCNC 2008 – IEEE Wireless Commun. and Networking Conf.*, pp. 1465–1470.

Necker M 2008 Interference coordination in cellular OFDMA Networks. *IEEE Network* **2**(6), 12–19.

Omiyi P, Haas H and Auer G 2007 Analysis of TDD cellular interference mitigation using busy-bursts. *IEEE Trans. Wireless Commun.* **6**(7), 2721–2731.

Ryu S, Ryu B, Seo H and Shi M 2005 Urgency and efficiency based wireless downlink packet scheduling algorithm in ofdma system *Proc. VTC 2005 Spring – IEEE 61st Vehicular Technology Conf.*

Saul A 2008 Wireless resource allocation with perceived quality fairness *Asilomar Conference on Signals, Systems and Computers*, pp. 1557–1561.

Saul A and Auer G 2009 Multiuser resource allocation maximizing the perceived quality. *EURASIP J. Wireless Commun. Networking*, pp. 1–15.

Schwarz H and Marpe D 2007 Overview of the Scalable Video Coding Extension of the H.264/AVC Standard. *Circuits and Systems for Video Technology, IEEE Transactions on* **17**(9), 1103–1120.

Sinanović S, Serafimovski N, Haas H and Auer G 2008 Maximising the System Spectral Efficiency in a Decentralised 2-link Wireless Network. *EURASIP J. Wireless Commun. Networking*, pp. 1–13.

Song G and Li Y 2005 Utility-based resource allocation and scheduling in ofdm-based wireless broadband networks. *IEEE Commun. Mag.* **43**(12), 127–134.

Song G, Li Y, Cimini L, Jr. and Zheng H 2004 Joint channel-aware and queue-aware data scheduling in multiple shared wireless channels *Proc. WCNC 2004 – IEEE Wireless Commun. and Networking Conf.*, pp. 1939–1944.

Sternad M, Ottosson T, Ahlen A and Svensson A 2003 Attaining both coverage and high spectral efficiency with adaptive OFDM downlinks *Proc. VTC 2003 Fall – IEEE 58th Vehicular Technology Conf.*, pp. 2486–2490.

Tobagi FA and Kleinrock L 1975 Packet switching in radio channels: Part II – The hidden terminal problem in carrier sense multiple-access and the busy-tone solution. *IEEE Trans. Commun.* **23**(12), 1417–1433.

WINNER+ 2009a Celtic Project CP5-026 First Set of Best Innovations in Advanced Radio Resource Management (ed. Saul A.). Public Deliverable D1.1, Wireless World Initiative New Radio – WINNER+, http://projects.celtic-initiative.org/winner+/index.html (accessed June 2011).

WINNER+ 2009b Celtic Project CP5-026 Intermediate Report on System Aspect of Advanced RRM (ed. Monserrat J. and Sroka P.). Public Deliverable D1.5, Wireless World Initiative New Radio – WINNER+, April 2009, http://projects.celtic-initiative.org/winner+/index.html (accessed June 2011).

WINNER+ 2009c Celtic Project CP5-026 Results of Y1 Proposed Candidate Proof-of-concept Evaluation (ed. Safjan K.). Public Deliverable D4.1, Wireless World Initiative New Radio – WINNER+, http://projects.celtic-initiative.org/winner+/index.html (accessed June 2011).

Zhao R, Walke B and Hiertz GR 2006 An efficient IEEE 802.11 ESS mesh network supporting quality-of-service. *IEEE J. Select. Areas Commun.* **24**(11), 2005–2017.

3

Carrier Aggregation

Jose F. Monserrat, Jorge Cabrejas, Xavier Gelabert and Miia Mustonen

3.1 Basic Concepts

International Mobile Telecommunications Advanced (IMT-Advanced) requirements establish a minimum support of 1 Gbps and 100 Mbps peak rates for low-mobility and high-mobility users, respectively. In order to fulfill these challenging requirements, wider channel bandwidth than legacy Third Generation (3G) systems has to be supported. However, the available spectrum resources of operators differ depending on the specific country, being spread out over different frequency bands and with different bandwidths. Therefore, all IMT-Advanced technologies foresee one of their key features as the aggregation of continuous or discontinuous spectrum in order to achieve wider bandwidth and consequently increase transmission capability. This concept is usually known as Carrier Aggregation (CA) or spectrum aggregation.

Discontinuous CA has the advantage of having spectral diversity gain, due to the use of different frequencies, which implies different types of fading channels. However, this requires having several physical layer processing chains – one per each of the aggregated carriers. On the other hand, contiguous CA can save much spectrum because many subcarriers used as guard bands can be employed for data and control information. In the absence of carrier aggregation some frequency chunks became unused whereas with carrier aggregation these guard bands could be removed.

Moreover, contiguous carrier aggregation could use only one Base Band (BB) processing chain (large Fast Fourier Transform (FFT) block), if the transmitter and receiver are suited to it, thus eliminating the need for parallel Radio Frequency (RF) transceivers.

There are some concepts defined in the International Telecommunication Union–Radiocommunication Sector (ITU-R) regarding CA that must be clarified to understand the main research challenges and their constraints:

- Carrier Aggregation: to transmit data on multiple carriers, or sub-bands, contiguous or non-contiguously located by using one or several parallel RF transceivers. The User Equipments (UEs) may adopt a single wideband-capable RF front-end and a single FFT, or multiple legacy RF

Mobile and Wireless Communications for IMT-Advanced and Beyond, First Edition.
Edited by Afif Osseiran, Jose F. Monserrat and Werner Mohr.
© 2011 Afif Osseiran, Jose F. Monserrat and Werner Mohr. Published 2011 by John Wiley & Sons, Ltd.

front-ends (<20 MHz) and FFT engines. The choice between single or multiple transceivers depends on power consumption, cost, size, and flexibility to support other aggregation types. Larger bandwidths than 20 MHz can be supported using carrier aggregation.

- Component Carrier (CC): the independent RF sub-band, or carrier, that is aggregated with other sub-bands to conform a larger bandwidth. Each component carrier maintains its original structure to support single-carrier-capable users, even if it is aggregated to a larger bandwidth.
- Guard Band: the guard subcarriers that cannot be used for transmission. In the case of contiguous carrier aggregation this guard band is located at the edge of the aggregated bandwidth. In case of noncontiguous aggregation, each sub-band will have its own guard bands, thus reducing efficiency.
- Center frequency of the aggregated bandwidth: the center of the total aggregated bandwidth.
- Center frequency of the component carrier.

In the IEEE 802.16m draft standard a further subclassification of component carriers is made. According to the transportation of signaling information, CCs are classified as:

- Primary Carrier: this is the carrier used by the Base Station (BS) to exchange control signals plus additional traffic data with the UE. Full control of UE mobility, state and context is made through the primary carrier. It is worth noting that each UE has only one primary carrier allocated. However, from the point of view of the BS all component carrier may act as primary carriers for some UEs.
- Secondary Carrier: one IMT-Advanced UE might have one or several secondary carriers allocated for traffic exchange. These secondary carriers are controlled by the signaling conveyed in the primary carrier.

Concerning their usage, CCs are further classified into:

- Fully configured carrier: a primary carrier is always fully configured since all control channels are transmitted. A secondary carrier for a UE, in case of being fully configured, may serve as primary carrier for others UEs.
- Partially configured carrier: this CC cannot operate on a stand-alone basis because it only carries a limited set of control channels. These partially configured carriers are mainly configured for downlink-only transmissions, such as multicast/broadcast traffic.

Finally, from the UE's operation perspective, the IEEE 802.16m draft standard also defines two different multicarrier operation modes:

- Multicarrier aggregation: in which the UE receives control information from primary carrier and processes data on the primary or secondary carriers. The UE must be able to receive simultaneously data from all component carriers since the Medium Access Control (MAC) layer scheduler may assign resources to the UE in the primary and multiple secondary carriers. Therefore, channel state reporting mechanisms should inform about all active component carriers.
- Multicarrier switching: in which the UE is connected to just one component carrier at a time switching from the primary to the secondary carrier according to the control information. The UE connects with the secondary carrier during a specified time period and then comes back to the primary carrier to receive further instructions.

Dealing with CA, several research issues need to be analyzed in depth to exploit fully the opportunity offered by bandwidth increase. Some of these research topics are treated in this chapter considering the investigation from several perspectives. Section 3.2 deals with regulatory aspects.

Section 3.3 makes a deep analysis of the state of the art focusing on the results derived from the Wireless World Initiative New Radio + (WINNER+) project. Section 3.4 focuses on the cognitive aspects related with CA and Section 3.5 addresses signalling issues. Finally, section 3.6 discusses about hardware and legal limitations around CA concept.

3.2 ITU-R Requirements and Implementation in Standards

The usage of radio frequencies is administered globally by the ITU-R via the Radio Regulations (RR). The RR is a binding agreement between governments. In ITU-R, the world is divided into three regions and the allowed spectrum use in RR differs among those regions. Region 1 consists of Europe, the Middle East and Africa, Region 2 is North- and South America and Region 3 is Asia Pacific. ITU-R specifies the spectrum use allowed mainly by allocating spectrum bands to different radio services. A service in ITU-R means a certain type of usage, characterized for example by some transmission characteristics or a kind of transmitter. There are approximately 40 ITU-R services, for example, mobile service, fixed service and fixed satellite service. The whole radio spectrum is allocated to different ITU-R services, thus, there is not unused available spectrum. Spectrum allocations can be either primary or secondary allocations. The services with primary allocations have a higher status and priorities than the services with secondary allocations. A band usually has more than one allocation in RR from which national administrations can choose. This gives considerable freedom for national administrations to decide how to use each band. National administrations are also in charge of dividing the portions of these spectrum bands, for example, in the case of mobile communications to different operators (Takagi and Walke 2008).

ITU-R has defined several requirements for IMT-Advanced systems. These requirements are related to technical performance, spectrum and supported services. Regarding the spectrum bands, an IMT-Advanced system has to support a scalable bandwidth up to and including 40 MHz. This may be supported by single or multiple RF carriers. The scalability is defined as an ability to support at least three bandwidth values. As an example, the Third Generation Partnership Project (3GPP) Long Term Evolution (LTE) Release 8 (Rel-8) system is able to support six different bandwidth values up to 20 MHz. In order for LTE Release 10 (Rel-10) (LTE-Advanced (LTE-A)) to meet the IMT-Advanced requirement for system bandwidth, it needs to aggregate at least two CCs to support wider transmission bandwidths effectively. Even if ITU-R has set the requirement for the supported transmission bandwidth to 40 MHz, the IMT-Advanced systems are encouraged to support operation in wider bandwidths – even up to 100 MHz – to enable systems to reach the research target peak data rates: 100 Mbps for high mobility and 1 Gbps for low mobility (ITU-R 2008b).

Another spectrum related requirement is that IMT-Advanced systems need to be able to utilize at least one band identified for IMT (ITU-R 2008a). In World Radio Conferences (WRCs) (World Administrative Radio Conference (WARC) in 1992, WRC in 2000 and in 2007), the following spectrum identifications for IMT or IMT-2000 have been made:

- 450–470 MHz
- 698–960 MHz
- 1710–2025 MHz
- 2110–2200 MHz
- 2300–2400 MHz
- 2500–2690 MHz
- 3400–3600 MHz

The list is simplified and the actual allocations vary considerably on different ITU-R regions. Actually, the widest contiguous spectrum allocated for IMT is 200 MHz on the 3.4-3.6 GHz band,

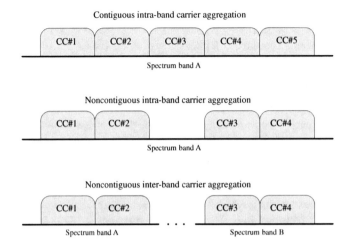

Figure 3.1   Carrier aggregation

which is available in a large number of countries in Regions 1 and 3. Therefore, the amount of contiguous transmission bandwidth for one operator in a certain geographical area is limited. In order to meet both the requirement on transmission bandwidth and the requirement for utilizing the IMT bands, the concept of CA is crucial. Therefore, CA has been identified as one of the major features in both 3GPP LTE-A and Institute of Electrical and Electronics Engineers (IEEE) 802.16m technologies to meet the IMT-Advanced requirements.

In 3GPP, three possible CA scenarios – also shown in Figure 3.1 – are considered:

- Intraband contiguous CA, where contiguous bands wider than 20 MHz are used.
- Intraband noncontiguous CA, where multiple CCs belonging to the same band are used in a non-contiguous manner.
- Interband noncontiguous CA, where multiple CCs belonging to different bands are aggregated.

In the Uplink (UL), the focus is on the intraband noncontiguous case, whereas in the Downlink (DL) the interband case is also considered. Four different deployments are considered for CA. In one of the most typical deployment scenario the antennas are collocated and have the same beam patterns for different components. If the CCs are at the same band or the frequency separation is small, this would lead to nearly the same coverage for all CCs. Large frequency separation between CCs would lead to the second scenario where the coverage of the CCs are different. In the third scenario, different beam directions or patterns are used for different CCs to shift the beams across carriers and by doing so improve throughput at sector boundaries. In the fourth scenario, one CC provides macrocell coverage and remote radio head cells are placed at traffic hotspots for additional throughput by other CCs. This last scenario is, however, not considered in UL (Iwamura et al. 2010).

On the other hand, IEEE has not clearly stated, in the IMT-Advanced candidate proposal, which are their choices for carrier aggregation. Only the following text can be found in (IEEE 802.16 2010) concerning the IEEE 802.16m requirement of the operating bandwidth: "[...] *a common MAC entity to control a PHY spanning over multiple frequency channels. The channels may be of different bandwidths (e.g. 5, 10 and 20 MHz) on contiguous or non-contiguous frequency bands. The channels may be of the same or different duplexing modes, for example, FDD, TDD, or a mix of bidirectional and broadcast only carriers [...].*"

The challenges that arise from CA are mainly due to the fact that UEs can transmit and receive from more than one carrier – and in theory from more than one cell. This requires some changes in the radio layer 1 and 2 and adds complexity. It changes Radio Resource Management (RRM) requirements because the UE will need to measure several carriers at the same time. The UE power consumption is increased with the number of CCs because UE will have to monitor more physical downlink control channels, which are detected by blind decodings. Two methods that are investigated to allow power savings are *discontinuous reception* and *component carrier activation/deactivation*. Moreover, the design and deployment of control signaling channels for multiple CCs is crucial for efficient data transmission control and the overall system performance. Because of different interference conditions on different carriers, LTE-A needs to support CC-specific UL power control for both contiguous and noncontiguous CA. Mobility management, more specifically radio resource control measurements and handovers, are also issues that require specific attention (WINNER+ 2010).

As a step from carrier aggregation towards more flexible spectrum use, cognitive radio introduces a novel approach to accessing spectrum – in an opportunistic way. This may increase the efficiency in spectrum use, however, the rights of current spectrum users need to also be preserved. The introduction of cognitive radio techniques therefore also requires acceptance from the regulatory domain. As a sign of awareness of this new emerging technology in the regulatory domain, the next WRC of ITU-R in 2012 will consider regulatory measures and their relevance in order to enable the introduction of Software Defined Radio (SDR) and Cognitive Radio System (CRS) (agenda item 1.19). The preparatory work for this agenda item at the European level in European Conference of Postal and Telecommunications Administrations (CEPT) Electronic Communications Committee (ECC) has been conducted in a Conference Preparatory Group (CPG) Project Team A (PT A), which has been responsible for all the aspects of the agenda item. The CEPT ECC response to the agenda item finished in June 2010 and it was sent to the ITU-R Working Party (WP) 1B "*Spectrum management methodologies and economic strategies*," which has the main responsibility for the preparations on this agenda item at the international level. As the first step towards the next WRC, ITU-R WP 1B developed the definitions for CRS and SDR in September 2009. The official International Telecommunication Union (ITU) definition for CRS is "[...] *a radio system employing technology that allows the system to obtain knowledge of its operational and geographical environment, established policies and its internal state; to dynamically and autonomously adjust its operational parameters and protocols according to its obtained knowledge in order to achieve predefined objectives; and to learn from the results obtained* [...]" (ITU-R 2009a). ITU-R WP 1B has also finished its work on this agenda item in June 2010 by finalizing the Conference Preparatory Meeting (CPM) text. ITU-R WP 5A "*Land mobile service excluding IMT; amateur and amateur-satellite service*" has been responsible for the technical work regarding the agenda item. ITU-R WP 5A is developing an ITU-R report "*Cognitive radio systems in the land mobile service*." The purpose of this report is to provide answers to the questions posed by the ITU Radiocommunication Assembly (RA) in 2007. The content of the report covers the following aspects of the CRS:

- the ITU definition;
- closely related radio technologies and their functionalities;
- key technical characteristics;
- requirements;
- performance;
- benefits;
- the potential applications;
- the operational implications;
- capabilities that facilitate coexistence with existing systems;
- possible spectrum-sharing techniques; and
- the effect on the efficient use of radio resources.

There are also various other ongoing activities regarding CRS at the European and international level. At the European level, in CEPT ECC Working Group Spectrum Engineering (SE) 43 is investigating the use of cognitive radio systems in the TV white spaces, that is, 470 - 790 MHz band. The outcome of these investigations is reported in the ECC report *"Technical and operational requirements for the operation of cognitive radio systems in the white spaces of the frequency band 470–790 MHz."* The report is mainly band specific, considering sharing scenarios and national deployment scenarios of systems or services to be protected. However, some of the issues may be applicable for other frequency bands. CEPT ECC Working Group Frequency Management (FM) has initiated work to monitor and investigate the Cognitive Radio (CR) technological and characteristics development and to study how to regulate the spectrum requirements of such technologies. The working group will receive information by monitoring the European Union research projects and the European Telecommunications Standards Institute (ETSI). At the international level, ITU-R WP 5D "IMT Systems" is also working on CRS and the outcome of their studies will be published in a report *"Cognitive Radio Systems Specific for International Mobile Telecommunications (IMT) Systems."*

3.3 Evolution Towards Future Technologies

The purpose of this section is not to clear up the current technology of spectrum aggregation in the future Fourth Generation (4G) standards of mobile communications such as LTE-A or Worldwide Interoperability for Microwave Access (WiMAX) IEEE 802.16m, but to present novel techniques, algorithms or proposals based on spectrum aggregation that could be employed in those future standards. Even though this section mainly focuses on LTE-A technology, some proposals could be used in future releases of WiMAX.

With the arrival of CA, a large number of open issues should be addressed in the future. Is it possible to use a different coding method that improves the performance with longer block sizes? Is it better to schedule resource in a contiguous or noncontiguous way? Which are current trends concerning scheduling in IMT-Advanced systems? What is the best reporting process in CA? Finally, this section presents current trends in CA for IMT-Advanced systems.

3.3.1 Channel Coding

Obviously, if larger bandwidths are used then transmission capability generally increases. Nevertheless, when a BS has to transmit data to several frequency bands the best choice is still not clear whether to encode data in separate transport blocks or to combine all information and distribute the bits in the physical layer. In the first case each CC has its own configuration parameters (for example, modulation, coding rate, antenna configuration) as well as a different Hybrid Automatic Repeat-reQuest (HARQ) entity. On the other hand, in the second case there is a unique HARQ entity for all the CCs. The difference between the alternatives is shown in Figure 3.2. Both schemes have advantages and disadvantages. Minimum changes in the specifications will be required if scheduling, Multiple-Input Multiple-Output (MIMO), link adaptation and HARQ are performed over groups of carriers of 20 MHz. For instance, a user receiving information in 100 MHz bandwidth will need five processing chains for reception in the physical layer, one per each 20 MHz block. Moreover, this alternative is backward-compatible with legacy technologies, as LTE or IEEE 802.16e. On the other hand the main benefit of using only one HARQ process is that the longer the transport block size the more efficient the channel coder is.

LTE-A and WiMAX IEEE 802.16m, use Turbo Codes (TCs) as a mandatory channel coding. However, the IEEE 802.16m standard proposes Low-Density Parity-Check (LDPC) codes as an additional alternative. Both coding methods are close to the Shannon limit and can achieve low bit error rates for low Signal to Interference plus Noise Ratio (SINR) applications. In general, turbo codes have good performance for all block sizes once a certain minimum length is surpassed. This, together with the high computational complexity of TCs, motivates that packet data units are segmented in several

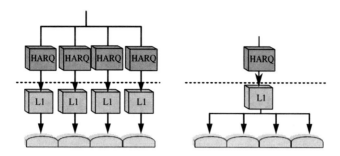

Figure 3.2 Two alternatives for the mapping to the components carriers

turbo blocks before coding. On the contrary, LDPC codes tend to behave better than TCs with larger blocks, most of all when increasing the SINR. This effect is due to the segmentation of the transport block in several turbo blocks. When the number of turbo blocks increases it is unlikely to complete successfully the transmission of all blocks, hence reducing the system's effective throughput. On the other hand, with a single LDPC block channel coding becomes more robust.

Given that 4G technologies will support bandwidths up to 100 MHz, the number of transmitted bits will increase. This section therefore compares the performance of LDPC codes and TCs with increasing bandwidths or, which is the same, with larger coding blocks. Another open question in case of usage of TCs is the convenience of the segmentation. Precisely, this segmentation has not been carried out in the analysis of this section. The maximum block size that the turbo coder can process is 6144 bits whereas with LDPC this restriction does not exist. Thus, in order to be totally fair in the comparison between these two schemes, segmentation should be taken into consideration.

In the comparison of LDPC codes and TCs, several coding rates and modulations have been simulated. However, for the sake of clarity only two types of simulation assumptions (QPSK with coding rate 1/2 and 16-QAM with coding rate 1/2) are depicted in Figure 3.3. This figure shows a difference of performance lower than 0.3 dB. The parameters of these simulations are listed in Table 3.1. Note that, for a given bandwidth, the guard bands have been accounted for. For instance, for a bandwidth of 100 MHz, 110 Resource Blocks (RBs) have been simulated instead of 100 RBs. Depending on the coding rate and the modulation, this difference is more or less significant. However, this difference

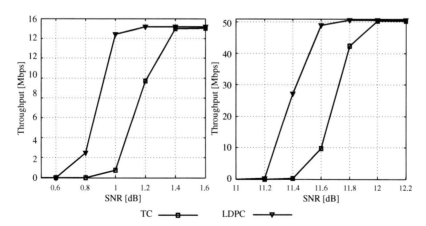

Figure 3.3 Performance comparison between LDPC codes and TCs for (15156 bits, rate = 1/2) (left) and for (50576 bits, rate = 5/6) (right)

Table 3.1 Basic parameters used in the performance assessment

Parameter	Value
Modulation	QPSK/16-QAM
Channel model	AWGN
Channel coding	LDPC/TC
Coding rates	0.5/0.83
Number of RBs	110 (Figure 3.3)
Matrix H	WiMAX (IEEE 802.16e 2005)
Decoding algorithm	Sum-product
Number of iterations	100 (LDPC)/8 (TC)
Segmentation	No (LDPC)/Yes (TC)

of performance has a relative importance due to its value. The most important conclusion is that both LDPC and TC perform very similarly similar for these block sizes.

Finally, as shown in Figure 3.4, when the encoded block increases its size there is a clear benefit in not segmenting the transport block. In this case, there is a difference of more than 2.5 dB in the SINR value required to reach the maximum throughput, that is, without any failure in the transmission. The reason is that when the number of turbo blocks increases it is unlikely to successfully complete the transmission of all blocks. Note that this maximum difference between TCs and LDPC codes occurs only in limited operating conditions.

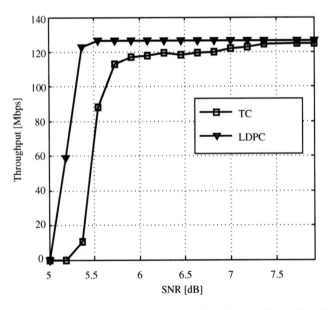

Figure 3.4 Performance comparison between turbo codes with packets no larger than 6144 and LDPC codes for (126476 bits, rate = 5/6) and QPSK

3.3.2 Scheduling

In IMT-Advanced systems two or more CCs belonging to the same of different frequency bands can be aggregated. It is therefore possible to schedule several CCs to a single user. In the assessment of the scheduling methods in CA, the Proportional Fair (PF) scheduler is going to be used due to its simplicity and good performance. This method allocates a RB to the user i^* who maximizes its instantaneous data rate over its average data rate (see Equation (4.8)). At slot s, $R_i(n, s)$ is the instantaneous transmission rate on the RB n for the user i and $\overline{R}_i(s)$ is its average data rate.

$$i^* = \arg \max_i \Gamma_{i,n} = \arg \max_i \frac{R_i(n, s)}{\overline{R}_i(s)} \tag{3.1}$$

PF will be compared with the Round Robin (RoR) scheduler, which allocates full resources to a single user in a circular sequence.

One of the more challenging topics of the resource scheduling in CA is to determine which aggregation strategy is preferable from the system performance point of view. In case of noncontiguous aggregation two scheduling strategies have been investigated in WINNER+ (2009): (1) joint PF scheduling of all RBs available in the aggregated bands and (2) separate and independent PF executed in any aggregated band. For contiguous aggregation, only joint PF scheduling of all available resource blocks has been employed. For the performance evaluation a system-level simulation in the DL of a single-cell scenario has been be considered, without any interference. Channel model corresponds to the Urban Macrocell (UMa) scenario specified in the IMT-Advanced guidelines (ITU-R 2009b). An aggregation of three component carriers of 20 MHz was assumed. A full buffer traffic model was considered. Link adaptation was used, where the modulation scheme was adaptively controlled based on the achieved SINR. In order to evaluate all the considered schemes, a simple PF scheduler was employed. Due to the lack of inter-cell interference, and with the aim of comparing the different aggregation approaches in low SINR regions, the total transmit power of the BS was tuned from − 6 up to 10 dBm. The comparison was carried out for 20 users uniformly distributed in the cell. Figure 3.5 present the total cell throughput obtained. The clear advantage of noncontiguous over contiguous aggregation can be noticed, especially when the transmitted power is low and the number of users is high. This is due to the higher spectral diversity experienced when noncontiguous aggregation is performed. However, this comes at the expense of hardware redundancy since more than one physical layer (and possibly MAC) processing chain should be available. Moreover, an interesting feature is that higher cell throughput has been achieved for separate CC scheduling than for joint scheduling of all aggregated bands.

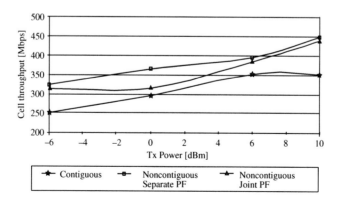

Figure 3.5 Cell throughput achieved with 20 users per cell

In a similar way to WINNER+ (2009) the QoS performance of two multiuser scheduling schemes called Separated Random User Scheduling (SRUS) and Joint User Scheduling (JUS) are compared in Zhang et al. (2009). At the BS the traffic to the users was buffered in different First Input First Output (FIFO) queues. The SRUS method allocates users to only one of the CCs in a random and uniform way and uses a different scheduler for each CC. Evidently, this scheme is not efficient because a CC could be unused while the other is still busy. On the other hand, the JUS method can assign resource of all CCs with a common scheduler. JUS increases the complexity of the scheduler but achieves the maximum spectral efficiency. In a full buffer scenario, the simulation results show that cell throughput of both schemes are similar using RoR, whereas in the case of PF the gain is between 5–7 % between both schemes. Assuming real traffic, the average user latency of SRUS is larger than JUS by around 90–100% when the arrival rate is small. However, as the traffic load increases the performance becomes increasingly similar.

In Chen et al. (2009), the authors made a similar study. In this case, the results show that CA always entails higher throughput than the independent use of CCs. The benefit of dynamic load balancing is also showed. Moreover, the fairness among users is only guaranteed when CA is used. In case of uncoordinated use of CCs this fairness is lost, most of all when users' allocation is not symmetrical.

In Wang et al. (2010b) the authors considered two schedulers based on the PF concept. The first uses Equation 3.1 independently for each CC. As for the second, user throughput is averaged over all aggregated CCs. This last method increases fairness among users. In a full buffer scenario the results show that there is no significant difference between both methods in terms of cell throughput. However, in terms of coverage, the second method has a meaningful gain, 50% when 20% of users can use CA and 90% when this percentage increases up to 50%.

CA and Coverage
Aggregation of lower frequency bands, for example, 450–470 MHz or 698–862 MHz bands, yields better results, as the signal attenuation is lower. It is known that path loss and doppler effect depend on the frequency.

The higher the frequency the higher the path loss is. Consequently, not all the frequency carriers have the same coverage. It will cause the users located at the cell edge to be scheduled on fewer carriers, while others can be scheduled on the entire aggregated spectrum. Fairness could therefore be lost when CA is used.

Figure 3.6 shows an example where two CCs are aggregated. Users near the BS (e.g. UE1) can use both CCs (f_1 and f_2) whereas users on the cell edge (e.g. UE2) can only use one frequency carrier f_1.

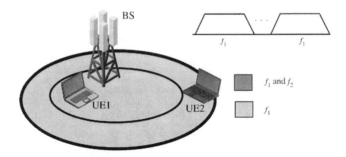

Figure 3.6 Coverage of a cell with two CCs

These challenging problems have attracted the attention of the scientist community. For instance, in Songsong et al. (2009) a scheduling method based on user grouping is proposed. Users are divided into groups according to the number of carriers they can be scheduled on. The path loss equations depend on the frequency carrier and the distance between the UE and the considered BS. It is therefore easy to obtain the coverage radius for a certain frequency carrier, assuming that the cell is circular. Hence, each CC will have a maximum radius in which it is possible to operate. After grouping, users are scheduled based on the PF method but taking into account the probability that users fall within the coverage of one or several carrier. In the proposed method, users' throughput is adjusted proportionally to the number of carriers on which they can be scheduled. Simulation results show higher fairness with the proposed method at the expenses of an increasing throughput degradation with the number of users.

So far, all analyzed studies are focused on the DL. However, although there are some similarities in CA between DL and UL, there are also some differences. Due to the limited transmission power of the UE, increasing the bandwidth does not always result in an increase of the user performance. In Wang et al. (2010a) an adaptive power control algorithm is proposed in order to compensate these variations in the allocated bandwidth. Firstly, the transmission power on each CC is independently estimated based on the allocated resources and the CC-specific power control parameters. Next, if the total estimated power is higher than the UE maximum power, then the transmission power in each CC is reduced by the same offset. Two types of users were considered in Wang et al. (2010a): Rel-8 users were only allocated to one CC while LTE-A users were assigned to either one CC or two CCs based on their path losses. The simulation results show that the cell edge user throughput of LTE-A and Rel-8 users is the same. However, the average user throughput in LTE-A is higher. The gain achieved with carrier aggregation depends upon the system load, being higher for lower loads.

3.3.3 Channel Quality Indicator

Orthogonal Frequency Division Multiplexing (OFDM) systems are particularly efficient when link adaptation and user multiplexing in the frequency domain are employed. However, it involves an instantaneous knowledge of the channel quality. Frequency Selective Scheduling (FSS) significantly improves system performance. Depending on the Channel Quality Indicator (CQI) bandwidth used, explicit CQI feedback for every RB can result in significant overhead and, therefore, reduced capacity. In case of reducing excessively the CQI bandwidth the FSS performance benefit could decrease. This is aggravated in LTE-A because multiple CCs can be located in different bands in which the doppler frequency is different. This fact is directly related to the required CQI reporting period.

An efficient and flexible technique for CQI reporting would optimize the tradeoff between the system performance of a FSS algorithm and the uplink bandwidth occupancy. It could therefore be useful to define a flexible CQI reporting method to select a certain level of granularity in the time domain and in the frequency domain depending on the radio channel conditions the UE experiences.

3GPP designed the Rel-8 CQI reporting method as follows:

- The CQI table comprises 16 entries (CQI is 4 bits) (3GPP 2010).
- The table is defined in terms of channel coding rate and modulation scheme.
- An UE reports a CQI index corresponding to a transport format with 10% BLock Error Rate (BLER) target at the first transmission, over the set of RB corresponding to the CQI value.

The eNodeB (eNB) controls time and frequency resources used by the UE. The UE transmits the CQI reports periodically on the Physical Uplink Control CHannel (PUCCH) for subframes without

Table 3.2 Look-up table

Parameter	Value
$(\sigma_{\tau 1}, f_{d1})$	(N_{f1}, N_{TTI1})
$(\sigma_{\tau 2}, f_{d2})$	(N_{f2}, N_{TTI2})
...	...
$(\sigma_{\tau n}, f_{dn})$	(N_{fn}, N_{TTIn})

Table 3.3 Simulation scenarios used in the CQI analysis

Deployment scenario	v[km/h]	f_d [Hz]	f_c [MHz]
UMi	3	7	2500
UMa	30	56	2000
RMa	120	89	800

the Physical Uplink Shared CHannel (PUSCH) transmission and on the PUSCH for those subframes with scheduled PUSCH transmissions.

The proposed innovation is based on building a lookup table mapping the doppler frequency f_d and the delay spread σ_τ that characterize the channel conditions with the required periodicity of CQI reports in frequency and time, expressed in number of RBs (N_f) and TTI (N_{TTI}) respectively. This lookup table was built using system level simulations WINNER+ (2009). Table 3.2 depicts this lookup table. The gain achieved when using variable reporting periods is assessed in terms of the tradeoff between the system performance and the uplink overhead.

Simulations were conducted over a dynamic system-level simulator that follows the indications of ITU-R (2009b) and using the scenarios in Table 3.3. The technology used in the assessment was LTE in its Frequency Division Duplex (FDD) version. Nevertheless, a nonstandard method of CQI reporting was used. The reporting period can be configured as a multiple of 1 ms. Moreover, it was assumed that the UE sends a CQI report with information about the whole bandwidth but with a configurable granularity in frequency. Frequency granularity is expressed in terms of number of LTE RBs. Specifically, the values used in this assessment were 1, 2, 5, 10, 25, 50 RBs for the reporting bandwidth and 1, 2, 4, 8, 16, 32, 64, 128 ms for the reporting period. More details on the simulation assumptions can be found in WINNER+ (2009).

From the obtained results it can be concluded that, in all scenarios when the reporting period increases, significant degradation of the spectral efficiency is observed. This degradation is more or less significant depending on the specific scenario. The results are shown in Figure 3.7 where each pair of numbers represents the reporting period and reporting bandwidth respectively. Based on the desired tradeoff (between the maximization of the cell efficiency and the minimization of the uplink overhead) one can identify, in the figure, the optimal pair (number of TTI – number of RBs).

3.3.4 Additional Research Directions

This section aims to sumarize additional research directions that represent open questions to be solved in the future. The research lines focus on fractional soft handover in CA (Chang et al. 2009), flexible carrier aggregation for the home base station (Li et al. 2009) and neighbor carrier signal-strength estimation (Yuan et al. 2010).

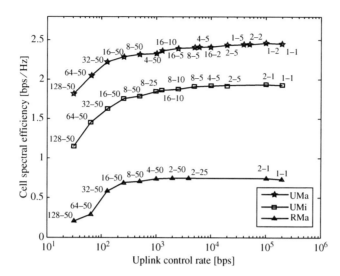

Figure 3.7 Cell spectral efficiency versus uplink overhead for UMa, UMi and RMa scenarios. Each pair of numbers represents the reporting period and reporting bandwidth, respectively

In Chang et al. (2009) the authors proposed a fractional soft handover scheme. In the LTE standard only hard handover is allowed. With the proposed soft handover, packets are forwarded from the original eNB to the target eNB before breaking the communication. In order to improve the handover scheme, the technique sets one CC as a special carrier for the user to establish a Radio Resource Control (RRC) connection with the target eNB. The authors compare the proposed soft handover method based on CA with the legacy hard handover mechanism and other conventional soft handover procedures. The new proposal reduces the outage probability while saving radio resources and power use.

In Li et al. (2009) the authors studied how to aggregate carriers for Home Base Station (H-BS). They proposed three types of CA allocation patterns: fixed, random and autonomous. In the former, the carrier aggregation is fixed as defined by the operators. In the second case, H-BS selects its CC randomly from the available set of RB when it is turned on or when the channel quality is unbearable. In the latter case, the bandwidth could be divided in components of certain size that H-BS can choose according to interference measurements. The results demonstrate that flexible carrier aggregation mechanisms improve system performance even when H-BSs have cochannel interference from the surrounding macrocells. A gain of between 2 to 8 dB is achieved as compared with fix assignment.

Finally, the UEs need to measure periodically the signal strength from all the CCs. Obviously, these measurements entail a rather high cost. In Yuan et al. (2010) the authors proposed to measure the center CC in each band and use correlation to extrapolate results to neighbor bands, thus reducing measurement costs. Evidently, this technique suffers from higher errors when increasing carriers distance. However, for distances around 40 MHz this error is less than 1 dB. On the other hand, in order to reduce the fast-fading impact, the final estimation is filtered, which reduces the estimation error.

3.4 Cognitive Radio Enabling Dynamic/Opportunistic Carrier Aggregation

As earlier stated, CA enables bandwidth resources to be increased and fully utilized in order to provide enhanced peak data rates to service-demanding users. In this context, the CR paradigm

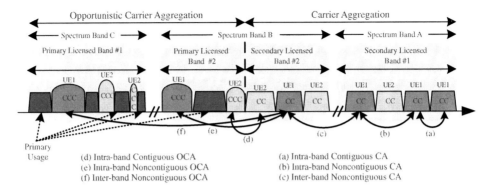

Figure 3.8   Carrier aggregation and opportunistic carrier aggregation

Akyildiz et al. (2006); Haykin (2005); Mitola and Maguire (1999), may further extend these benefits by adopting the Dynamic Spectrum Access (DySA) concept (Zhao and Sadler 2007). As a result, Cognitive Component Carriers (CCCs) can be aggregated, in addition to CCs, on an opportunistic and noninterfering basis. In this case, we may appropriately coin the term Opportunistic Carrier Aggregation (OCA) as a method that efficiently uses the spectrum resources by aggregating both CCs and CCCs when available. Accordingly, OCA provides extended capabilities and improved flexibility in the aggregation of spectrum resources, therefore enhancing both data rates and spectrum utilization requirements.

In the following, we will refer to an OFDM-based cognitive radio system, which is able to aggregate carriers in bands belonging to licensed legacy systems (i.e. primary systems), in addition to those bands being licensed to the cognitive (or secondary) system. In this case, the cognitive system may enhance its spectrum access, and consequently its performance, by means of aggregating unused primary spectrum resources. Accordingly, Figure 3.8 represents the case where contiguous and noncontiguous spectrum bands are aggregated within a same band (intraband) or between different bands (interband). Note that "traditional" carrier aggregation is carried out within licensed bands to the secondary system (i.e. spectrum bands A and part of B). On the contrary, opportunistic carrier aggregation is performed within licensed bands to the primary system when these are available – that is, spectrum bands C and part of B.

3.4.1 Spectrum Sharing and Opportunistic Carrier Aggregation

It is widely held that traditional Fixed Spectrum Assignment (FSA), where spectrum is licensed for exclusive use to a licensee, leads to inefficient and poor use of expensive and scarce spectrum resources Akyildiz et al. (2006); Haykin (2005). In this sense, the concept of Dynamic Spectrum Access (DySA) has gained increased momentum in recent years. DySA advocates a more flexible use of the spectrum owned by the licensee (also known as the primary system) through well defined spectrum-sharing mechanisms with a cognitive (or secondary) system (Döttling et al. 2009, Chapter 11).

In this context, the suitability of OFDM-based systems has been widely suggested and recommended given its demonstrated spectral flexibility (Weiss and Jondral 2004). This property enables it to exploit spectrum opportunities fully during primary idle periods by selecting an adequate set of subcarriers for transmission, that is, those falling in detected unused bands, while preventing the use of those subcarriers that fall within detected occupied bands. Hence, OFDM access provides an adaptive transmit filter enabling flexible and opportunistic access to the licensed spectrum.

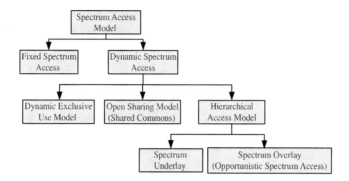

Figure 3.9   Spectrum access taxonomy

The use of OFDM-based systems by the secondary system for OCA, in addition to the standard licensed use, comes at a reduced cost because a particular set of subcarriers may be fed by zeroes in the corresponding transmitter Inverse Fast Fourier Transform (IFFT) input to prevent interference with the primary system. At the receiver, the FFT operation implemented to recover the transmitted data will still be valid for the OCA operation mode, thus coming at no extra cost. Implementation issues for DySA using Orthogonal Frequency Division Multiple Access (OFDMA) schemes can be found in Poston and Horne (2005); Rajbanshi et al. (2006) and references therein. Particularly in Poston and Horne (2005), the authors propose the adoption of Discontiguous OFDM (DOFDM) for DySA in idle television channels. They provide a preliminary description of a DOFDM prototype and describe some special considerations for its implementation. Similarly in Rajbanshi et al. (2006), an efficient implementation of a Non-Contiguous OFDMA (NC-OFDMA) transceiver is presented for cognitive radio applications. In this work, due to the deactivation of several carriers during the IFFT procedure at transmission, the authors propose a pruning mechanism for the FFT at the receiver in order to reduce the execution time. Results show improved performance with respect to other pruning methods when medium to large number of subcarriers have been deactivated.

Several DySA implementations arise according to the way in which spectrum is shared between two or more network entities. These implementations may be broadly categorized into three models Zhao and Sadler (2007): the *Dynamic Exclusive Use Model*, the *Open Sharing Model* and the *Hierarchical Access Model*, which are briefly described below (see Figure 3.9).

- The Dynamic Exclusive Use model allows a spectrum owner (i.e. the licensee) to grant spectrum access rights over a certain spectrum band to a cognitive system. This cognitive system can make exclusive use of the spectrum as long as it follows the rules provided by the licensee.
- The Open Sharing Model advocates for an open access to shared resources with no exclusive or priority rights over the considered spectrum. In this case, fairness must be guaranteed among users accessing the shared spectrum.
- The Hierarchical Access Model provides an access structure based on primary (or licensed) and secondary (or cognitive) users. In this case, the primary system opens its spectrum so that secondary users may access it in a noninterfering manner.

Based on the above descriptions, OCA will inherit the same DySA paradigms and implementations. In the case of OCA, shared spectrum resources in licensed bands will be identified in order to accommodate so-called CCCs for subsequent CA. This process will not only require the successful detection of available spectrum (i.e. *white spaces*) but also the determination of the suitability of such bands for CA. In addition, each UE in the considered scenario may, in turn, observe different spectrum availability conditions, thus adding increased complexity to the CCC selection process regarding whether or not to aggregate a particular CCCs to a particular UE. Furthermore, DySA scenarios are dynamic in nature meaning that available spectrum conditions may suffer from high variability. Then, OCA procedures should be spectrum agile in the sense that they should vacate spectrum suddenly occupied by the licensed system. All these issues call for efficient functionalities to be implemented, such as spectrum awareness, CCCs identification and selection, along with spectrum mobility. These functionalities will be addressed in the subsequent sections.

3.4.2 Spectrum Awareness

Spectrum awareness is a key concept enabling nonintrusive DySA, extensible to nonintrusive OCA, to shared spectrum resources. It essentially provides a set of mechanisms capturing information on the occupancy of a particular spectrum band in a particular geographical area and at a given instant in time.

Several methods and mechanisms providing spectrum awareness information have been identified in the literature, among them: *Spectrum Sensing*, *Geo-Location Databases* and *Beacon Signaling* have attracted major attention. A short overview of these methods is given in the following. For detailed information, the reader is referred to Appendix B.

Spectrum Sensing

Spectrum sensing is the task of obtaining awareness about the spectrum usage and existence of primary users in a particular geographical area. This awareness is usually undertaken locally by the secondary user, which performs measurements over a particular radio frequency band. While spectrum sensing usually refers to the energy measurement over a given spectrum band, it actually encompasses broader connotations on obtaining spectrum usage characteristics across multiple dimensions such as time, space, frequency and code (Yucek and Arslan 2009). It also involves determining what types of signals are occupying the spectrum including the modulation, waveform, bandwidth, carrier frequency, etc.

Geo-Location Databases

Spectrum awareness by means of geo-location databases operates as follows. An unlicensed (secondary) device estimates its position and checks a database in order to identify those spectrum bands that are being used by licensed (primary) services in the vicinity of the unlicensed device. This implies that the device must have an estimate of its position, an estimate of its position error, an estimate of its interference range, and access to a database of potential licensed service areas (Brown 2005).

Beacon Signaling

In the context of spectrum awareness, beacon signals can be utilized to indicate if a particular licensed spectrum band is occupied, and thus available for secondary access (Hulbert 2005). Beacon signaling would ease the performance requirements of spectrum sensing by secondary devices and also facilitate the delivery of database information in absence of Internet connectivity. Beacons may advertise either permission (grants) for cognitive radios to access spectrum, or alternatively, denials of spectrum access for cognitive radios, see, for example, CEPT-ECC (2010); Hulbert (2005). However, as shown in Mangold et al. (2006), a dual beaconing approach may offer greater reliability than using a single beacon either for grant or for denial. Furthermore, beacon broadcasting can be used within a network to inform neighbors about available communication channels (Zhao et al. 2005).

3.4.3 Cognitive Component Carrier Identification, Selection and Mobility

The selection of bandwidth-limited CCCs may impose high constraints in terms of signaling and increased complexity while not providing substantial aggregated spectrum and thus limiting increased throughput (see for example, in Figure 3.8, the potential CCCs identified in spectrum band C for UE2). Moreover, particular primary bands may present high variability in terms of usage, meaning that potential CCCs lying in these bands must be frequently vacated and reassigned, which, again, increases the burden of signaling. For the same reasons, the cognitive system will preferably aggregate intraband contiguous spectrum (refer to Figure 3.8 case (d)) rather than intraband noncontiguous (Figure 3.8 case (e)) or interband noncontiguous (Figure 3.8 case (f)) aggregations.

Upon service demand by a particular UE, radio and spectrum resource management mechanisms will provide efficient algorithms for the assignment of CCs and CCCs to the existing users. In this sense, we may identify carrier selection policies that operate upon session establishment (intersession carrier selection) and during the session's duration (intrasession carrier selection). For the case of intrasession carrier selection, existing mobility mechanisms of traditional networks should be extended in order to support spectrum-agile intraband and interband handoffs between CCs and CCCs in both contiguous and noncontiguous bands. Different UEs may be configured with a set of different CCs and/or CCCs, according to their hardware capabilities, channel conditions, QoS requirements and spectrum availability among other issues. In this way, OCA adds a new dimension, that is, spectrum availability, into the already complex process of CA.

3.5 Implications for Signaling and Architecture

When using carrier aggregation, some signaling information related to the allocation of the resources is provided. As previously mentioned, the 3GPP is currently addressing this specific topic where backward compatibility and the reduction of complexity are being considered. However, with the objective of having maximum system performance, all cases must be studied.

There are five key aspects that must be established with respect to signaling in CA:

1. CC discovery and accessibility. Among the configured CCs, at least one downlink CC should transmit the required broadcast information to enable UEs to access the system. In order to increase flexibility and guarantee backward compatibility, the most reasonable proposal is that all CCs are accessible to all UEs, being Release 8 or 10. Those UEs with the capability to support multiple CCs will detect the stronger CC in the initial access and receive system information from all CCs, as this system information will be common among aggregated CCs. However, if any CC system information exists, this should be transmitted by each CC in a different signaling channel. It must be noted that, for the sake of simplicity, the synchronization of all of the aggregated carriers is required, with respect to both time and frequency.
2. Once synchronized, the UE will use a random access channel to connect to the system. If the CA is symmetric between UL and DL then the DL CC will coincide with its paired UL carrier. However, in the asymmetric carrier aggregation case, in which multiple DL CCs may be associated with only one UL CC, the configuration between UL CC and DL CC may cause a serious ambiguity problem as the BS may not know which DL CC is selected by the UE. In order to solve this problem CCs must be identified, transmitting this information in the corresponding signaling broadcast channel. Besides, this identifier simplifies the scheduling process and allows future handovers between CCs.
3. Another important aspect is related to the downlink signaling control for the scheduling. In LTE-A UEs must be able to decode control channel DL information on multiple DL CCs. Besides, any control channel on a CC can assign resource blocks in both DL and UL in one of multiple CCs.

Therefore, the CC identifier is needed to specify to the UE on which component carrier the control channel assigns resources. The BS could decide on transmitting the scheduling information of the whole aggregated band several times, one per CC, or could limit this transmission to just the best CC. In IEEE 802.16m the situation is simpler, since each user has only one primary carrier with all control information. However, data can be transmitted in the primary plus in one or several secondary carriers. Again CC identifiers are needed to identify the exact allocation of resources.

4. In LTE-A and additional control channel reports on the HARQ status. Again, in case of asymmetry it will be required to select the DL CC on which this channel is transmitted. The most reasonable alternative is to select only the best CC, although this issue requires further research. In IEEE 802.16m all control information is transmitted through the primary CC and therefore this problem does not exist.

5. In the uplink, in case of a multiple CC assignment, the UE may have multiple HARQ processes in parallel, one per CC. This would mean that multiple ACK/NACK messages corresponding to the DL CC transport blocks should be transmitted using the UL control channel. Moreover, the UE will also feedback multiple channel state indicators, including optimum MIMO precoding matrices, one per CC. This is the information used for CC handover and hence it is important to guarantee its proper reception. Again one UL control channel could report about all the UL CCs using the CC identifiers. The decision on transmitting several parallel UL control channels will lie again on the BS design.

With regards to the channel state reports, it has been confirmed in WINNER+ (2009) that significant overhead reduction can be achieved if the carrier and scenario are known. Flexibility in the reporting mechanism is therefore recommended. The Channel State Information (CSI) reporting procedure can be performed automatically by the UE depending on the frequency carrier that UE is using to communicate with the base station and depending on the speed that characterizes its radio channel. In this case no specific requirement on signaling and specific measurement would be expected. Nevertheless, assuming that the BS is able to know the lookup table used by the UE and/or assuming that the BS can set its own lookup table in the UE, the BS could decide to force the UE to change the granularity of the CSI reporting referring to another pair (number of frequencies, number of time slots) contained in the lookup table in order to optimize its radio resources allocation strategy. In such a scenario, a new signal is expected from the UE to the BS where the UE communicates the lookup table to the base station every time that the UE is going to camp on a new cell or a new signal is expected from the BS to the UE where the BS communicates its own lookup table every time that a new UE is going to be served.

3.6 Hardware and Legal Limitations

Due to hardware limitations at the UE, in a short time it is expected that no more than 8192 FFTs (2^{13}) will be implemented by vendors. Considering LTE subcarrier spacing, this results in no more than 120 MHz. For IEEE 802.16m with component carriers of 10 MHz, the maximum bandwidth is reduced to 80 MHz. Lower bandwidth can be configured by transmitting zeros in the unused subcarriers. Another approach to aggregate spectrum is not to increase FFT size but to use multiple RF transceivers. In this case, although the guard bands of each component carrier cannot be utilized, the hardware complexity is lower and backward compatibility is easily reached.

Another issue is the existence of such a large continuous bandwidth. In fact, the result of WRC 2007 may not allow such wide carrier bandwidth for new radio systems. Given the current spectrum distribution, the only valid alternative to process the entire frequency band from about 400 MHz to about 6 GHz seems to be the use of parallel receivers for each different band.

This may lead to another hardware limitation, due to antennas, at least on the mobile side. Indeed, the antenna should cover the whole set of bands decided by WRC 07, especially the lower band (698–960 MHz) (not to mention 400 MHz). If we assume its size should remain compatible with that of a smartphone, then its electrical dimension (dimension over wavelength) in the 698–960 MHz band will be about 0.25, and its relative bandwidth (bandwidth over central frequency) should be about 0.3. For these values, electromagnetic theory predicts a maximum possible radiation efficiency (ratio of power actually radiated to the power put into the antenna terminals) of about 0.25, which is quite poor. Indeed, radiation efficiency decreases as the antenna length decreases and for larger relative bandwidths. Therefore, alternative solutions should be found, such as frequency reconfigurable antennas.

The last hardware limitation is the difficulty for the RF part when aggregation is performed on noncontiguous bands. As explained in (Shukla et al. 2006) there is a research challenge related to the mitigation of intermodulation distortion, especially if fragments share a transmit/receiver chain, or chains need combining to share an antenna or amplifier, to reduce component count and overall size.

References

3GPP 2010 Physical layer procedures. Technical Specification Group Radio Access Network 36.212 v9.2.0, 3rd Generation Partnership Project (3GPP), http://www.3gpp.org/ftp/Specs/html-info/.

Akyildiz I, Lee W, Vuran M and Mohanty S 2006 Next generation/dynamic spectrum access/cognitive radio wireless networks: A survey. *Computer Networks* **50**(13), 2127–2159.

Brown T 2005 An analysis of unlicensed device operation in licensed broadcast service bands *Proc. First IEEE International Symposium on New Frontiers in Dynamic Spectrum Access Networks*, pp. 11–29.

CEPT-ECC 2010 Technical and Operational Requirements for the Possible Operation of Cognitive Radio Systems in the "White Spaces" of the Frequency Band 470–790 Mhz (Annex 3 to Doc. SE43(10)103). Technical report, Conférence Européenne des Post et Telecommunications, http://www.ict-qosmos.eu/project/standardisation-and-regulation/ceptecc-se43.

Chang J, Li Y, Feng S, Wang H, Sun C and Zhang P 2009 A fractional soft handover scheme for 3GPP LTE-advanced system *Proc. ICC 2009 – IEEE Int. Conf. Commun.*, pp. 1–5.

Chen L, Chen W, Zhang X and Yang D 2009 Analysis and simulation for spectrum aggregation in LTE-Advanced system *Proc. VTC 2009 Fall – IEEE 70th Vehicular Technology Conf.*, pp. 1–6.

Döttling M, Mohr W and Osseiran A 2009 *Radio Technologies and Concepts for IMT-Advanced*. John Wiley & Sons, Ltd., Chichester.

Haykin S 2005 Cognitive radio: brain-empowered wireless communications. *IEEE Journal on Selected Areas in Communications* **23**(2), 201–220.

Hulbert A 2005 Spectrum Sharing Through Beacons *Proc. PIMRC 2005 – IEEE 16th Int. Symp. on Pers., Indoor and Mobile Radio Commun.*, pp. 989–993, Berlin, Germany.

IEEE 802.16 2010 System Description Document (SDD). Document IEEE 802.16m-09/0034r3, IEEE 802.16 Broadband Wireless Access Working Group, Task Group m, http://www.ieee802.org/16/tgm /core.html.

IEEE 802.16e 2005 Draft IEEE Standard for Local and Metropolitan Area Networks. Part 16: Air Interface for Fixed and Mobile Broadband Wireless Access Systems. Amendment for physical and Medium Access Control Layers for Combined Fixed and Mobile Operation in Licensed Bands. Document IEEE P802.16e/D12, Technical Specification Group Services and System Aspects, http://ieeexplore.ieee.org/xpl/mostRecentIssue.jsp?punumber=4039740.

ITU-R 2008a Requirements Evaluation Criteria and Submission Templates for the Development of IMT-Advanced. Report ITU-R M.2133, International Telecommunications Union Radio (ITU-R), http://www.itu.int/publ/R-REP/en.

ITU-R 2008b Requirements related to Technical Performance for IMT-Advanced Radio Interface(s). Report ITU-R M.2134, International Telecommunications Union Radio (ITU-R), http://www.itu.int/publ/R-REP/en.

ITU-R 2009a Definitions of Software Defined Radio (SDR) and Cognitive Radio System (CRS). Report ITU-R SM.2152, International Telecommunications Union Radio (ITU-R), http://www.itu.int/publ/R-REP/en.

ITU-R 2009b Guidelines for Evaluation of Radio Interface Technologies for IMT-Advanced. Report ITU-R M.2135-1, International Telecommunications Union Radio (ITU-R), http://www.itu.int/publ/R-REP/en.

Iwamura M, Etemad K, Fong MH, Nory R and Love R 2010 Carrier aggregation framework in 3GPP LTE-Advanced. *IEEE Commun. Mag.* **48**(8), 60–67.

Li J, Liu Y, Duan J and Liang X 2009 Flexible carrier aggregation for home base station in IMT-Advanced System *Proc. 5th International Conference on Wireless Communications, Networking and Mobile Computing*, pp. 1–4.

Mangold S, Jarosch A and Monney C 2006 Operator assisted cognitive radio and dynamic spectrum assignment with dual beacons – detailed evaluation *Proc. IEEE 1st International Conference on Communication Systems Software & Middleware*, pp. 1–6, New Delhi, India.

Mitola J and Maguire G 1999 Cognitive radio: making software radios more personal. *IEEE Personal Communications* **6**(4), 13–18.

Poston J and Horne W 2005 Discontiguous OFDM considerations for dynamic spectrum access in idle TV channels *Proc. First IEEE International Symposium on New Frontiers in Dynamic Spectrum Access Networks*, pp. 607–610.

Rajbanshi R, Wyglinski AM and Minden GJ 2006 An efficient implementation of NC-OFDM transceivers for cognitive radios *Proc. 1st International Conference on Cognitive Radio Oriented Wireless Networks and Communications*, pp. 1–5.

Shukla A, Willamson B, Burns J, Burbidge E, Taylor A and Robinson D 2006 A Study for the Provision of Aggregation of Frequency to Provide Wider Bandwidth Services. Technical Report R/06/01773, QINETIQ, http://www.aegis-systems.co.uk/download/1722/aggregation.pdf.

Songsong S, Chunyan F and Caili G 2009 A resource scheduling algorithm based on user grouping for LTE-Advanced system with carrier aggregation *Proc. International Symposium on Computer Network and Multimedia Technology*, pp. 1–4.

Takagi H and Walke BH 2008 *Spectrum Requirement Planning in Wireless Communications* first edn. John Wiley & Sons, Ltd., Chichester.

Wang H, Rosa C and Pedersen K 2010a Performance of uplink carrier aggregation in lte-advanced systems *Proc. VTC 2010 Fall – IEEE 72st Vehicular Technology Conf.*, pp. 1–5.

Wang Y, Pedersen K, SĂ andrensen T and Mogensen P 2010b Carrier load balancing and packet scheduling for multi-carrier systems. *IEEE Trans. Wireless Commun.* **9**(5), 1780–1789.

Weiss T and Jondral F 2004 Spectrum pooling: an innovative strategy for the enhancement of spectrum efficiency. *IEEE Communications Magazine* **42**(3), 8–14.

WINNER+ 2009 Celtic Project CP5-026 Intermediate Report on System Aspect of Advanced RRM (ed. Monserrat J. and Sroka P.). Public Deliverable D1.5, Wireless World Initiative New Radio – WINNER+, April 2009, http://projects.celtic-initiative.org/winner+/index.html (accessed June 2011).

WINNER+ 2010 Celtic Project CP5-026 Strategies and technologies for spectrum utilisation and sharing aspects of IMT (ed. Siebert M.). Public Deliverable D3.3, Wireless World Initiative New Radio – WINNER+, http://projects.celtic-initiative.org/winner+/index.html (accessed June 2011).

Yuan P, Xiao D, Han J and Jing X 2010 Neighbor carrier signal strength estimation for carrier aggregation in LTE-A *Proc. WASE International Conference on Information Engineering*, vol. 1, pp. 284–287.

Yucek T and Arslan H 2009 A survey of spectrum sensing algorithms for cognitive radio applications. *IEEE Communications Surveys and Tutorials* **11**(1), 116–130.

Zhang L, Wang YY, Huang L, Wang HL and Wang WB 2009 QoS performance analysis on carrier aggregation based LTE-A systems *Proc. IET International Communication Conference on Wireless Mobile and Computing*, pp. 253–256.

Zhao J, Zheng H and Yang GH 2005 Distributed coordination in dynamic spectrum allocation networks *Proc. First IEEE International Symposium on New Frontiers in Dynamic Spectrum Access Networks*, pp. 259–268.

Zhao Q and Sadler B 2007 A Survey of Dynamic Spectrum Access. *IEEE Signal Processing Magazine* **24**(3), 79–89.

4

Spectrum Sharing

Jaakko Vihriälä and Mehdi Bennis

4.1 Introduction

Spectrum sharing refers to a wide variety of techniques aiming at using spectral resources in geographical locations where any particular part of the spectrum can be used simultaneously by at least two systems. Signals from other systems can be separated by detecting unused time or frequency slots, for instance. This closely resembles cognitive radio, where white-space detection is an essential feature.

A very well-known example is the use of the frequency band reserved for fixed satellite communications. Satellite stations transmit beacons providing information about spectrum usage. If the cognitive system does not receive this beacon then it is free to use the spectrum. A second example is femtocells. A common requirement with femtocells is that they do not cause interference to the primary system, which has rights to use the band. Especially for intersystem sharing, where the primary system (cellular network) and the femtocell belong to different operators, it is important to somehow guarantee that the femto-to-macro interference is minimal. In intrasystem sharing we have more options: by allowing controlled amount of interference to the macro system, we can maximize the sum rate of the macro and femto layers.

Spectrum sharing has become a high-priority research topic over the past few years. The motivation behind this lies in the fact that the limited spectrum is currently inefficiently utilized. As recognized by the World Radio Conference (WRC)'07, the amount of identified spectrum is not large enough to support large bandwidths for a substantial number of operators. On the other hand, existing cellular architectures are designed to cater large coverage areas, which does not achieve the expected throughput to ensure seamless mobile broadband in the Uplink (UL) as User Equipments (UEs) move far from the Base Station (BS). This is due to the increased intercell interference, as well as to the constraints on the transmit power of the UEs. Therefore, it is of paramount importance for future mobile cellular systems to share the frequency spectrum and coexist in a more efficient manner. This does not mean that different operators would share spectrum, but one single operator uses spectrum-sharing techniques with, for example, femtocells. However, we present one technique in which more than one operator trades the spectrum in an auction. This is not possible in existing cellular architectures, but future enhancements in the architecture may allow it.

Mobile and Wireless Communications for IMT-Advanced and Beyond, First Edition.
Edited by Afif Osseiran, Jose F. Monserrat and Werner Mohr.
© 2011 Afif Osseiran, Jose F. Monserrat and Werner Mohr. Published 2011 by John Wiley & Sons, Ltd.

This chapter is organized as follows. Section 4.2 gives an overview of the literature on spectrum sharing. Section 4.3 analyzes a proposal of spectrum sharing using Game Theory (GT). In section 4.4, spectrum auctioning between several operators is studied, aiming at optimizing bandwidth usage. Section 4.5 presents several femtocell techniques. Both practical and theoretical solutions are investigated to solve the interference problem inherent to femtocells. Section 4.5 also reviews the standardization activity in the femtocell area, including the Third Generation Partnership Project (3GPP), 3GPP2, and the FemtoForum[1]. Finally, in section 4.6 the main conclusions on spectrum sharing and femtocells are drawn, discussing the techniques and challenges, and giving a short insight into the future research of the spectrum-sharing area.

4.2 Literature Overview

Over the course of the last two decades, many approaches have been proposed to address the issue of *dynamic spectrum allocation, which consists of dynamically allocating the spectrum. The objective is to allow adaptive allocation of spectrum over space and time to various UEs in order to increase overall the spectrum efficiency.* In the mid-to-late 1990s, some of the initial foundational work on spectrum sharing was done in (Satapathay and Peha 1996, 1997, 1998). It was shown that, from a queuing theory perspective, independent networks could achieve an overall better capacity through cooperation. Examining the Federal Communications Commission (FCC) proposed listen-before-talk approach, they showed that such greedy algorithms could lead to very poor spectrum utilization and hence developed a system that imposed artificial penalties on greedy algorithms to keep them constrained. In his PhD dissertation, Mitola (2001) proposed the concept of Cognitive Radio (CR), which is essentially a software-defined radio with artificial intelligence capable of sensing and reacting to its environment. Later on, some work was conducted as a part of the Dynamic Radio for IP-Services in Vehicular Environments (DRiVE) and Spectrum Efficient Unicast and Multicast Services over Dynamic multi-Radio Networks in Vehicular Environments (OverDRiVE) projects aiming at providing video content delivery to vehicles (Leaves et al. 2001a,b). The OverDRiVE architecture involves partitioning the spectrum in space, frequency and time where certain blocks would be allocated by a central authority to each Radio Access Network (RAN) in response to their predicted capacity needs. The CORVUS project used similar ideas (Cabric et al. 2005) creating a channelized spectrum pool from unused licensed spectrum and proposing algorithms to allocate it efficiently. A Dynamic Intelligent Management of Spectrum for Ubiquitous Mobile access Networks (DIMSUM-Net) architecture was proposed in Buddhikot (2007) providing opportunistic access to large parts of under utilized spectrum. A simpler pragmatic approach offering coordinated, spatially aggregated spectrum access via a regional spectrum broker is argued to be more attractive in the immediate future. These architectures are centralized, therefore requiring someone to decide who should use which spectral resources at what time, while guaranteeing minimal interference to licensed UEs. While achieving good results, current politics involved in frequency licensing would make adopting these approaches unlikely because the need for a central authority hampers feasible deployment. As a result, more recent research has mainly focused on distributed techniques for dynamic spectrum allocation, where no central spectrum authority is required.

From a system-level perspective, several works have been carried out on resource sharing between operators, commonly referred to as channel borrowing (Katzela and Naghshineh 1996). When channels are borrowed between cells, networks have more flexibility, which improves performance compared to traditional channel assignment schemes as well as preventing congestion in some cells. In Zhang and Yum (1989), simulations related to channel-borrowing schemes are examined depicting

---

[1] http://www.femtoforum.org/femto/

the achievable gains in terms of blocking probability. In Pereirasamy et al. (2004) the performance of spectrum sharing is investigated for the Universal Mobile Telecommunication System (UMTS) Frequency Division Duplex (FDD) Downlink (DL). The obtained capacity gain is due to the increased trunking[2] efficiency when both subnetworks share radio resources. It is seen that the lower the number of nominal channel is, the higher the trunking gain is obtained. This study is further extended in Pereirasamy et al. (2005), showing that spectrum sharing offers capacity gains under restricted and cooperative conditions, thanks to the trunking gain where resources are pooled. Moreover, the cooperation in terms of interoperator signaling to maintain the grouped BS's orthogonality is shown to be necessary to ensure the full benefit obtained from resource pooling.

There exist two scenarios of practical interest for the study of Dynamic Spectrum Allocation (D.S.A) (Zhao and Sadler 2007): Hierarchical Spectrum Access (HSA) and Open Spectrum Access (OSA). In HSA, CRs coexist with legacy systems if and only if the additional interference overcome by the preexisting systems is below a specific threshold. These thresholds can be predefined by network operators, manufacturers or regulators to ensure certain QoS for primary systems. Moreover, CRs only transmit using radio resources left unused by the primary systems, hence CRs are often called opportunistic or secondary radio devices. Such unused radio resources are called Spectrum Access Opportunities (SAO) or available channels. Typically, an available channel consists of nonoccupied time slots in Time Division Multiple Access (TDMA), frequency bands in Frequency Division Multiple Access (FDMA), spatial directions in Space Division Multiple Access (SDMA), tones in Orthogonal Frequency Division Multiple Access (OFDMA), or spreading codes in Code Division Multiple Access (CDMA).

It is worth noting that interference management is done differently in a CDMA-based system in which the simultaneous intracell and intercell transmissions, cause interference due to universal frequency reuse. Nevertheless, in OFDMA-based systems, since UE information is spread by hopping in the time-frequency grid, the transmissions within a cell can be kept orthogonal but adjacent cells share the same bandwidth and intercell interference still exists. Hence, interference management is more challenging in this case and, as a result, remains an open issue. Finally, OFDMA system has the advantage of the full frequency reuse of CDMA while retaining the benefits of narrow band system without intracell interference. In HSA, once the SAO have been identified each CR decides whether to transmit or not based on their own performance criterion. Furthermore, the HSA model considers two different spectrum sharing scenarios, the underlay and the overlay approaches (Buddhikot 2007; Zhao and Sadler 2007). With underlay, opportunistic devices have to meet a certain power constraint in order to keep the interference level they generate on the primary systems always below their noise floor. Here, opportunistic and primary transmitters can transmit simultaneously without generating harmful interference to the primary receivers. On the other hand, the overlay approach, which targets the spectral white spaces, does not impose any limit on the transmit power. It only requires opportunistic radio devices to identify radio resources left unused by the primary network and exploit them subject to the constraint that those resources can be required by the primary system at any time. In this approach, the opportunistic and primary players do not transmit simultaneously.

In Open Spectrum Access (OSA), the notion of primary and secondary systems does not exist. In this scenario, each UE has the same rights to access the spectrum at any time. OSA typically includes the case of unlicensed bands. Radio devices operating in these bands include cordless telephones, wireless sensors and devices operating under the standards of Wi-Fi (IEEE 802.11), Zig-Bee (IEEE 802.15.4), and Bluetooth (IEEE 802.15.1). Here, each technology implements different Modulation and Coding Schemes (MCSs), so there exists neither a common multiple access technique nor a

---

[2] In a trunked system, channels are pooled for common use on a need basis. Trunking gain is the improvement obtained (in offered traffic intensity) when channels are merged into a pool.

signaling system to harmonize the use of these bands. Different governmental agencies, such as the European Radio Communication Office (ERO) in Europe and the FCC in the United States of America, have defined a set of rules either in terms of power spectral density masks or in terms of time duty cycles, depending on the application. The power spectral density mask defines the limits on the Peak-to-Average Power Ratio (PAPR) as a function of the frequency offset around the central frequency. The time duty cycle defines the longest cumulative period a specific device is allowed to transmit within a time unit.

4.2.1 Spectrum Sharing from a Game Theoretic Perspective

Studying dynamic spectrum sharing from a game theoretical perspective is important in many ways. First, by modeling dynamic spectrum sharing among network users (primary and secondary users) as games, the network users' behaviors and actions can be analyzed in a formalized game structure, by which the theoretical achievements in GT can be fully utilized. Second, GT equips us with various optimality criteria for the spectrum-sharing problem. These criteria, such as price of anarchy and price of stability (Fudenberg and Tirole 1991), are of high concern in the case of multiplicity of equilibria. Third, the application of GT to spectrum sharing enables us to derive efficient distributed approaches for dynamic spectrum access using local information. This aspect is useful when centralized control is not available and flexible self-organized approaches are deemed necessary. An overview of game theoretic dynamic spectrum sharing is found in Ji and Liu (2007). It is worth noting that various types of games exist depending on the considered scenario along with its underlying assumptions. In a noncooperative game, each player is selfish and unconcerned about all the other players' performance. In contrast, in a cooperative approach, each player maximizes a common benefit for the set of players assuming that all the other players have adopted the same cooperative behavior.

In Scutari et al. (2006), a unified framework based on potential games is proposed with an application to power control in CDMA, in which sufficient conditions are given for the existence and uniqueness of the Nash Equilibrium (N.E). Moreover, different distributed algorithms are provided along with their convergence properties. In Perlaza et al. (2009a) potential games (Fudenberg and Tirole 1991) were studied for BS selection and sharing in self-configuring networks. It is shown that depending on the number of transmitters, BS selection might perform better than BS sharing. This results in a Braess paradox (Perlaza et al. 2009a), where increasing the strategy space of each player leads to a degeneration of the global network spectral efficiency. On the other hand, spectrum sharing was studied using congestion games (Liu et al. 2009; Liu and Wu 2008) modeling the competition for resources among multiple selfish players. Similar to the potential games approach, when a player unilaterally switches its strategy, the change in its own payoff is the same as the change in a global objective, known as the potential function. In (Etkin et al. 2007), the spectrum-sharing problem in an unlicensed band is studied where a noncooperative game is shown to yield inefficient (Dubey 1986) outcomes. In Scutari et al. (2008a,b) the authors study the problem of maximizing mutual information subject to spectrum mask constraints and transmit power, for both simultaneous and asynchronous cases. The existence of the N.E is proven and sufficient conditions are given for the uniqueness along with distributed iterative algorithms to reach the N.Es In (Perlaza et al. 2009b) the benefits of bandwidth limiting in decentralized vector multiple access channels are examined. First, the authors study the case when transmitters use nonintersecting sets of bands, whereas in the second case all available frequency bands are freely exploited using Successive Interference Cancellation (SIC). A closed-form expression is given for the optimal number of accessible bands in the former case, while bandwidth limiting does not bring any significant improvement to the network spectral efficiency in the latter. In (Han and Liu 2005) a game theory approach for distributed resource allocation is adopted, where a power-control game at the user level and a throughput game at the system level are studied. The proposed games are shown to converge to near-optimal solutions. In Han et al. (2004) the authors propose a noncooperative game approach for distributed subchannel assignment, adaptive modulation and power control for multicell Orthogonal Frequency Division Multiplexing (OFDM) networks, where each user water-fills its power on different subchannels, regarding other users powers as interference.

Compared to Nash games where all players take their moves simultaneously, Stackelberg games (Fudenberg and Tirole 1991; Stackelberg 1934) model the scenario when a hierarchy exists between players. The players of a Stackelberg game include a leader and a follower/followers in which a leader commits to a strategy first, and then a follower selfishly optimizes his own reward considering the strategy selected by the leader.

The game leader perfectly knows the set of strategies and the utilities of the followers. Similarly, it is guaranteed that the followers can observe the actions of their leader(s). The framework of Stackelberg games is readily applied to the problem of dynamic spectrum access in which primary and secondary systems interact. Here, two different types of players (primary and secondary systems) exist whose priority is different upon accessing the radio resources. This concept naturally arises in multiple practical situations: (1) when primary and secondary systems share the spectrum, (2) when users have access to the medium in an asynchronous manner, (3) when operators deploy their networks at different times and (4) when some network entities have more power than others, such as BSs. In a two-level Stackelberg game, the game leader moves first and the other players follow and play simultaneously. In the recent literature, an application of Stackelberg games in dynamic spectrum access was presented in Simeone et al. (2008) where the design of a multiantenna access point using a noncooperative Stackelberg game model is investigated. A primary system might offer a set of channels to the opportunistic UEs in exchange for cooperation in the form of distributed space-time coding. It is shown that spectrum leasing based on trading secondary spectrum access for cooperation is a promising framework for cognitive radio. Moreover, in Hayel et al. (2009) the authors analyze the effect of hierarchy in energy-efficient power-control games for multiple access channel for both the individual user and overall network performance. It is shown that both the leader and followers benefit from hierarchy and following is more energy-efficient than leading. In Bloem et al. (2007), a Stackelberg formulation is applied to power control and channel allocation in the context of cognitive radio networks where a Stackelberg leader charges a virtual price for using the licensed frequency band that impacts the utilities of the player. Two price update algorithms were proposed. Simulations show that the total utility of the network increases when this virtual pricing is played compared with the traditional Stackelberg game. Finally, a survey of hierarchical power allocation games can be found in Bennis et al. (2009a).

4.2.2 Femtocells

Femtocells (Chandrasekhar and Andrews 2008; Xia et al. 2010) have recently gained attention in both academic and standardization forums due to their ability to increase capacity and coverage in indoor hotspots and cell-edge areas. The inherent problem with femtocells, that is interference, has probably been the most studied topic. Power Control (PC) is an evident means to reduce interference. In Arulselvan et al. (2009) the authors study power-control mechanisms for High Speed Downlink Packet Access (HSDPA) femtocells. Several algorithms are studied in the paper, taking into account the throughput of the femto UEs and the degradation in the macrocell performance. In Chandrasekhar et al. (2009) the authors investigated PC and provide a link budget analysis enabling simple and accurate insights to two-tier networks. The PC is based on a utility-based femtocell

Signal-to-Interference-plus-Noise Ratio (SINR) adaptation. Joint self-optimization of spectrum assignment and transmission power for OFDMA femtocells is given in Bernardo et al. (2010). The method is distributed and each femtocell acts as an autonomous entity. Besides, spectrum and power assignment are based on UEs' reported measurements (sensing intercell interference from other femtocells and the from the macrocells).

Interference management (in UMTS/High-Speed Packet Access (HSPA) networks) in a wider sense is discussed in Yavuz et al. (2009). This paper studies Home NB (HNB) autonomous carrier selection, DL power self-calibration, as well as UL interference management. Coverage and capacity are also analyzed. Garcia et al. (2010) examines dynamic interference coordination for open and closed access femtocells in LTE-Advanced (LTE-A) networks. This paper assumes dedicated carriers for femtocell operations. Therefore, there is no interference to/from the macro layer. The problem addressed in the paper is the interference between randomly deployed femtocells. Numerical results suggest that open access is preferable, at least in the DL case.

4.3 Spectrum Sharing with Game Theory

In this section, the problem of spectrum sharing where transmitter and receiver pairs share the same spectrum is examined. First the noncooperative spectrum sharing game is presented followed by the hierarchical approach.

4.3.1 Noncooperative Case

We consider the following interoperator spectrum game: K transmitter-receiver pairs share a spectrum band composed of N carriers. This study focuses on the DL. Let $p_i^{(n)}$ denote the transmit power of transmitter $i \in \{1, ..., K\}$ on carrier $n \in \{1, ..., N\}$, and $|h_{i,j}^{(n)}|^2$ the channel gain between transmitter i and receiver j in carrier n. Finally, σ_n^2 is the noise variance, assumed to be constant over all carriers. In a noncooperative approach, operator $i \in \{1, ..., K\}$ *selfishly* maximizes his utility function defined as the achievable rate R_i subject to the total transmit power constraint $\bar{P}_i$:

$$\max_{p_i^{(1)},...,p_i^{(N)}} R_i = \max_{p_i^{(1)},...,p_i^{(N)}} \sum_{n=1}^{N} \log_2 \left(1 + \frac{|h_{i,i}^{(n)}|^2 p_i^{(n)}}{\sigma_n^2 + \sum_{j \neq i}^{K} |h_{j,i}^{(n)}|^2 p_j^{(n)}} \right) \qquad (4.1)$$

$$s.t. \sum_{n=1}^{N} p_i^{(n)} \leq \bar{P}_i, \text{ for } p_i^{(1)}, ..., p_i^{(N)} \geq 0$$

The solutions to Equations (4.1) are given by the Water-Filling power allocation solutions (Yu et al. 2004):

$$p_i^{(n)} = \left(\frac{1}{\mu_i} - \frac{\sigma_n^2 + \sum_i |h_{j \neq i,i}^{(n)}|^2 p_{j \neq i}^{(n)}}{|h_{i,i}^{(n)}|^2} \right)^+ \qquad \text{for } i = 1, ..., K \qquad \text{and } n = 1, ..., N \qquad (4.2)$$

where $(x)^+ = \max\{x, 0\}$ and $\mu_i > 0$ is the Lagrangian multiplier chosen to satisfy the power constraint $\sum_{n=1}^{N} p_i^{(n)} = \bar{P}_i$.

In order to gain insight into the properties of the N.E of the interoperator spectrum game, let us take the case where two operators share two carriers ($K = N = 2$). It turns out that after solving Equation (4.1), there exists at least one N.E depending on the channel conditions. A full characterization of the N.E regions for the two-operators two-carriers case (but also holds for the general case) can be found in Bennis et al. (2009b), in which it was shown that, under certain channel realizations, the spectrum-sharing game is predictable with a *unique* N.E when:

$$\sum_{i=1, j\neq i}^{K} \frac{|h_{ji}^{(n)}|^2}{|h_{ii}^{(n)}|^2} < 1, \quad n = 1, ..., N \tag{4.3}$$

The physical meaning of Equation (4.3) is that the uniqueness of the N.E is ensured if the transmitter-receiver pairs are sufficiently far from each other.

4.3.2 Hierarchical Case

Here, the interoperator spectrum-sharing problem is extended to the case where the concept of hierarchy among players is taken into account. This situation is inherent when primary and secondary operators share the same spectrum. In the following, the interoperator spectrum-sharing problem with hierarchy is described for the two-operators two-carriers case. This case is subsequently extended to the general case with an arbitrary number of operators.

A *Stackelberg* game is proposed to model the spectrum-sharing problem where one of the two operators is chosen to be the leader (primary operator). The Stackelberg Equilibrium (S.E.) (Stackelberg 1934) is the best response when a hierarchy of actions exists between players. *Backward induction* (Fudenberg and Tirole 1991) is used to find the S.E., assuming that players can reliably forecast the behavior of other players, believing that other players can do similarly. First, in the high-level problem, primary operator 1 maximizes his own utility function (achievable rate). Then, in the low-level problem, secondary operator 2 (follower) maximizes his own utility taking into account $\mathbf{p}_1^{SE}$, the optimal power allocation of operator 1. By denoting $(\mathbf{p}_1^{SE}, \mathbf{p}_2^{SE})$ as the S.E., where $\mathbf{p}_i^{SE} = (p_i^{(1)^{SE}}, \ldots, p_i^{(N)^{SE}})$ for $i \in \{1, 2\}$. Then, the rate optimization problem for operator 1 (leader) can be written as:

$$\max_{p_1^{(1)}, ..., p_1^{(N)}} \sum_{n=1}^{N} \log_2 \left(1 + \frac{|h_{11}^{(n)}|^2 p_1^{(n)}}{\sigma_n^2 + |h_{21}^{(n)}|^2 p_2^{(n)} \{p_1^{(n)}\}} \right) \tag{4.4}$$

$$s.t. \sum_{n=1}^{N} p_1^{(n)} \leq \bar{P}_1, \text{ for } p_1^{(1)}, ..., p_1^{(N)} \geq 0$$

The rate optimization problem for operator 2 (follower) is:

$$\max_{p_2^{(1)}, ..., p_2^{(N)}} \sum_{n=1}^{N} \log_2 \left(1 + \frac{|h_{22}^{(n)}|^2 p_2^{(n)}}{\sigma_n^2 + |h_{12}^{(n)}|^2 p_1^{(n)^{SE}}} \right) \tag{4.5}$$

$$s.t. \sum_{n=1}^{N} p_2^{(n)} \leq \bar{P}_1, \text{ for } p_2^{(1)}, ..., p_2^{(N)} \geq 0$$

Using backward induction, and given the best response of the follower, Equation (4.4) can be rewritten as:

$$\max_{p_1^{(1)},...,p_1^{(N)}} \sum_{n=1}^{N} \log_2 \left(1 + \frac{|h_{11}^{(n)}|^2 p_1^{(n)}}{\sigma_n^2 + |h_{21}^{(n)}|^2 \left(\frac{1}{\mu_2} - \frac{\sigma_n^2 + |h_{12}^{(n)}|^2 p_1^{(n)}}{|h_{22}^{(n)}|^2} \right)^+} \right) \quad (4.6)$$

$$s.t. \sum_{n=1}^{N} p_1^{(n)} \leq \bar{P}_1, \text{ for } p_1^{(1)}, ..., p_1^{(N)} \geq 0$$

Therefore, the hierarchical sharing game boils down to solving Equation (4.6) where several cases are considered. In Bennis et al. (2009b) a detailed solution for the special case with two operators and two carriers was presented.

The interoperator spectrum sharing in the context of two operators can be extended to the more general case with $K > 2$ operators sharing the same spectrum. The problem is formulated in the same way where the leader's optimization problem is written as:

$$\max_{p_1^{(1)},...,p_1^{(N)}} \sum_{n=1}^{N} \log_2 \left(1 + \frac{|h_{11}^{(n)}|^2 p_1^{(n)}}{\sigma_n^2 + \sum_{j \neq 1}^{K} |h_{j1}^{(n)}|^2 p_j^{(n)} \{p_1^{(n)}\}} \right) \quad (4.7)$$

$$s.t. \sum_{n=1}^{N} p_1^{(n)} \leq \bar{P}_1, \text{ for } p_1^{(1)}, ..., p_1^{(N)} \geq 0$$

Solving Equation (4.7) becomes more challenging in the general case, in which the utility function of the primary operator is nonconvex ($p_j^{(n)}$ is function of $p_1^{(n)}$). Nevertheless, there exist suboptimal and low-complexity methods to solve the problem. To this end, and motivated by the work of Yu et al. (2006), Lagrangian duality theory is used, wherein the *duality gap* (Boyd and Vandenberghe 2004) provides an effective tool for solving the nonconvex optimization problem (Bennis et al. 2009b).

Figure 4.1 depicts the performance of the best and worst N.E in terms of average user throughput. Five carriers and two players are considered. The best N.E refers to the equilibrium maximizing the sumrate of both operators whereas the worst N.E case minimizes it. It is worth noting that the worst N.E acts like a lower bound for the N.E Furthermore, the Stackelberg approach is closer to the centralized approach (sum-rate optimization over all transmitter and carriers) as compared to the selfish case. This is due to the fact that, in the Stackelberg approach, operators take into account other operators' strategies whereas in the selfish case, operators behave carelessly by using the water-filling solution. On the other hand, Figure 4.2 depicts the Cumulative Density Function (CDF) of the ratio between the achievable rates of the hierarchical and noncooperative approaches. In this scenario, it is assumed that $K = 4$ operators of which one primary operator and three secondary operators share $N = 5$ carriers. As shown, the primary operator (i.e. operator 1) always improves its achievable rate compared to the purely selfish approach.

4.4 Spectrum Trading

Recently, there has been a huge interest in flexible and dynamic spectrum access aiming at an efficient radio resource allocation. Clearly, future radio systems need to have high spectrum efficiency and be able to share and coexist with other systems in a more efficient manner. This coexistence issue has brought up several new technical challenges. In this section, radio resources are traded between several operators. Primary operators refer to systems with priority in accessing the spectrum whereas

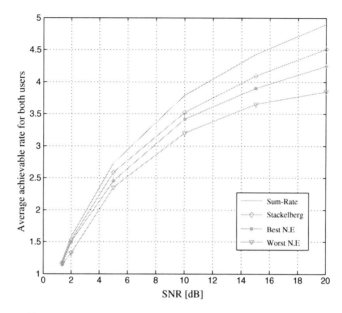

Figure 4.1 Average achievable rate for both users versus the signal-to-noise ratio for the centralized, Stackelberg approach, best and worst N.E for the noncooperative game

secondary operators utilize the spectrum when the primary system is idle. In particular, if primary systems do not use their resources during a certain period of time, these resources can be sold or traded to other operators. As a result, radio resources are efficiently utilized, satisfying both sellers (primary) and buyers (secondary), increasing thereby the total network spectral efficiency.

The problem of spectrum trading becomes more challenging with an increasing number of operators trying to gain more customers. This brings up the notion of pricing, that is, how operators

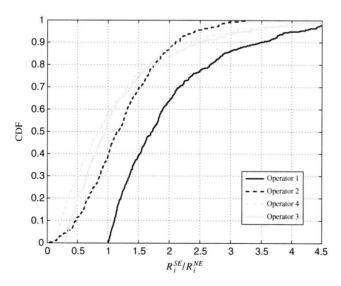

Figure 4.2 CDF of the ratio of the rates between the hierarchical (Stackelberg) and noncooperative (selfish) approaches for different operators

optimize their pricing strategy to satisfy their customers. Clearly, pricing is a very critical aspect that ought to be properly tackled so that both operators and customers are satisfied. In this section, the spectrum trading problem is investigated from a game theory perspective, where operators compete for the offered price and achieve the maximum profit. Two cases are herein considered: (1) the primary system does not use its own radio resources, which can be reused by secondary systems paying the requested price; (2) the cost of a secondary system sharing a primary system spectrum band is accounted for, where both primary and secondary systems share the same frequency band.

The spectrum demand function (also referred to as utility function) for the secondary system can be defined as follows (Hossain et al. 2009):

$$U(\mathbf{q}) = \sum_{i=1}^{n} q_i \alpha_i - \frac{1}{2} \left(\sum_{i=1}^{n} q_i^2 + 2\gamma \sum_{i \neq j} q_i q_j \right) - \sum_{i=1}^{n} p_i q_i \tag{4.8}$$

where $\mathbf{q} = (q_1, ..., q_n)$ is the set of spectral resources (also referred to as products), p_i is the price per product, and γ measures the substitutability among products. If $\gamma = 0$, each operator has monopolistic market power, while if $\gamma = 1$ the products are perfect substitutes. A negative γ implies that the products are complementary. Finally, α_i measures the quality of the spectral resource (for instance channel quality or spectral efficiency). Thus, an increase in α_i augments the marginal utility of consuming product q_i. The rationale behind using the quadratic utility function defined in Equation (4.8) is its concavity, which makes it easier for use in further analysis and is of particular interest in GT because it constitutes a second-order approximation to other types of nonlinear cost functions.

In order to derive the optimum spectrum demand, the first-order condition is used to determine the optimal consumption of products q_k, $1 \leq k \leq n$:

$$\frac{\partial U(q)}{\partial q_k} = \alpha_k - q_k - \gamma \sum_{j \neq k} q_j - p_k = 0 \tag{4.9}$$

which yields:

$$p_k = \alpha_k - q_k - \gamma \sum_{j \neq k} q_j \tag{4.10}$$

Equation (4.10) is generally used in Cournot oligopoly situations (Fudenberg and Tirole 1991) when there are n operators with identical products operating in a market with known market demand function. Moreover, every spectrum seller (primary operator) rationally assumes that the production of other operators will not be modified as a response to their moves and vice-versa (best response). The profit of every primary operator k is given by:

$$\Pi(q_k, q_{j \neq k}) = p_k \times \sum_{k=1}^{n} q_k = \left(\alpha_k - q_k - \gamma \sum_{j \neq k} q_j \right) \times \sum_{k=1}^{n} q_k \tag{4.11}$$

where $q_{j \neq k}$ denotes the spectrum usage of operators other than operator k. The first-order derivative of the profit with respect to the quantity q_k results in:

$$\frac{\partial \Pi(q_k, q_{j \neq k})}{\partial q_k} = 0 \tag{4.12}$$

$$q_k\{q_{j \neq k}\} = \frac{\alpha_k - \gamma \sum_{j \neq k} q_j}{2} \tag{4.13}$$

Summing over all operators, the Cournot equilibrium q_k^C for demand and price is

$$q_k^C = p_k^C = \frac{\alpha_k\left(\gamma(n-2)+2\right) - \gamma \sum_{j \neq k} \alpha_j}{(2-\gamma)(\gamma(n-1)+2)} \tag{4.14}$$

Another oligopoly model known as the Bertrand model (Fudenberg and Tirole 1991) is a model of price-setting oligopoly. In this case, prices are differentiated (unlike in the Cournot approach) and secondary operators seek the cheapest price. Summing over all operators, Equation (4.12) can be written as:

$$\sum_i \alpha_i - \sum_i q_i - \gamma(n-1) \sum_i q_i - \sum_i p_i = 0 \tag{4.15}$$

After some mathematical derivations (summing over all operators), the spectrum demand function for operator k can be expressed as follows (Hossain and Han 2009):

$$q_k^B(p_k, p_{j \neq k}) = \frac{(\alpha_k - p_k)(\gamma(n-2)) - \gamma \sum_{j \neq k}(\alpha_j - p_j)}{(1-\gamma)(\gamma(n-1)+1))} \tag{4.16}$$

4.4.1 Revenue and Cost Function for the Offering Operator

In order to develop a cost function, the Quality of Service (QoS) of the primary system needs to be considered. Therefore, two cases are considered here. In the first case, a spectrum band is shared with the secondary system, whereas in the second case the portion band is dedicated to the secondary system. The revenue R_i and costs C_i for primary operator i are calculated as follows:

$$R_i = c_1 M_i, \tag{4.17}$$

$$C_{q_i} = c_2 M_i \left(B_i^r - \alpha_i \frac{W_i - q_i}{M_i} \right)^2 \tag{4.18}$$

where c_1 and c_2 are the weights for the revenue and cost functions, respectively, B_i^r and α_i denote the bandwidth requirement and the spectral efficiency for primary operator i. Finally, M_i is the number of primary users, W_i is the total bandwidth and q_i is the spectrum demand of operator i.

Based on the aforementioned model, the players are the primary operators offering spectrum whose strategy is based on maximizing the price per unit of spectrum p and the payoff is the profit revenue – costs after the spectrum transaction. The solution of this game is a Nash Equilibrium.

The profit for every operator i is calculated as:

$$Pr_i\{p_i\} = q_i p_i + R_i - C_i, \tag{4.19}$$

where p_i denotes the set of prices offered to all players in the game by operator i. By definition, the N.E of a game is a strategy profile where no player can increase his payoff by choosing a different action, given other players' actions. In this case, the N.E is obtained using the best response function defined as the best strategy of one player given others' strategies. Hence, the best response function of operator i, given a set of prices offered by other primary operators, is:

$$BR_i(p_i) = \arg \max_{p_i} Pr_i(p_i) \tag{4.20}$$

The N.E is calculated by taking the first order of the profit: $\frac{\partial Pr_i(p)}{p_i} = 0$ for all operators i. It is worth noting that the spectrum demand is taken into account in order to compute the primary operator profit. Finally, the corresponding p_i^* is calculated, which is the N.E of the spectrum trading game.

4.4.2   Numerical Results

We assume a scenario with three operators of which two are primary. The bandwidth allocated to operators is $W_i = 20$ MHz. Ten primary UEs are served per operator. The target Bit Error Rate (BER) for the secondary operator is $BER^t = 10^{-4}$. The bandwidth requirement is set to be 2 Mbps and $c_1 = c_2 = 2$. SNR = 9 dB and the substitutability factor varies between 0.1 to 0.6.

Figure 4.3 depicts the profit of primary operator 1 as a function of the offered price. Clearly, when the offered price increases so does the profit, because more revenue is generated due to the higher price. However, for a certain value p_1, the revenue starts to dwindle as the secondary

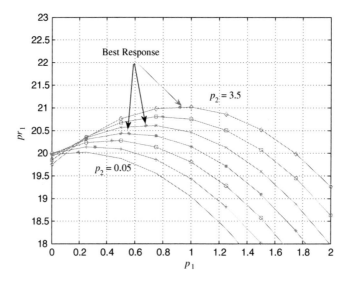

Figure 4.3   Profit of primary operator 1 as a function of the offered price to secondary operator 2. This is plotted for several values of the offered price from secondary operator p_2

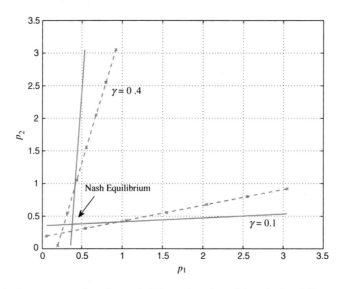

Figure 4.4   Best response function and N.E as a function of the substitutability parameter γ

operator decreases its spectrum demand due to the expensive prices charged by the primary operator. Furthermore, it is seen that the price that results in the highest profit for the operator is the best response: given the prices of operator 2, operator 1 maximizes its profit accordingly. Besides, in Figure 4.3 the best response for operator 1 increases with increasing p_2. This goes in line with the intuition since the primary operator has to increase its prices as well.

On the other hand, Figure 4.4 gives some insights about the N.E with respect to p_1 and p_2. The best response functions for both primary operators are shown, as a function of the substitutability values ($\gamma = 0.1$ and $\gamma = 0.4$). The N.E point gives less prices for both operators meaning that the secondary operator has the freedom of changing and picking up the best primary operator. In this way, both primary operators have no choice but to decrease their respective prices. Otherwise, the secondary system will not seek any further spectrum.

4.5 Femtocells and Opportunistic Spectrum Usage

Femtocells (Chandrasekhar and Andrews 2008) are low-cost, low-power, short-range, plug-and-play BSs that can be used for offloading traffic from the macro cells. This entails a boost in the spectral efficiency and improving indoor coverage. Since femtocells are deployed by the end user, the operator does not have any control in the deployment.

Femtocells can be divided into categories based on the access type. In open access, any UE may connect to the femtocell. In closed access, only a subset of UEs is allowed to connect to the femtocell. These UEs form a Closed Subscriber Group (CSG). The third access type is hybrid access, where all UEs have limited connectivity. Only the UEs in the CSG have full access rights.

The uncoordinated deployment of femtocells creates obvious problems with interference. The Home eNB (HeNB) may be located anywhere, for example very close to an eNodeB (eNB). This would cause severe interference in both femtocell and in the eNB. In coordinated deployment this would never happen, because network planning would select the locations of BSs (either eNB or HeNB far away from each other).

Examples of femto-to-macro and macro-to-femto interference are given in Figure 4.5, where useful signals are represented with solid lines and interference with dashed lines. Nominal femtocell coverage area is shown in gray. Figure 4.5(a) illustrates the interference in the DL. The HeNB transmits to its UE, generating interference to the Macro UE (MUE) nearby. This situation is especially difficult since the MUE is close to the cell edge and receives only a weak signal from the eNB. The same situation is shown in Figure 4.5(b) but for the UL case. A UE connected to a femtocell is close to the macro eNB, generating interference in the eNB receiver. On the other hand, Figure 4.5(c) shows a scenario with macro-to-femto interference in the DL. A femtocell close to an eNB is experiences interference from the macro network's DL signal. In Figure 4.5(d) a MUE is close to a femtocell, causing interference to the HeNB.

Since femtocells can be close to each other (due to uncoordinated deployment), interference occurs between adjacent femtocells. Figure 4.6(a) shows the DL case. Two HeNBs (HeNB 1 and HeNB 2) are located close to each other, and their DL signals cause interference to the other HeNBs. The corresponding problem in the UL is shown in Figure 4.6(b). Uplink transmissions of the UEs interfere other HeNBs' UL reception.

Ideally, a solution should be found to all of the interference problems described in Figures 4.5 and 4.6. This is a very challenging task due to the uncoordinated nature of the femtocells. However, as shown in this section, some of the problems can be solved.

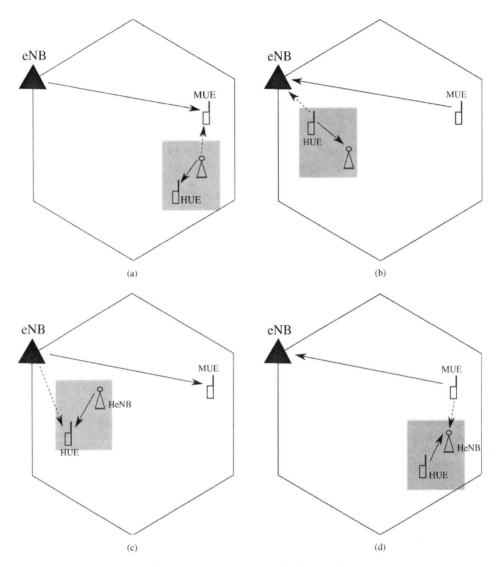

Figure 4.5   Interference with femtocells

In this chapter it is assumed that femtocells operate on the same frequency as the macro network. Therefore, the most important and most detrimental type of interference is from the femtocells to the macro network. Moreover, the interference from the macro network to the femtocells and the interference between adjacent femtocells have to be considered to obtain the best possible performance. Several solutions to this problem are provided:

- Self-organized femtocells: femtocells operate only if they do not cause significant degradation to the macro network. This method is blind and there is no control between macro and femto layers. This method does not require changes in the macro network.
- Femtocells and beacons: beacons (broadcast channels from the HeNB) are used to deliver interference information which used by the HeNBs in Physical Resource Block (PRB) assignment to avoid interference.

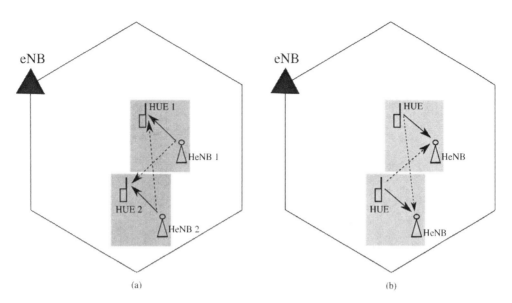

Figure 4.6   Femto-to-femto interference

• Femtocells with intercell interference coordination: femtocells avoid using those PRBs that might interfere with the macro network.
• Femtocells and GT: GT can be used to allocate PRBs and transmission powers to both MUE and Femto-UEs (FUEs).

In the following subsections we will describe these methods. Beforehand, a short overview of the femtocells mainly for Long-Term Evolution (LTE) is given.

4.5.1   Femtocells and Standardization

Basic concepts of several of the envisaged innovations of femtocell networks deployment are currently under discussion or specification by standardization bodies. In particular, the status of 3GPP is of special interest, because this is the reference system. Therefore, in what follows, an overview of the state of the art in the 3GPP standardization is provided, mainly up to LTE Release 10 (Rel-10), focusing on HeNBs.

During Long Term Evolution (LTE) standardization, the concept of femtocells was taken into consideration from the beginning. Together with the all-IP architecture of the evolved packet core with reduced number of nodes and interfaces, a tighter integration of femtocells into the network compared with Third Generation (3G) femtocells was achieved by reusing the same interfaces as macrocell eNBs to communicate with the core network.

The macro eNB is connected to the Evolved Packet Core (EPC) via the S1 interface, which consists of an interface to the Mobile Management Entity (MME) (S1-MME) handling the control plane and an interface to the Serving Gateway (S-GW) (S1-U) handling the user plane data. The macro eNBs are connected via the X2 interface, which allows efficient mobility management handled in a decentralized way. Figure 4.7 illustrates the interfaces used for a macro eNB in LTE. The main concern for femtocell integration into the Evolved Universal Terrestrial Radio Access Network (E-UTRAN) architecture during standardization was the large number of expected femtocells and, consequently, the large number of S1 Stream Control Transmission Protocol (SCTP) and

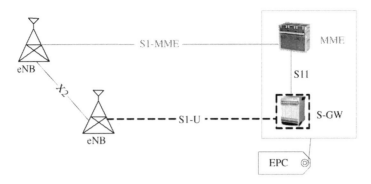

Figure 4.7   An example of LTE architecture for Macro eNB

GPRS Tunneling Protocol (GTP) connections to the MME/S-Gateway (GW), which could easily overload the nodes in the core network. In order to avoid this and improve the scalability of the core network, an operator can deploy a HeNB Gateway HeNB-GW, which combines the traffic of multiple femtocells reducing the number of S1 connections (S1-MME and S1-U) to the core network.

There are three variants, illustrated in Figure 4.8, of how femtocells can be integrated into the core network: (1) the HeNB-GW combines both the control and user plane traffic and distributes the control traffic to the MME and the bearer traffic to the S-GW; (2) the HeNB-GW only combines the control plane traffic from multiple HeNBs to the MME, while user plane data is routed directly to the S-GW; (3) the HeNBs are connected directly to the MME and S-GW. Another aspect of the large number of expected femtocells is that the number of peer-to-peer connections between macro and

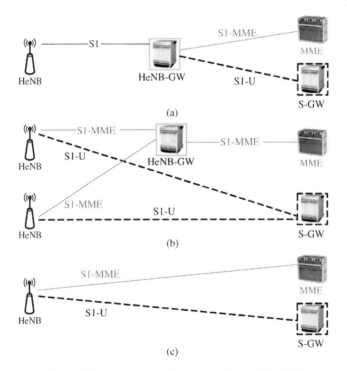

Figure 4.8   Femtocell architecture variants within LTE

femtocells and between femtocells can be very large. Thus, this interface has not been defined for femtocells up to Release 9 (Rel-9) and is under discussion for Rel-10.

It is worth noting that there is no standardized mechanism for cooperative/networked femtocells and between femtocells and macrocells yet. The architecture for mobile femtocells has also not yet been studied in 3GPP, Next Generation Mobile Network (NGMN), and FemtoForum.

4.5.2   Self-Organized Femtocells

The objective of self-organized femtocells is to avoid coordination between the femto and macro layers (WINNER+ 2009a), even if the macro network and femtocells belong to the same operator. The bandwidth is shared between the two layers and the challenge is to use interference avoidance techniques to minimize interference between layers.

We propose the following method for interference avoidance: the femtocell (both HeNB and Home User Equipment (HUE)) measures the Path Loss (PL) from the nearest eNB. If this PL is large enough we can be reasonably sure that using the femtocell in the UL band in Time Division Duplex (TDD) mode does not cause measurable interference to the macro networks UL band (see Figure 4.9), provided that the transmission powers in the femtocell are low enough. Note that the macro system is in FDD mode. This is especially true because, typically, the HeNB and HUEs are indoors, and penetration losses due to walls attenuate the signal to the eNB.

Since the PL measurement is done by the HeNB and HUEs, combining these measurement provides accurate estimate of the PL. Simulation results WINNER+ (2009a,b) show that taking the minimum of the PL measurements provides a robust PL estimate, with fast fading averaged out.

If the minimum of the PLs is above the threshold (nearest eNB is far away), the femtocell operation is allowed. Simulations suggest that a threshold of 85 dB is a good compromise (WINNER+ 2009a).

The problem is that if the femtocell is close to an eNB, the PL is low, and the femtocell cannot operate on the cellular UL band. On the other hand, femtocells are purchased by the EUs in situations where the coverage is not acceptable. In practice this is not an important problem because if the eNB–HeNB distance – and hence PL – is small then there is no need for the EU to purchase a HeNB since there are no coverage problems.

The concept of measuring the PL to allow secondary spectrum use is not limited to femtocells. Also Device-to-Device (D2D) communications are feasible using the same principle, see Chapter 9 for more information on D2D communications.

This method allows also interoperator sharing. Femtocells could use a different UL band from operators, provided that the interference from the femtocell to the macro network is negligible. Also D2D communications are possible with interoperator sharing. Note that, in the interoperator case,

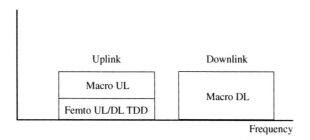

Figure 4.9   Spectrum arrangement for self-organized femtocells. The entire UL band is shared in time and frequency domains

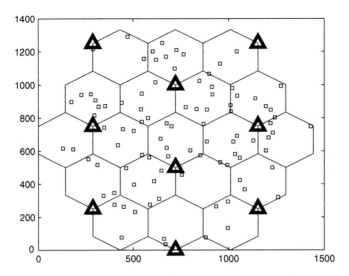

Figure 4.10   An example of randomly dropped femtocells in a macrocell network

only femtocells and D2D communications use self-organization. Macro networks cannot share spectrum using this technique.

Femtocells (and D2D using this technique) are understood as the secondary UEs of the spectrum. Therefore, it is important not to interfere with the macro network's operation. However, we can accept some performance degradation in the femtocells, such as increased outage.

The performance is assessed with a system-level simulator. An example network is depicted in Figure 4.10 with 19 cells, and a number of femtocells randomly dropped in the network area. eNBs are marked with a triangle, and femtocells with a square. Simulation parameters are summarized in Table 4.1.

Figure 4.11 shows the CDF of UE throughputs (i.e. both FUEs and MUEs) for different number of femtocells per macrocell area. Only the UL was simulated since DL is unaffected.

There are several factors that contribute to the large gains. The most important one is the small distance between FUE and HeNB. In the simulations, HeNBs are assumed to be indoors. Secondly, UEs connected to a femtocell do not take any resources from the macro system. Thus, these PRBs can be given to MUEs. Thirdly, there is a mechanism to reduce transmission power of the FUEs and HeNBs to 10 dBm. This results in a smaller overall interference level in the network.

Table 4.1   Simulation parameters

Parameter	Value
System bandwidth	10 MHz
Channel	Typical Urban
Number of macrocells	19
Number of UEs/macrocell	15
Number of UEs in a CSG	2
Number of femtocells/macrocell	0, 1, 3, 5
PL Threshold	85 dB
Femtocell transmission power	10 dBm

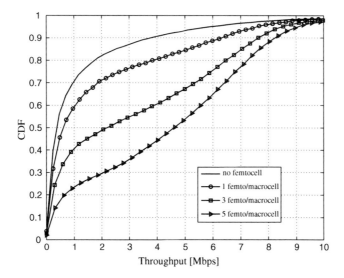

Figure 4.11   CDF of UE throughput

4.5.3   Beacon-Based Femtocells

This concept uses broadcast signaling to transmit control information to locally monitor the spectrum usage situation (WINNER+ 2009a). The broadcast messages (beacons) are received by UEs from their own network and also by UEs connected to other HeNBs. Then, the UEs process the measurements to understand the current spectrum usage situation. The UEs then report to their own HeNB. This way the HeNB acquires information about spectrum usage in the area, including spectrum used by other networks. This is illustrated in Figure 4.12. There are two HeNBs close to each other causing interference and, in the worst case, they might prevent communication in both cells.

FUE1a and FUE1b are connected to HeNB1, and FUE2 to HeNB2. FUE2 receives beacon information from HeNB1 and forwards this information to its own access point (HeNB2). Similarly, FUE1b receives beacon from HeNB2 and sends the information HeNB1. Consequently, both HeNBs in the area can have a clear picture of the spectrum usage in the local area.

The beacons may include information about the neighbor HeNBs (cell identities), which PRBs they are using, how much interference they are causing, their locations, information about their reference signals, bandwidth demand estimates, priority values, and TDD UL/DL switching point.

Each HeNB is required to assess its resource usage. If the HeNB has more resources than needed, it should release the unused PRBs and modify its beacon accordingly. These PRBs can then be used by other HeNBs. Besides, if there are free resources available – indicated by the reports sent by UEs – and a HeNB needs more PRBs, it should indicate this on its beacon. If no free PRBs are available, the HeNB should wait until resources become available. This may create a conflict. There may be two or more HeNBs that reserve a resource at the same time. This can be solved by using priority-based fairness mechanisms. Each HeNB is associated with a priority value, which is used to solve the conflict. The priority value should be known by the HeNBs in the local area. The priorities are dynamically updated based on commonly agreed rules to ensure fairness.

The following rules are used for priority management:

- Priorities are broadcasted using the beacons. The UEs receive the beacons from other HeNBs and relay the information to their own HeNB.
- Higher priority HeNBs can take resources from lower priority HeNBs.

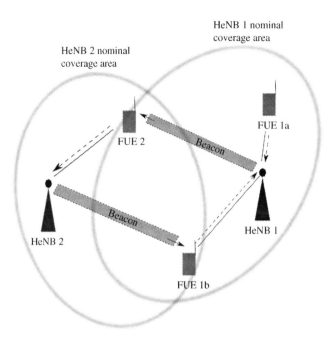

Figure 4.12 Example of beacon usage. Control signaling (UE→HeNB) is marked with dashed lines, data links (UE↔HeNB) with solid lines, and beacons with thick gray lines

- Free or unreserved PRBs can be used regardless of priority.
- Priorities are defined so that there is a penalty for reserving more resources than needed.

One example of priority calculation is based on the filtered average amount of PRBs reserved in the past:

$$p(t+1) = \alpha p(t) + (1-\alpha)\left(1 - \frac{N_{\text{tot}}}{N_{\text{res}}+1}\right) \tag{4.21}$$

where p is the priority value, t is the time instant, α is the filtering parameter ("forgetting factor"). Finally, N_{tot} and N_{res} are the number of total and reserved PRBs, respectively. More details and simulation results can be found in WINNER+ (2009a).

4.5.4 Femtocells with Intercell Interference Coordination

Consider the interference scenario of Figure 4.5(a), where the femtocell causes interference to a MUE close to the HeNB. The DL transmission could completely block the MUE's DL reception. If the HeNB uses the same resource blocks as the macro system for the MUE, this UE will be completely blocked due to the interference from the HeNB. The problem would not exist if the UE could connect to the femtocell. However, if the UE is not part of the CSG, this is not possible.

The concept of Inter-Cell Interference Coordination (ICIC) was extended to the femto-layer in WINNER+ (2009a,b), and is described briefly here. Consider the situation in Figure 4.13, where the femtocell and MUE1 are close to each other in cell edge area 1. The MUE detects the interference originating from the femtocell and sends a High Interference Indicator (HII) message to the serving

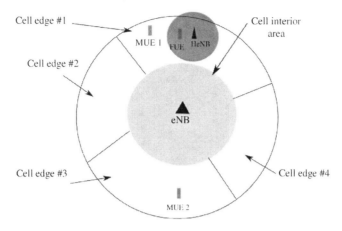

Figure 4.13 Division of macrocell into regions

eNB, which sends the message through the femto gateway to the HeNB causing the interference. Note that the HeNB can be identified by the DL broadcast channel. Then, the HeNB refrains from using the PRBs in question. Instead, resources used by MUE2 (far away from the femtocell) can be reused in the femtocell.

This method requires that the eNB knows the locations of the UEs. This requirement is reasonable because sectors are used in most cells and the distance can be estimated from the round trip delay or path losses. However, this approach requires a lot of signaling messages.

4.5.5   Femtocells with Game Theory

Using the game theoretic tools of section 4.3, and due to the strategic interaction between the macro and femtocell layers, GT is seen as a natural mathematical framework to analyze coexistence between the macro and femtocells. Here, the problem of spectrum sharing is investigated where a eNB is underlaid with multiple HeNB. This problem is considered from a *game theory* perspective where two games are investigated. First, in the noncooperative case, the eNB and HeNB – that is, players – behave selfishly aiming at improving their respective payoffs – that is, achievable rates. In the second case, a hierarchy is introduced among players as a means for improving the overall network efficiency. This problem is cast as a hierarchical game with a *leader-follower* approach in which the eNB is designated as the game's leader whereas HeNB are the game's followers.

Figure 4.14 depicts a cell with one eNB and two underlaid femtocells. This will be the scenario used in this section to validate the proposed method against simulation results. The maximum power constraint for each player i is assumed to be identical and normalized as $\bar{P}_i = 1$.

Figure 4.15 shows the average total sum rate of the network for the noncooperative sharing game $\mathcal{G}_1$ after solving (4.4), considering a channel noise variance $\sigma^2 = 0.1$. A number of observations can be made: the total sum rate of the network increases with increasing number of eNBs before leveling off. Moreover, it may be seen that for a given geographical area and an arbitrary number of carriers N, there is an *optimal*[3] number of BSs $M^\star$ that can be deployed in the network, after which the total sum rate of the network starts decreasing due to the interference-limited regime. The latter is essentially due to the selfishness nature of the HeNBs.

---

[3] This is referred to as the Braess paradox (Braess 1969), which implies that increasing the space of strategies of each player, that is, the number of eNBs each player can use, ends up degenerating the global performance of the network.

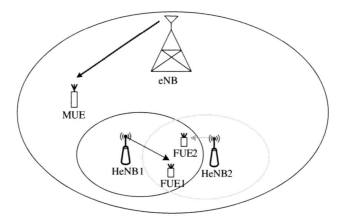

Figure 4.14   An example of a eNB and two underlaid femtocells

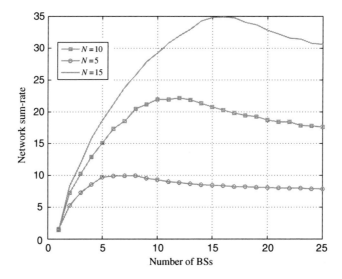

Figure 4.15   Sum-rate of the network as a function of the number of BSs for the noncooperative sharing game

Figure 4.16 illustrates the total sum rate of the network for the noncooperative, hierarchical and centralized approach for the case of $N = 5$ carriers. It is clear that through adopting a hierarchy the performance of the network outperforms the selfish approach and bridges the gap between both the noncooperative and the fully centralized one. In other words, it is possible to deploy more HeNBs through the concept of hierarchy. Finally, as expected, the centralized approach outperforms both approaches.

4.6   Conclusion, Discussion and Future Research

With the upsurge in cognitive radio studies for next wireless generations, pricing has received a lot of interest recently. We have presented a game theory model to obtain the optimal pricing for dynamic spectrum sharing, where several operators compete with each other to offer spectrum to secondary systems.

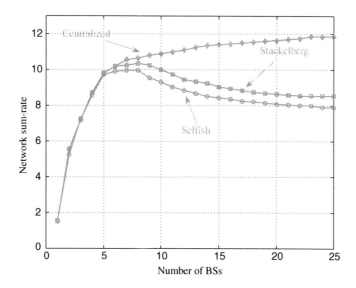

Figure 4.16 Sum-rate of the network as a function of the number of BSs, for the selfish, hierarchical and centralized approaches

The strategic interaction among wireless operators was modeled using specific tools from GT (hierarchical games). It was shown that through adopting hierarchy, operators can improve their respective payoffs, as opposed to the noncooperative case that yields a lower payoff, that is, N.E A natural extension to femtocells was carried out in which the macrocell was cast as a leader and the femtocells into followers. It was shown that the total sum-rate of the network increases by adding more femtocells into the network. However, there is an optimal number for these deployments after which interference becomes too detrimental, and hence the overall performance trickles down.

Hampered by the limitation of conventional macrocell networks due to poor indoor penetrations and the presence of dead spots, the indoor coverage is drastically altered. As a consequence, strategic means to improve the spatial reuse and coverage, such femtocells technologies, are required.

Femtocells are a promising technique for improving network performance and increasing spectral efficiency in the system level. From an individual UE point of view, higher QoS, improvement in cell edge performance, and better system throughput are expected without affecting macro network operation. The techniques presented in this chapter show that the most detrimental type of interference (femto-to-macro) can be solved. Moreover, solutions to femto-to-femto interference were given.

4.6.1 Future Research

Future research in the femtocell area will focus on the interference problem. This is a challenging task because the control signaling between the two tiers is limited or nonexistent. A possible solution to the femto-to-femto interference problem should use spectrum sensing to identify unused PRBs. A solution to macro-to-femto interference can also use spectrum sensing; the HeNB would choose only those resources where there is no interference. The concept of spectrum sensing is not new, but the characteristics of the femtocells probably requires a different approach from sensing and utilizing unused spectrum blocks.

An important concept for femtocells is the use of Multiple-Input Multiple-Output (MIMO) techniques, as is the case in macro networks. These can be used to to increase capacity and suppress the interference. If a certain level of coordination between femtocells is allowed,

Coordinated Multipoint transmission or reception (CoMP) can also be used. For instance, CoMP can also be used between tiers, that is, between femtocells and macrocells.

Even though we have concentrated on interference avoidance and management problems, there are still other issues that have to be dealt with before femtocell attain their full potential. These include security, mobility and handovers (femto-to-femto and macro-to-femto), just to name a few.

References

Arulselvan N, Ramachandran V, Kalyanasundaram S and Han G 2009 Distributed power control mechanisms for HSDPA femtocells *Proc. VTC 2009 Spring – IEEE 69th Vehicular Technology Conf.*

Bennis M, Lasaulce S and Debbah M 2009a *Hierarchy in Power Allocation Games*. Auerbach Publications, Taylor and Francis Group, CRC Press.

Bennis M, Letreust M, Lilleberg J, Lasaulce S and Debbah M 2009b Spectrum sharing games on the interference channel *Proc. IEEE GAMENET.*

Bernardo F, Agusti R, Cordero J and Crespo C 2010 Self-optimization of spectrum assignment and transmission power in OFDMA femtocells *Proc. Sixth Advanced International Conference on Telecommunications.*

Bloem M, Alpcan T and Basar T 2007 A Stackelberg game for power control and channel allocation in cognitive radio networks *Proc. The Second International Conference on Performance Evaluation Methodologies and Tools.*

Boyd S and Vandenberghe L 2004 *Convex Optimization*. Cambridge University Press, Cambridge.

Braess D 1969 Uber ein Paradoxon aus der Verkehrsplanung. *Unternehmensforschung* **24**(5), 257-268.

Buddhikot M 2007 Understanding dynamic spectrum access: models, taxonomy and challenges *Proc. IEEE International Symposium on New Frontiers in Dynamic Spectrum Access Networks*, pp. 649–663.

Cabric D, Mishra SM, Willkomm D, Brodersen R and Wolisz A 2005 A cognitive radio approach for usage of virtual unlicensed spectrum *Proc. IEEE International Symposium on Information Theory.*

Chandrasekhar V and Andrews JG 2008 Femtocell networks: a survey. *IEEE Commun. Mag.* **46**(9), 59–67.

Chandrasekhar V, Andrews JG, Muharemovic T, Shen Z and Gatherer A 2009 Power Control in Two-Tier Femtocell Networks. *IEEE Trans. Wireless Commun.* **8**(8), 4316–4328.

Dubey P 1986 Inefficiency of Nash equilibria. *Mathematics of Operations Research* **11**(1), 1–8.

Etkin R, Parekh A and Tse D 2007 Spectrum sharing for unlicensed bands. *IEEE J. Select. Areas Commun.* **25**(3), 517–528.

Fudenberg D and Tirole J 1991 *Game Theory*. MIT Press, Cambridge MA.

Garcia LGU, Pedersen KI and Mogensen PE 2010 On Open versus closed LTE-Advanced femtocells and dynamic interference coordination *Proc. WCNC 2010 – IEEE Wireless Commun. and Networking Conf.*

Han Z, Ji Z and Liu KJR 2004 Power minimization for multi-cell OFDM networks using distributed non-cooperative game approach *Proc. Globecom 2004 – IEEE Global Telecommunications Conf.*, vol. 6, pp. 3742–3747.

Han Z and Liu KJR 2005 Noncooperative power-control game and throughput game over wireless network. *IEEE Trans. Commun.* **53**(10), 1625–1629.

Hayel Y, Lasaulce S, El-Azouzi R and Debbah M 2009 Introducing hierarchy in energy games. *IEEE Trans. Commun.* **8**(7), 3833–3843.

Hossain E, Niyato D and Han Z 2009 *Dynamic Spectrum Access and Management in Cognitive Radio Networks* first edn. Cambridge University Press, Cambridge.

Ji Z and Liu KJR 2007 Dynamic spectrum sharing: a game theoretical overview. *IEEE Commun. Mag.* **45**(5), 88–94.

Katzela I and Naghshineh M 1996 Channel assignment schemes for cellular mobile telecommunication Systems: a Comprehensive Survey *Proc. IEEE Personal Communications.*

Leaves P, Ghaheri-Niri S, Tafazolli R, Christodoulides L, Sammur T, Stahl W and Huschke J 2001a Dynamic spectrum allocation in a multi-radio environment: concept and algorithm *Proc. Conference on 3G Mobile Communication Technologies*, pp. 53–57.

Leaves P, Ghaheri-Niri S, Tafazolli R, Christodoulides L, Sammut T, Staht W and Huschke J 2001b Dynamic spectrum allocation in a multi-radio environment: Concept and algorithm *Proc. IEEE Second International Conference on 3G Mobile Communication Technologies*, pp. 53–57.

Liu M. Ahmad S and Wu Y 2009 Congestion games with resource reuse and applications in spectrum sharing *Proc. IEEE GAMENET*, pp. 171–179.

Liu M and Wu W 2008 Spectrum sharing as congestion games *Proc. IEEE Allerton*, pp. 1146–1153.

Mitola J 2001 Cognitive radio for flexible mobile multimedia communications. *Mobile Networks and Applications* **6**(5), 435–441.

Pereirasamy M, Luo J, Dillinger JM and Hartmann C 2004 An approach for inter-operator spectrum sharing for 3G systems and beyond *Proc. PIMRC 2004 – IEEE 15th Int. Symp. on Pers., Indoor and Mobile Radio Commun.*, pp. 1952–1956.

Pereirasamy M, Luo J, Dillinger JM and Hartmann C 2005 Dynamic inter-operator spectrum sharing for UMTS FDD with displaced cellular networks *Proc. PIMRC 2005 – IEEE 16th Int. Symp. on Pers., Indoor and Mobile Radio Commun.*, pp. 1952–1956.

Perlaza SM, Belmega EV, Lasaulce S and Debbah M 2009a On the base station selection and base station sharing in self-configuring networks *Proc. Gamecomm*.

Perlaza SM, Lasaulce S, Debbah M and Bogucka H 2009b On the benefits of bandwidth limiting in decentralized vector multiple access channels *Proc. CROWNCOM*.

Satapathay D and Peha J 1996 Spectrum sharing without licenses: opportunities and dangers *Proc. Globecom 1996 – IEEE Global Telecommunications Conf.*, pp. 414–418.

Satapathay D and Peha J 1997 Performance of unlicensed devices with a spectrum etiquette *Proc. Globecom 1997 – IEEE Global Telecommunications Conf.*, pp. 414–418.

Satapathay D and Peha J 1998 Etiquette modification for unlicensed spectrum: approach and impact *Proc. VTC 1998 Spring – IEEE 48th Vehicular Technology Conf.*, pp. 272–276.

Scutari G, Barbarossa S and Palomar DP 2006 Potential games: a framework for vector power control problems with coupled constraints *Proc. ICASSP 2006 – IEEE Int. Conf. Acoust. Speech and Signal Processing*.

Scutari G, Palomar DP and Barbarossa S 2008a Competitive design of multiuser MIMO systems based on game theory: a unified view. *IEEE J. Select. Areas Commun.* **26**(7), 1089–1103.

Scutari G, Palomar DP and Barbarossa S 2008b Optimal linear precoding strategies for wideband noncooperative systems based on game theory – Part II: Algorithms. *IEEE Trans. Signal Processing* **56**(3), 1250–1267.

Simeone O, Stanojev I, Savazzi S, Bar-Ness Y, Spagnolini U and Pickholtz R 2008 Spectrum leasing to cooperating secondary ad hoc networks. *IEEE J. Select. Areas Commun.* **26**(1), 203–213.

Stackelberg VH 1934 *Marketform und Gleichgewicht*. Julius Springer, Vienna.

WINNER+ 2009a Celtic Project CP5-026 *Initial Report on System Aspects of Flexible Spectrum Use* (ed. Vihriälä J.). Public Deliverable D1.2, Wireless World Initiative New Radio – WINNER+, January 2009, http://projects.celtic-initiative.org/winner+/index.html (accessed June 2011).

WINNER+ 2009b Celtic Project CP5-026 Intermediate Report on System Aspects of Flexible Spectrum Use (ed. Saul A. and Vihriälä J.). Public Deliverable D1.6, Wireless World Initiative New Radio - WINNER+, October 2009, http://projects.celtic-initiative.org/winner+/index.html (accessed June 2011).

Xia P, Chandrasekhar V and Andrews JG 2010 Open vs. Closed Access Femtocells in the Uplink. IEEE Trans. Wireless Commun **9**(12), 3798–3809.

Yavuz M, Meshkati F, Nanda S, Pokhariyal A, Johnson N, Raghothaman B and Richardson A 2009 Interference management and performance analysis of UMTS/HSPA+ femtocells. *IEEE Commun. Mag.* **47**(9), 102.

Yu W, Ginis G and Cioffi JM 2006 Dual methods for nonconvex spectrum optimization of multicarrier systems. *IEEE Trans. Commun.* **54**(7), 1310–1322.

Yu W, Rhee W, Boyd S and Cioffi J 2004 Iterative water-filling for Gaussian vector multiple-access channels. *IEEE Trans. Inform. Theory* **50**(1), 145–152.

Zhang M and Yum T 1989 Comparison of channel-assignment strategies in cellular mobile telephone systems. *IEEE Trans. Veh. Technol.* **38**(4), 211–215.

Zhao Q and Sadler BM 2007 A survey of dynamic spectrum access. *IEEE Signal Processing Mag.* **24**(3), 79–89.

5

Multiuser MIMO Systems

Antti Tölli, Petri Komulainen, Federico Boccardi, Mats Bengtsson and Afif Osseiran

In Multiple-Input Multiple-Output (MIMO) communications, multiple antenna elements are employed both in the transmitter and the receiver, in order to obtain increased data rates or improved reliability compared to single-antenna transmission. In cellular systems, the Base Station (BS) may operate in Single-User (SU)-MIMO mode (see Figure 5.1 (a)), that is, to employ point-to-point transmission of data just for one multi-antenna User Equipment (UE) at a time. Alternatively, the BS can perform Multi-User (MU)-MIMO transmission (see Figure 5.1 (b)), that is, spatially multiplex data streams intended for different UEs that may be equipped with arbitrary numbers of antenna elements. The use of MU-MIMO mode offers potential system capacity gains especially when the BS employs a large antenna array.

The theory and practice of point-to-point MIMO communications are well-known, and the SU-MIMO mode has been adopted into wireless system standards such as Long Term Evolution (LTE), as one of the core methods to increase spectral efficiency. On the other hand, while the theory of the MU-MIMO transmission has also been relatively well investigated, the corresponding practical methods are less mature. Thus, the main research challenges reside in the development of MU-MIMO transmission concepts. Nevertheless, LTE Release 8 (LTE Rel-8) already supports a basic form of MU-MIMO mode.

This chapter discusses various aspects of MIMO communications, with a special focus on the downlink direction of cellular systems. First, an overview of the fundamentals and existing techniques is given in section 5.1. Then, the MIMO related standards and standardization activities are described in section 5.2. In section 5.3, system optimization methods for the case of nearly perfect Channel State Information at the Transmitter (CSIT) are addressed in detail. Some recent advances concerning limited CSI feedback and sounding are briefly presented in section 5.4. Finally, the future directions and challenges of MIMO techniques are briefly discussed in section 5.5.

5.1 MIMO Fundamentals

The use of the term MIMO is meant to differentiate it from a classical wireless system where a single antenna is used both at the transmitting and at the receiving end, called Single Input Single Output (SISO). More generally, it is common to define an antenna system related to the number of antennas at the receiver and transmitter. Figure 5.2 shows four antenna configurations that can

Mobile and Wireless Communications for IMT-Advanced and Beyond, First Edition.
Edited by Afif Osseiran, Jose F. Monserrat and Werner Mohr.
© 2011 Afif Osseiran, Jose F. Monserrat and Werner Mohr. Published 2011 by John Wiley & Sons, Ltd.

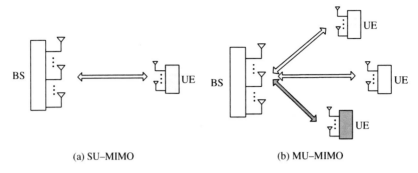

(a) SU–MIMO   (b) MU–MIMO

Figure 5.1   SU- and MU-MIMO transmissions

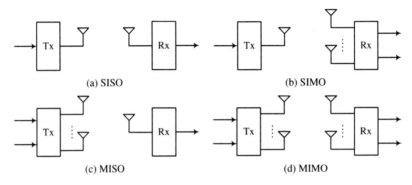

(a) SISO   (b) SIMO

(c) MISO   (d) MIMO

Figure 5.2   Antenna configurations

characterize any wireless radio communication system: SISO, Single Input Multiple Output (SIMO) (which is characterized by a single transmit antenna and multiple receive antennas), Multiple-Input Single-Output (MISO) (which is characterized by multiple transmit antennas and a single receive antenna), and MIMO, where multiple antennas are used at both ends of the transmission. Multiantenna techniques are used in order to achieve

- diversity gain;
- array gain; or
- multiplexing gain.

Roughly speaking, the role of diversity is to increase the reliability of the radio link. The diversity gain is the gain obtained from receiving independently (or partially correlated) faded replicas of the signal. Diversity can be achieved by transmitting or receiving the signal over multiple independently fading paths (in time, frequency or space). Transmit diversity techniques attain exclusively diversity gain, while space-time-coding methods aim for a beneficial tradeoff between diversity and multiplexing. These techniques do not require knowledge of the channel matrix, that is, CSIT. Formally, the diversity gain for a MIMO system designates the negative slope of the average error probability, P_e, versus the Signal to Noise Ratio (SNR) and is given by

$$d = -\lim_{\text{SNR}\to\infty} \frac{\log(P_e(\text{SNR}))}{\log(\text{SNR})} \tag{5.1}$$

Array gain, also called beamforming gain, is the average increase in the receive SNR due to a coherent combining from multiple antennas at the receiver or transmitter or both. Array gain requires partial or full channel knowledge at the transmitter or receiver and depends on the number of transmit and receive antennas.

Spatial multiplexing gain is the increase of data rate at no additional power consumption, obtained for example by transmitting multiple data streams or layers that are spatially multiplexed into the same time and frequency slot. Formally, the multiplexing gain of a MIMO transmission strategy, achieving $C(\text{SNR})$ as the information rate at an arbitrary SNR, is defined as

$$r = \lim_{\text{SNR}\to\infty} \frac{C(\text{SNR})}{\log(\text{SNR})} \tag{5.2}$$

A simple approach to realize spatial multiplexing is to apply linear transmit precoding (beamforming) so that each data stream is conveyed by a spatial beam that can be resolved by the receiver. Spatial multiplexing is feasible even without CSIT, as long as the receiver is capable of resolving the different streams. Here, each stream may be transmitted by one transmit antenna element, or alternatively precoded over multiple antennas with a predefined transmit precoder, or with a predefined (time-varying) sequence of precoders. This technique is called open-loop precoding. On the other hand, closed-loop transmit precoding takes advantage of CSIT in one form or another. Typically, CSIT means either direct knowledge of the channel matrix or its statistics. Alternatively, the receiver may report directly to the transmitter informing it of the desired transmit precoding matrix.

While CSIT is beneficial in the SU-MIMO case, it is essential for MU-MIMO transmission. This is because the BS transmitter needs to ensure that the signals intended for one UE do not severely interfere the reception of another UE. Insufficient CSIT will soon render the MU-MIMO system interference-limited, as the UEs – each equipped with a limited number of antennas – cannot cancel the multiple access interference. Therefore, acquisition of the CSIT is one of the most important research challenges for cellular MIMO systems. One further important research issue is scheduling, or UE selection and grouping. That is, before multiuser precoding, a set of spatially compatible UEs should be selected so that the precoder design for this set becomes relatively easy.

This section provides an overview of the fundamentals and the existing literature related to MIMO communication. More precisely, section 5.1.2 focuses on point-to-point MIMO communication and section 5.1.3 concentrates on MU-MIMO communications. Section 5.1.3 considers the problem of allocating resources among multiple UEs across the space, frequency and time dimensions while strategies for designing a MIMO system with co-channel interference are introduced in section 5.1.4.

5.1.1 System Model

Let us assume a narrowband frequency flat fading downlink MU-MIMO channel (at some time instant) with K UEs, each equipped with N_{R_k} antennas, and a single BS having N_T antennas. The input-output relation for the channel of UE k is described as

$$\mathbf{y}_k = \mathbf{H}_k \mathbf{W} \mathbf{x} + \mathbf{n}_k \tag{5.3}$$

where $\mathbf{y}_k \in \mathbb{C}^{N_{R_k}}$ is the signal vector received by UE k, $\mathbf{H}_k \in \mathbb{C}^{N_{R_k} \times N_T}$ is the channel matrix for UE k, $\mathbf{W} = [\mathbf{w}_1, \ldots, \mathbf{w}_S] \in \mathbb{C}^{N_T \times S}$ is the precoder matrix of the BS, $\mathbf{x} = [x_1, \ldots, x_S]^T \in \mathbb{C}^S$ is the vector of normalized data symbols so that $E[|x_s|^2] = 1$, and $\mathbf{n}_k$ represents noise and interference. In wideband systems, Equation (5.3) can denote a single carrier out of a wideband multicarrier channel.

Let $\mathcal{U}$ denote the set of active or scheduled UE indexes. The BS transmits S independent data streams,

$$S \leq \min(N_T, \sum_{k \in \mathcal{U}} N_{R_k}) \tag{5.4}$$

and for each data stream s, $s = 1, \ldots, S$, the scheduler unit associates an intended UE k_s. The transmit power of stream s is $p_s = \|\mathbf{w}_s\|_2^2$, where $\|.\|_2$ denotes the Euclidian norm operator. Note that in general more than one stream can be assigned to one UE, that is, the cardinality of the set of scheduled UEs, $\mathcal{U} = \{k_s | s = 1, \ldots, S\}$, is less than or equal to S.

In the special case of SU-MIMO transmission, there is only one scheduled UE, and the dependence of the UE index k in Equation (5.3) can be dropped. The same mathematical relationship and several of the standard MIMO results, can also be applied to model other situations, such as time-dispersive single antenna channels, relay communication, and other so-called virtual MIMO scenarios.

At the receiver side, the desired data may be recovered by using linear techniques such as Zero Forcing (ZF) and linear Minimum Mean Square Error (MMSE) (which is also called LMMSE) estimation, or nonlinear techniques such as Maximum Likelihood (ML) detection or Successive Interference Cancellation (SIC) (especially when spatial multiplexing is used).

Let $\boldsymbol{v}_s$ be a linear receive beamformer (or antenna combining vector) for data stream s, used by UE k_s to generate the decision variable, $\hat{x}_s = \boldsymbol{v}_s^H \mathbf{y}_{k_s}$, where $(.)^H$ denotes the Hermitian operator. For arbitrary linear transmit and receive beamformers, the SINR of data stream s can be expressed as

$$\Gamma_s = \frac{\left|\boldsymbol{v}_s^H \mathbf{H}_{k_s} \mathbf{w}_s\right|^2}{N_0 \|\boldsymbol{v}_s\|_2^2 + \sum_{i=1, i \neq s}^{S} \left|\boldsymbol{v}_s^H \mathbf{H}_{k_s} \mathbf{w}_i\right|^2} \tag{5.5}$$

where the noise-plus-interference was assumed to be white, that is, $\mathbf{n}_k \sim \mathcal{CN}(0, N_0 \mathbf{I})$. A commonly used linear MMSE estimator that maximizes the Signal to Interference plus Noise Ratio (SINR) at the receiver is given by

$$\boldsymbol{v}_s = \left(\mathbf{H}_{k_s} \mathbf{W} \mathbf{W}^H \mathbf{H}_{k_s}^H + N_0 \mathbf{I}\right)^{-1} \mathbf{H}_{k_s} \mathbf{w}_s \tag{5.6}$$

Example

In a SU-MIMO scenario, the expression of the receiver combining vectors $\boldsymbol{v}_s$, $s = 1, \ldots, S$ is more compact. Let $\boldsymbol{\Upsilon} = [\boldsymbol{v}_1, \ldots, \boldsymbol{v}_S]$ be the receiver combining matrix of a UE k receiving S streams. Then, for ZF and linear MMSE receivers, $\boldsymbol{\Upsilon}$ are given by

$$\boldsymbol{\Upsilon}^{MMSE} = \mathbf{H}_k \mathbf{W} \left(\mathbf{W}^H \mathbf{H}_k^H \mathbf{H}_k \mathbf{W} + N_0 \mathbf{I}\right)^{-1} \tag{5.7}$$

$$\boldsymbol{\Upsilon}^{ZF} = \mathbf{H}_k \mathbf{W} \left(\mathbf{W}^H \mathbf{H}_k^H \mathbf{H}_k \mathbf{W}\right)^{-1} \tag{5.8}$$

In general, ZF reception may not be feasible in the multiuser case since the interstream interference can be nulled by the receiver k only if $S \leq N_{R_k}$, unless ZF transmit precoding was used.

5.1.2   Point-to-Point MIMO Communications

Transmission Methods and Capacity

The pioneering work by Winters (1984, 1987) proposed the use of Space Division Multiple Access (SDMA) to boost up the capacity of wireless communication systems. Multiple receive antennas in the Uplink (UL) enable spatial separation of the signals from different UEs, and hence allow several UEs to simultaneously communicate with the BS. By the mid-1990s the same idea was transferred to point-to-point communications with multiple antennas at both the transmitter and the receiver. It was noticed that the uplink SDMA with single-antenna transmitters is, in fact, similar to point-to-point MIMO communication without Channel State Information (CSI) at the transmitter. This resulted in a large potential for the use of the angular or space domain to convey multiple independent data streams by a single multiantenna transmitter to one multiantenna receiver. In a MIMO system, the signals transmitted from colocated antennas can still be separated at the receiver provided that the scattering environment is rich enough. A MMSE receiver with SIC was shown to be a theoretically optimal solution for both SDMA with single transmit antennas and MIMO without transmitter CSI (Tse and Viswanath 2005).

The research on point-to-point MIMO communications was pioneered by Telatar (1999, 1995) and Foschini (1996) and Foschini and Gans (1998). Foschini considered the case where the CSI of the MIMO channel is only available at the receiver and not at the transmitter. For such a case, he also proposed a capacity-achieving transmission architecture called the Bell Labs Space-Time Architecture (BLAST). Both Foschini and Telatar also provided an outage capacity analysis for the slow-fading MIMO channel and demonstrated that, in ideal conditions, the capacity can grow linearly with the minimum number of transmit and receive antennas. Telatar also showed that, *the MIMO channel matrix, with full CSI at the transmitter, can be converted into parallel, noninterfering SISO subchannels through a Singular Value Decomposition (SVD) so that the number of parallel data streams is dictated by the rank of the matrix.* The optimal capacity-achieving transmission is then carried out by pre and post combining each stream with the right and left singular vectors of the channel matrix, respectively. Denoting the SVD of the channel matrix by $\mathbf{H} = \mathbf{U}\boldsymbol{\Lambda}\mathbf{V}^H$, the signal transmitted over the antennas is $\mathbf{V}\mathbf{d}$, where vector $\mathbf{d}$ contains the data signal. $\boldsymbol{\Lambda}$ is a diagonal matrix with the singular values $\{\lambda_i\}$ for $i = 1, \ldots, N_{\min}$ on the main diagonal, where $N_{\min} = \min(N_T, N_R)$. The receiver estimates the data using $\hat{\mathbf{d}} = \mathbf{U}^H\mathbf{y}$, resulting in

$$\hat{\mathbf{d}} = \boldsymbol{\Lambda}\mathbf{d} + \mathbf{U}^H\mathbf{n} = \boldsymbol{\Lambda}\mathbf{d} + \tilde{\mathbf{n}} \tag{5.9}$$

that is, a number of equivalent parallel scalar channels, as illustrated in Figure 5.3. The effective noise $\tilde{\mathbf{n}}$ is uncorrelated between the scalar channels if $\mathbf{n}$ is spatially white, because $\mathbf{U}$ is unitary. The channel capacity is obtained by power allocation, using water-filling over the parallel SISO subchannels with gains corresponding to the singular values of the channel matrix. Here, each parallel channel i is allocated with transmit power p_i so that the data signal becomes $\mathbf{d} = \text{diag}(\sqrt{p_1}, \ldots, \sqrt{p_{N_{\min}}}) \cdot \mathbf{x}$, where $\mathbf{x}$ is a vector of normalized data symbols. The resulting information capacity is given by Tse and Viswanath (2005)

$$C = \sum_{i=1}^{N_{\min}} \log\left(1 + \frac{p_i\lambda_i^2}{N_0}\right) \tag{5.10}$$

where N_0 is the noise variance, and the power levels are obtained by the water-filling principle. Furthermore, the capacity-achieving transmission strategy requires a Gaussian codebook with continuous rate allocation among the parallel subchannels.

Extensions of the MIMO capacity to frequency selective channels were provided in Raleigh and Cioffi (1998). The SVD-based transmission has a clear similarity to the Orthogonal Frequency Division Multiplexing (OFDM) system, where the channel is transformed into a set of parallel independent subchannels. Capacity-optimal power allocation for a MIMO-OFDM system is achieved via simultaneous water-filling over space and frequency.

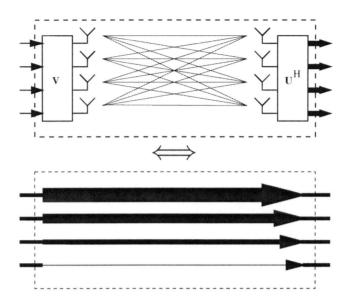

Figure 5.3 Illustration of the SVD based solution for full CSI at the transmitter

The capacity analysis of the MIMO channels promises large gains, compared to conventional SISO channels. However, these gains are based on the assumption that each transmit-receive antenna pair experiences independent identically distributed (i.i.d.) fading. This can only be achieved in a rich scattering environment (Goldsmith et al. 2003). In general, the capacity depends on the channel correlation between the antenna elements.

In addition to the throughput optimal design, several other design criteria have been proposed in the recent literature in order to consider the quality of the communication link between the transmitter and the receiver, given in terms of Mean Square Error (MSE), SINR, or Bit Error Rate (BER). In Lee and Petersen (1976) Salz (1985) and Yang and Roy (1994), the minimization of the sum of the MSE of all the channel substreams (the trace of the MSE matrix) under a sum power constraint was considered in the context of cross-coupled wired communication, that is, Digital Subscriber Lines (DSL). This criterion was later extended to the wireless MIMO system in Scaglione et al. (1999, 2002), and generalized to weighted MSE in Sampath et al. (2001). It was shown that the optimum structure for the linear precoder and decoder diagonalizes the MIMO channel into eigen subchannels, for any set of error weights. Thus, it differs from the capacity optimal solution only by the power allocated to each subchannel. In fact, the capacity optimal and the MMSE solutions are connected through the MSE matrix; the first design minimizes the determinant and the second minimizes the trace of the MSE matrix. In addition to the aforementioned criteria, several other optimization criteria have been considered for SU-MIMO communication (Ding et al. 2003; Palomar 2003, 2005; Palomar and Barbarossa 2005; Palomar et al. 2005, 2003, 2004). Palomar *et al.* provided a general framework based on convex optimization for a large variety of MSE-, SINR- and BER-based design criteria subject to different power constraints. Moreover, they considered the power minimization problem subject to Quality of Service (QoS) constraints given in terms of MSE, SINR, or BER for fixed modulation and coding schemes.

While optimal transceiver design with ideal CSI at the transmitter is rather simple, the case without CSI at the transmitter is less straightforward. In general, multiple antennas can be used for increasing the amount of diversity or the number of spatial multiplexing dimensions in wireless communication systems (Tse and Viswanath 2005). Based on the original BLAST scheme, several other spatial multiplexing schemes have been introduced. Multiantenna schemes aiming at maximizing the available diversity have been proposed (Alamouti 1998; Naguib et al. 1998; Tarokh et al. 1998). Alamouti proposed a simple but elegant space-time coding technique, which turned out to be optimal from both the diversity and multiplexing perspectives for the case with two transmit antennas and one receive antenna. The same idea was generalized to orthogonal designs with any number of transmit antennas in Tarokh et al. (1999). The coding structure from the orthogonal designs achieves the full diversity order as well but, unlike the Alamouti scheme, it reduces the achievable multiplexing gain. Zheng and Tse (2003) proved that a part of the diversity and multiplexing gains can be obtained simultaneously. Furthermore, they characterized the optimal tradeoff between the type of gain achievable by any coding scheme.

CSI Acquisition

Perfect CSI at the transmitter is often difficult to achieve in real networks. Ideally, full CSI at the transmitter can be achieved in Time Division Duplex (TDD) systems, where the reciprocal UL and Downlink (DL) channels are time-multiplexed on the same physical wireless channel. The transceiver can obtain the CSI while receiving information in the current time slot and can use the same CSI to transmit in the next time slot, as long as the TDD frame length is shorter than the channel coherence time and the transmit and receive radio frontends are well calibrated. In general, this can be guaranteed in low mobility environments for practical system parameters (Lebrun et al. 2005). On the other hand, in high-mobility environments the CSI becomes quickly outdated as the UE velocity increases due to the time delay between the estimation of the channel and the transmission of the data. CSI is obtained via channel estimation, and a channel estimate is always a distorted version of the real channel. Transmitter optimization with noisy channel estimates remains a largely unresolved research problem. It has been proposed that worst-case design criteria be used to guarantee robust performance for any realization of the actual and estimated channels, that is, worst-case MSE precoder design (Guo and Levy 2006; Vorobyov et al. 2003) or to combine robust beamforming with space-time coding (Pascual-Iserte et al. 2006).

In Frequency Division Duplex (FDD) systems, full CSI knowledge at the transmitter requires an ideal feedback channel from the UE(s). This, however, results in an enormous overhead due to the number of channel coefficients that need to be quantized and sent back to the transmitter over a limited bandwidth feedback channel. Hence, an ideal feedback is impractical for FDD systems with any mobility. When only some statistical channel information (distribution, mean, covariance) is available at the transmitter, the transmission strategy may be designed based on the statistics instead of the instantaneous information. It was discovered in Goldsmith et al. (2003); Jorswieck and Boche (2004) and Visotsky and Madhow (2001) that the capacity-achieving eigenvectors should correspond to the eigenvectors of the statistical covariance matrix of the channel. Not surprisingly, this finding is very similar to the solution with instantaneous channel knowledge at the transmitter. However, the optimal power allocation among the transmit directions is not achieved by water-filling. Rather, it requires numerical solutions that resemble water-filling where the interstream interference due to nonorthogonal transmission is considered. In order to improve the reliability of systems with statistical or noisy channel information, some authors have proposed coupling the statistical beamforming with space-time block codes (Jongren et al. 2002; Zhou and Giannakis 2003).

A multitude of solutions have also been proposed in the literature for FDD systems utilizing a low-rate CSI feedback from the receiver. A simple solution is to select a subset of transmit antennas to maximize the available rate, to minimize the transmit power with fixed rate requirements (Heath and Love 2005; Heath et al. 2001) or to switch between spatial multiplexing and transmit diversity depending on the instantaneous channel characteristics (rank, correlation between antennas, etc.) (Heath and Paulraj 2005; Love and Heath 2005). Another solution is to report an index of a transmission strategy (precoder) that has the best match with the instantaneous channel state. The transmit precoder matrix is chosen from a finite predefined codebook known by both receiver and transmitter (Love et al. 2003; Mukkavilli et al. 2003). Due to its practical nature, such codebook-based precoding has been adopted to be one of the MIMO transmission techniques in the LTE standard.

5.1.3 Multiuser MIMO Communications

Capacity

In information theory, the UL channel with multiple UEs/transmitters and a single receiver is called the Multiple Access Channel (M.A.C). The capacity region for the Gaussian M.A.C has been known for quite a while (Gallager 1985). Following the pioneering work on the use of multiple receive antennas in the UL by Winters, the scalar Gaussian M.A.C capacity region was extended to Inter-Symbol Interference (ISI) in (Cheng and Verdú 1993) and MIMO channels in (Telatar 1999; Yu et al. 2004).

In information theory, the downlink channel with a single transmitter and multiple UEs/receivers is called the Broadcast Channel (BC). The capacity region for the Gaussian degraded BC has also been known for more than 30 years (Cover 1972). The degraded BC implies that the Gaussian channel has a scalar input and scalar outputs, that is, a single-antenna transmitter and several receivers. For such a case, the capacity region is achieved by using superposition coding at the transmitter and interference subtraction at the receivers. However, for the nondegraded BC, where the transmitter has a vector input, that is, multiple transmit antennas, superposition coding no longer achieves the capacity.

The foundation for the information theory studies on the capacity of the nondegraded Gaussian MIMO BC was laid in Costa's (1983) landmark paper, where he studied the link capacity with both additive Gaussian noise and additive Gaussian interference, and where the interference is noncausally known at the transmitter but not at the receiver. Costa concluded that the effect of the *interference can be completely canceled out at the transmitter by using specific precoding called Dirty Paper Coding (DPC)*. Consequently, the capacity is identical to the case when the interference is known also at the receiver.

Caire and Shamai (2003) proposed to use Costa's precoding for transmitting over the multiantenna BC. By comparing DPC to Sato's cooperative upper bound (Sato 1978) in a simple case of two UEs with single-antenna receivers, they showed that DPC achieves the sum capacity. For the case with more than two UEs, they proposed a suboptimal precoder design for DPC based on ZF. For any given UE, the ZF part completely suppresses interference caused by subsequently encoded UEs while the DPC coding is applied to the previously encoded UEs. This strategy was shown to be asymptotically optimal for both high and low SNR regions. These results sparked off a number of new studies (Vishwanath et al. 2003; Viswanath and Tse 2003; Yu and Cioffi 2004), which generalized the results to the case of any number of UEs and an arbitrary number of receive antennas.

The duality between the DPC region for the MIMO BC and the capacity region of the MIMO M.A.C was established in (Vishwanath et al. 2003; Viswanath and Tse 2003). Both papers utilized the reciprocity of the UL and DL channels, combining it with the previously established duality between the scalar Gaussian BC and M.A.C (Jindal et al. 2004). They showed that the DPC region is exactly equal to the M.A.C capacity region, where the UEs have the same sum power constraint as in MIMO BC. Unlike in MIMO BC, in MIMO M.A.C the rate maximization can be formulated as a concave function of the transmit covariance matrices. Therefore, the DPC-M.A.C duality allows the DPC region to be found using a standard convex optimization problem solver. A somewhat different approach to prove the same result was taken in Yu and Cioffi (2004), where the sum-rate optimal precoding structure was shown to correspond to a decision-feedback equalizer.

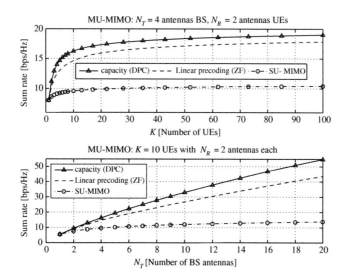

Figure 5.4   Achievable sum rates for MU-MIMO transmission, SNR $= 10$ dB

The DPC region was finally shown to be the capacity region of the entire nondegraded Gaussian MIMO BC in Weingarten et al. (2006), which proved the optimality of DPC. The MIMO BC capacity region was also characterized under various constraints on input (transmit) covariance matrices, including, as special cases, the sum and per-antenna power constraints.

The DPC rate region establishes the fundamental performance limits for MU-MIMO communication in the downlink. Several authors have analyzed the asymptotic gains from the optimal DPC compared to the SU capacity with Time Division Multiple Access (TDMA) and to linear beamforming, in a multiple-antenna BC (Jindal and Goldsmith 2005; Sharif and Hassibi 2007). They showed that, for Rayleigh fading channels, the gain from DPC largely depends on the ratio between the number of transmit and receive antennas, and the total number of UEs. As a conclusion, *the highest DPC gain is achieved with a large number of UEs, and when the ratio between the number of antennas is high.* On the other hand, the DPC gain converges to unity for both low and high SNRs when the number of receive antennas is higher than or equal to the number of transmit antennas. One of the main results is that the sum rate of both DPC and beamforming grows asymptotically as $N_T \log(\log(K N_R))$ where N_R and N_T denote the number of receive antennas per UE and transmit antennas, respectively. The corresponding result for TDMA is $\min\{N_R, N_T\} \log(\log(K))$.

The positive effect on the system sum rate of having a multitude of UEs with independent channels is called multiuser diversity. Figure 5.4 illustrates how multiuser diversity as well as the number of transmit antennas affect the average achievable sum rates in uncorrelated Rayleigh fading channels.

Resource Allocation and Linear Precoding

A major challenge for wireless communication systems is how to allocate resources among UEs across the space (including different cells), frequency and time dimensions with different system optimization objectives. In MIMO-Orthogonal Frequency Division Multiple Access (OFDMA) systems, this leads to a three-dimensional subchannel and power allocation problem, that is, how many subchannels should be allocated to each UE in different dimensions.

For a single-cell MU-MIMO system, the computation of the optimal sum capacity achieving allocation and the transmit covariance matrices for the frequency-selective MIMO BC requires solving the corresponding convex M.A.C dual optimization problem, and transforming the solution back to BC (Vishwanath et al. 2003). The computational complexity becomes very high for an increasing number of subcarriers, UEs and antennas. Therefore, suboptimal but less complex allocation techniques have been proposed (Caire and Shamai 2003; Tejera et al. 2006; Tu and Blum 2003).

Multi-user diversity for the dirty paper approach with ZF-DPC was studied in Tu and Blum (2003), where a greedy scheduling algorithm was proposed for the selection of UEs and their encoding order for maximizing the sum rate. The greedy UE selection and ordering algorithm combined with ZF-DPC was shown to have a sum rate very close to capacity. Moreover, Tejera et al. (2006) investigated different subchannel allocation methods aiming at maximizing the sum rate. They extended the suboptimal ZF-DPC with single-antenna UEs from Caire and Shamai (2003) to a more general case with multiple receive antennas per UE, and extended the greedy approach to the multidimensional allocation case.

The sum rate maximizing solutions can occasionally result in a very nonuniform rate allocation between UEs. Therefore, other transmitter design criteria should be considered in order to guarantee, for example, the QoS for all UEs. The symmetric (balanced) capacity providing absolute fairness between UEs becomes an important performance metric for delay-constrained applications (Lee and Jindal 2006; Sartenaer et al. 2005). A commonly used compromise between these two strategies is so-called *proportional fair scheduling (Viswanath et al. 2002), where the relative rates, normalized by the historic rate of the corresponding UE*, are considered. The weighted symmetric capacity refers to the situation where the weighted UE rates are equal, while their rates belong to the boundary of the capacity region. This enables the system to control the rates assigned to UEs that belong to distinct service priority classes.

The capacity-achieving schemes generally require very complex nonlinear precoding based on DPC. Therefore, research on suboptimal, but less complex transmission techniques is justified. Linear precoding or beamforming (Godara 1997) is a suboptimal transmission strategy that enables the spatial separation of several concurrently served UEs. Each UE stream is encoded independently and spread over multiple antennas by a beamforming weight vector. Mutual interference between multiple streams is controlled or even completely eliminated by the proper selection of weight vectors. Unlike the sum-rate capacity of MIMO BC using the DPC, the sum rate achieved by optimal beamforming cannot be written as a convex optimization problem (Sharif and Hassibi 2007). Despite its sub-optimality, beamforming combined with a proper selection of UEs has been shown to have the same asymptotic sum-rate as the DPC, when the number of UEs approaches infinity (Sharif and Hassibi 2005, 2007; Yoo and Goldsmith 2006). *This is due to the multi-user diversity effect, that is, the probability of finding a set of close-to-orthogonal UEs with large channel gains increases for a large number of UEs.*

In general, any sub-optimal allocation problem can be divided into two phases.

1. A set of UEs is selected for each orthogonal dimension (frequency/sub-carrier, time).
2. The transceivers are optimized for the selected set of UEs per orthogonal dimension.

The optimal UE selection per each orthogonal dimension is a difficult non-convex combinatorial problem with integer constraints (Sharif and Hassibi 2005; Yoo and Goldsmith 2006). Consequently, finding the optimal solution requires an exhaustive search over the entire UE set, which is computationally prohibitive for a large number of UEs. A more detailed treatment on the design of the linear transceiver with different optimization criteria is given in Section 5.3.

Several scheduling algorithms based on, for example best UE selection, largest eigenvalue selection and greedy UE/beam selection, have been proposed for DL beamforming. In (Dimic and

Sidiropoulos 2005) authors utilized the sub-optimal greedy UE selection algorithm from (Tu and Blum 2003) for the ZF beamforming with single antenna receivers. (Yoo and Goldsmith 2006) also used the greedy algorithm with an additional semi-orthogonality test between UEs and showed that the performance of the ZF beamforming with sub-optimal UE selection is still asymptotically optimal. Since the performance of the ZF beamforming is always inferior to optimal linear beamforming, this result proved the asymptotic optimality of linear beamforming in general.

Often, the allocation problems have been addressed for systems with UEs having a single receive antenna. When the UEs are equipped with multiple receive antennas, receiver beamforming further enhances the data rates. The signal space of each UE has multiple dimensions, allowing for multiple beams to be allocated per UE. Therefore, the receiver signal space has to be considered when selecting the optimal sets of UEs, as well as the dimension and orientation of the signal subspace used by each selected UE, for each orthogonal dimension. This further complicates the optimization problem. Since the transmitter vectors, and thus the corresponding receiver vectors at each UE, are affected by the set of selected UEs, it is impossible to know the actual receiver structure at the transmitter before the final beam allocation. An obvious candidate for an intelligent initial guess of the receiver matrix is the optimum single UE receiver, that is, the left singular vectors of the channel matrix. This decomposes the system into a MIMO BC with virtual single-antenna UEs with corresponding channel gains. This type of approach has been taken in (Primolevo et al. 2005; Tölli and Juntti 2005). However, it was shown in (Yoo and Goldsmith 2006) that the performance penalty for not performing receive beamforming decreases as the number of UEs approaches infinity.

Several suboptimal solutions for the general MIMO-OFDMA resource allocation problem have been proposed. One way to simplify the problem is to restrict to ZF transmission, which allows separating the beamformer design from the power allocation, making it much easier to solve (Chan and Cheng 2007). Another idea aiming at simplifying the problem is to group UEs according to the spatial separability or comparability of the UE channels for maximizing the system throughput (Fuchs et al. 2007; Shen et al. 2006; Spencer and Swindlehurst 2004a; Zhang and Letaief 2005).

The availability and quality of transmitter CSI in MU-MIMO BC have a profound effect on the achievable rates.

In contrast to the point-to-point MIMO capacity, where the transmitter CSI availability only affects the SNR offset to the capacity and not the multiplexing gain (Telatar 1999, 1995), both the multiplexing gain and the rate achievable from the MU-MIMO BC are greatly affected by the level of CSI available at the transmitter. Jindal (Jindal 2006) demonstrated that the throughput of the MU-MIMO DL with linear transmit beamforming becomes saturated with imperfect or noisy transmitter CSI. This is due to increased multi-user interference. However, full multiplexing gain can be achieved if the quality of CSI is increased linearly as a function of SNR.

Random opportunistic beamforming and nulling is a simple but remarkable limited feedback strategy for MIMO DL channels (Sharif and Hassibi 2005; Viswanath et al. 2002). Multiple random orthonormal beams are formed at the BS, and multiple UEs are simultaneously scheduled on these beams. Each UE reports the channel quality metric, that is, SINR, for the strongest beam(s), and the UEs with the highest instantaneous metric value are scheduled at a time. The opportunistic transmission strategy relies on the fact that with a large number of UEs, the probability of finding a set of nearly orthogonal UEs with high channel gains is high. This has been shown to achieve asymptotically the performance of linear beamforming (Viswanath et al. 2002) and to have the same capacity scaling obtained with perfect CSI using the DPC (Sharif and Hassibi 2005), as the number of UEs approaches infinity. However, it may result in very poor performance from both the capacity and fairness point of views, when applied in a system with a low or medium number of UEs. Therefore, the

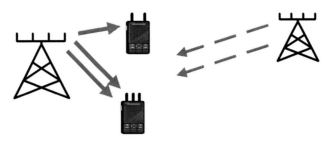

Figure 5.5 Multi-user MIMO system with interference

transmitter CSI is important for systems with a low to moderate number of UEs, and especially with a large number of transmit antennas (Jindal 2006).

In a realistic network with multiple UEs, the assumption of having full CSI from all UEs may be overly optimistic. This is due to the excessive overhead required for providing the transmitter with instantaneous CSI. The combination of opportunistic beamforming for the initial UE selection from a finite UE set and the use of supplementary CSI feedback for the selected UEs has been proposed in Codreanu et al. (2007) and Kountouris and Gesbert (2005) allowing for improved optimization of linear transmit and receive beamformers at the BS and UEs, respectively. The supplementary CSI for the selected UEs can greatly improve the performance of the opportunistic beamforming, especially for a low number of UEs.

5.1.4 MIMO with Interference

In cellular MIMO systems with interference, there are multiple UEs located in different cells that suffer from cochannel interference (see Figure 5.5). A similar scenario may arise also in *ad hoc* networks where each transmitter–receiver pair suffers from the interference originated from other UEs using the same channel. In general, the interfering signals are assumed to be unknown in systems where cooperation between transmitters is not possible. In such a case, neither joint transmission from several transmitters nor multiuser detection at the receivers is possible. In the theoretical analysis, both the interfering signals as well as the desired signal are generally modeled as Gaussian distributed, which is the usual form of the optimal MIMO signalling (Cover and Thomas 1991; Telatar 1995).

As discussed in Section 5.1.2, the optimum transmit strategy for a single-link MIMO system with transmitter CSI but without interference is water-filling over noninterfering SISO subchannels, which is achieved through a SVD of the channel matrix. Assuming that the interference structure is known at the transmitter, the same result is generalized to a single-link MIMO system with a fixed interference structure, that is, covariance matrix. Consequently, the parallel subchannels are obtained by the SVD of the prewhitened channel matrix (Bliss et al. 2002; Farrokhi et al. 2001; Ye and Blum 2003). However, when the interference is not fully known at the transmitter, the optimal transceiver design is considerably more challenging.

In cellular systems, the intracell and intercell interference can be controlled with optimum frequency, time, and space allocation executed by a centralized unit in charge of all resources, provided that the CSI and the interference structure are known at the transmitters. However, the transmit strategy selected at one transmitter affects the interference experienced by the receivers associated to other transmitters, and hence, their receive strategies. This in turn affects the transmit strategies of the other transmitters. The system capacity is neither a convex nor a concave function with respect to the transmit covariance matrices of the UEs (Ye and Blum 2003). Thus, analytical or even numerical solutions are difficult to find. However, in a sufficiently low interference regime, the UEs are decoupled and the optimization problem becomes a concave function of the transmit covariance matrices. On the other hand, the optimum transmission strategy in the large interference regime is to assign a single beam for all UEs (Ye and Blum 2003).

In practical applications, one may not always assume that the intercell interference structure seen at the receiver is perfectly known at the transmitter. The required signaling feedback would make this assumption infeasible. For example, in practical adaptive MIMO-OFDM cellular systems, the ideal feedback would require that the interference covariance matrix is reported to the transmitter for each subcarrier and for each transmitted data frame. In Cepeda et al. (2002), the problem of nonreciprocal interference in the case of adaptive modulation in general was recognized. A simple feedback method was developed to compensate it in a flat fading single-antenna transmission scenario. A simple and bandwidth efficient closed-loop compensation algorithm (Tölli et al. 2007), which is applied together with linear MMSE filter at the receiver, was proposed to compensate the nonreciprocity between UL and DL interference structure. The proposed compensation method was shown to achieve nearly the same performance as the ideal case where the interference structure per subcarrier is perfectly known at the transmitter.

Interference alignment has emerged as a very popular topic recently, with several important contributions shedding some light on optimal precoding strategies to approach the capacity of general interference networks with multiple transmitter–receiver pairs (Cadambe and Jafar 2008; Maddah-Ali et al. 2008).

The key idea of MIMO interference alignment is that transmitters construct signals such that they overlap over half of the signal space observed by each receiver where they constitute interference, leaving the other half of the signal space free of interference for the desired signal (Cadambe and Jafar 2008). Every UE is thereby able to access half of the spectrum free from other UEs assuming that all nodes (BSs, UEs) are equipped with equal number of antennas (Cadambe and Jafar 2008).

5.2   MIMO in LTE-Advanced and 802.16m

Standardization efforts to construct two International Mobile Telecommunications Advanced (IMT-Advanced) compliant systems – the Third Generation Partnership Project (3GPP), LTE Release 10 (LTE Rel-10) called LTE-Advanced (LTE-A) (3GPP 2010) and Worldwide Interoperability for Microwave Access (WiMAX) evolution IEEE 802.16m (IEEE 802.16 2010) – were finalized in the first half of 2011. In terms of MIMO transmission, both standards are evolving in the same direction: up to eight transmit antennas (2, 4 or 8) at the BS and up to four transmit antennas (1, 2 or 4) at the UE are supported. While simple transmit diversity modes are still supported, the emphasis has moved to spatial multiplexing and beamforming. Both open-loop and closed-loop SU-MIMO and MU-MIMO modes with adaptive rank are supported.

Reference signals, also known as pilot signals, are modulation symbols known to the receiver and they are used for MIMO channel estimation. In both upcoming systems, the reference symbol arrangement is essentially the same: there are two classes of pilots. The first class is the demodulation pilots that facilitate coherent reception, and they are precoded the same way as data. The second class is the sounding signals, that is, reference signals targeting CSI estimation, the purpose of which is to support for example transmit precoder matrix selection. These pilots are typically not spatially precoded, and in the downlink they are many times called "common pilots". Note that the reference signal strategy is different from that in LTE Rel-8, where no dedicated demodulation pilots are employed (apart from the case of closed-loop rank-1 transmission). In principle the new strategy facilitates any codebook- or noncodebook-based spatial precoding method, provided that CSI feedback is sufficiently accurate.

In both upcoming systems, support for a relatively broad range of advanced linear MU-MIMO schemes will be available. Backward compatibility is provided so that LTE-A can support legacy LTE Rel-8 UEs. Likewise, 802.16m will support legacy WiMAX UEs. In 802.16m the support can be created by so-called "time zones" so that some of the subframes comply with the old standard.

5.2.1 LTE-Advanced

Average spectral efficiency and cell-edge UE throughput are the main challenging indicators for LTE-A to fulfill the International Telecommunication Union – Radiocommunication Sector (ITU-R) requirements for IMT-Advanced systems. To address these challenges, multi-antenna techniques in LTE-A have been improved targeting efficient SU- and MU-MIMO transmissions.

The main driver in designing MIMO schemes for LTE Rel-10 comes from the need to consider different antenna configurations at the eNodeB (eNB), where it will be possible to deploy 2, 4 and 8 antennas. The main antenna deployment configurations have been identified as:

- Closely spaced (e.g. 0.5 to 0.7 times carrier wavelength) cross-polarized antenna scenario (closely-spaced Clustered Linear Array (CLA)).
- Widely spaced (e.g. 4 or even 10 times carrier wavelength) cross-polarized antenna scenario (widely spaced CLA).
- Uniform Linear Array (ULA) scenario.

Operators indicate the case of closely spaced CLA as the one with the highest priority. However, they also emphasize the need for optimizing the other possible scenarios.

Another driver is the availability (starting from Release 9 (Rel-9)) of dedicated reference signals that allow the use of nonpredefined beamforming weights for downlink transmissions. For example, beamforming based on ZF or on grid-of-beams could both be implemented, and different eNB vendors could choose different design criteria for the precoders.

Dynamic SU/MU-MIMO switching has been recognized to be an important feature, for which codebook design and feedback reporting should be optimized. Low computational complexity at the UE side, is another reason behind some of the choices made in 3GPP.

In the following a summary of reference signal design and multiantenna transmission schemes is given, for downlink and uplink transmissions. A more detailed description can be found in Akyildiz et al. (2010) and Sesia et al. (2011).

Downlink Reference Signal Design

In LTE Rel-10 the UE-specific Demodulation Reference Signal (DM-RS) (also referred to as dedicated pilots) is extended to allow data demodulation for up to eight layers. Recall that the UE-specific Reference Signal (RS) undergoes the same precoding as data symbols. It therefore allows for the possibility of not explicitly sending back the specific matrix index used for precoding. The Release 10 (Rel-10) RS pattern has been designed by extending the Rel-9 rank-2 UE-specific RS pattern using a frequency/time/code division multiplexing technique.

The Channel State Information Reference Signal (CSI-RS) (also referred to as common pilot) has been extended from Release 8 (Rel-8) to enable channel-state estimation for up to eight antenna ports. Moreover, CSI-RS also enables the UE to estimate the CSI for multiple cells, to allow future multi-cell cooperative transmission schemes. The total number of supported antenna ports is 40, which allows for CSI estimation of five cells with eight antenna ports per cell, or 20 cells in case of two antenna ports.

Uplink Reference Signal Design

Uplink DM-RSs are used to allow the demodulation of up to four spatial layers. They are precoded using the same precoder as the Physical Uplink Shared CHannel (PUSCH) data transmission in the same resource blocks when multiple antenna transmission is used. Rel-10 uses different Cyclic Shifts of the DM-RS base sequence to multiplex the DM-RS for different layers. In addition to this, two length-2 code division multiplexing codes can be applied to the two DM-RS symbols in the two slots of one subframe to further separate the multiplexed DM-RS.

The Sounding Reference Signals (SRSs) have been extended from Rel-8 in order to allow the sounding of up to 4 transmit antennas. The mechanism to separate the different SRSs is similar to the one used in Rel-8 and is based on the use of different cyclic shifts of Zadoff-Chu sequences and by assigning different frequency shifts. Other enhancements to the SRS, are represented by the possibility of allowing channel state estimation of the same UE in different cells to enable Coordinated Multipoint transmission or reception (CoMP) schemes (see Chapter 6 for more information).

Downlink Multiantenna Transmission Schemes

For the case of eight antennas at the eNB side, the precoding matrix is calculated by combining two instants of Precoding Matrix Indicator (PMI), one targeting wideband and/or long-term channel properties and the other targeting frequency-selective and/or short-term channel properties. The Channel Quality Indicator (CQI) is calculated assuming that the overall precoder is obtained as a product of the two PMI components. Conversely, the way of combining the two PMI is not specified in transmission, and a vendor-specific precoder implementation is possible by exploiting the UE-specific RS. The long-term PMI is designed to match the spatial covariance of cross-polarized antennas and single-polarized setups with different spacing and is based on Discrete Fourier Transform (DFT) beams. The frequency-selective and/or short-term PMI targets selection and cophasing. SU-MIMO can be used with up to eight independent streams, even if the total number of codewords is still two, as in Rel-8. In a MU-MIMO setup, up to four streams to each UE can be sent. For the case of two and four antennas the precoding codebook is the same as in Rel-8.

In order to allow efficient scheduling, precoding and rate adaptation, UEs send periodic and aperiodic channel status reports to the eNB. Aperiodic reports are sent through the PUSCH, whereas periodic reports are sent through the Physical Uplink Control CHannel (PUCCH). How to combine periodic and aperiodic reports is a matter of vendor implementation. In Rel-8, periodic reports are designed to provide a first approximated CSI before a UE is scheduled for transmission whereas aperiodic reports are designed to provide a more detailed CSI only for the UEs that have already been scheduled. Nevertheless, due to the adoption of more sophisticated MU-MIMO techniques, where an efficient UE scheduling requires a more refined CSI, 3GPP decided to introduce a more refined type of periodic reporting for Rel-10. This reporting is a form of hierarchical/differential feedback, technique that was initially studied under the framework of WINNER II project. The main idea behind the original hierarchical feedback was the possibility of allowing successive feedback refinement in time, for UEs experiencing a stable channel condition. In Rel-10, successive feedback refinement is designed to enable a detailed UE-triggered sub-band report only for the UEs observing a stable channel condition.

Uplink Multi-antenna Transmission Schemes

LTE Rel-10 UEs can support up to four independent streams, carrying up to two transport blocks, each associated with a Hybrid Automatic Repeat-reQuest (HARQ) indicator. Closed-loop precoding is used for the PUSCH in a similar way to Rel-8 Physical Downlink Shared CHannel (PDSCH): rank and precoding matrix are fixed by the eNB and signalled to the UE by means of a grant over the Physical Downlink Control CHannel (PDCCH). There are different codebooks for two and four transmit antennas, requiring respectively three and six bits to signal over the PDCCH the codeword to be used for transmission over the PUSCH. Some of the codewords are designed to turn off one of the antennas, for power-saving reasons. The rank-1 codewords are designed also to improve the MU-MIMO operations performed at the eNB side. Transmit diversity is not used for the PUSCH. This decision was based on simulation studies showing that transmission diversity-based multiantenna schemes were outperformed by long-term closed-loop precoding, in the meaningful SNR region (Sesia et al. 2011). On the other hand, in Rel-10 transmit diversity is used for the PUCCH. Additional tutorial details on UL MIMO for LTE-A can be found in Park et al. (2011).

5.2.2   WiMAX Evolution

The WiMAX evolution, IEEE 802.16m (IEEE 802.16 2010), supports two, four or eight transmit antennas at the BS. In the downlink SU-MIMO mode, a maximum of eight spatial streams per UE are allowed. However, in the downlink MU-MIMO mode, a maximum of four simultaneous streams – maximum two per UE – are allowed, even when the BS has eight transmit antennas. It is mandatory for the UE to employ at least two receive antennas.

At the UE, one, two or four transmit antennas are supported. Consequently, the maximum number of streams in the uplink SU-MIMO mode is four. The uplink MU-MIMO mode also allows multiple streams per UE, but the total number of simultaneous streams is restricted to four.

For downlink closed-loop codebook based precoding, IEEE 802.16m does not explicitly define the derivation of the precoding matrices, but allows it to be vendor-specific. The codebook may be optimized for correlated channels, and advanced options for PMI feedback are suggested: PMI may be transformed with long-term CSI or PMI may provide differential knowledge of the short-term CSI. The long-term CSI, including the channel covariance matrix or its eigenvectors, may be reported by the UE separately as well. In TDD systems, uplink sounding based adaptive downlink precoding is supported. The sounding signals may be transmitted by one or multiple UE antennas.

Furthermore, in the case of uplink closed-loop codebook based precoding, the BS signals the PMI to be used in the uplink transmission. The codebook may be created by the BS, or by the UE as instructed by the BS. In TDD mode, adaptive precoding based on the measurements of downlink reference signals is allowed. In this case, the form of the uplink precoding matrix need not be known by the BS.

5.3   Generic Linear Precoding with CSIT

Recall from section 5.1 that the optimal sum capacity achieving schemes for the downlink require non-linear precoding based on dirty paper coding (Viswanath and Tse 2003; Weingarten et al. 2006). On the other hand, linear precoding/beamforming is usually remarkably simpler to implement, and, hence, it is an important solution in practical system design. In this section, we focus on the transmitter and receiver optimization for a fixed set of UEs selected from a possibly larger group of UEs for each orthogonal dimension prior to the transceiver optimization. Furthermore, perfect CSIT knowledge is assumed.

The precoder design solutions described in this section are often solved by reformulating them as convex optimization problems. Convex optimization methods (Boyd and Vandenberghe 2004), such as second-order cone programming (SOCP) (Lobo et al. 1998), semidefinite programming (SDP) (Vandenberghe and Boyd 1996) and geometric programming (GP) (Boyd et al. 2007; Chiang 2005), are very important and powerful tools that allow for efficient numerical solutions to many signal-processing and communications problems – especially linear transceiver design problems. Some communications problems solved via convex optimization can be found in the tutorial papers (Boyd et al. 2007; Chiang 2005; Gershman et al. 2010; Luo 2003; Luo and Yu 2006).

5.3.1 Transmitter–Receiver Design

In general, the joint design of a linear transceiver for a MU-MIMO scenario is a much more demanding optimization problem than the point-to-point scenario considered in section 5.1.2. Due to the noncooperative UEs , it is often difficult to simplify the optimization problems to an easily solvable form unlike in the SU case (Palomar et al. 2003). The optimization problems employed in the linear beamformer design are not convex in general. Therefore, the problem of finding the global optimum for most of the optimization criteria is intrinsically intractable.

Consider the system model defined in section 5.1.1, and let $\mathcal{A}_n$ be an arbitrary subset of transmit antennas. If P_n is the maximum transmit power for the antenna set $\mathcal{A}_n$, then the sum power constraint for $\mathcal{A}_n$ can be expressed as

$$\sum_{s=1}^{S} \left\| \mathbf{w}_s^{[n]} \right\|_2^2 \leq P_n \tag{5.11}$$

where $\mathbf{w}_s^{[n]} \in \mathbb{C}^{|\mathcal{A}_n|}$ are the parts of $\mathbf{w}_s \in \mathbb{C}^{N_T}$ that contain the weights of the antennas belonging to $\mathcal{A}_n$. The transmit power constraint model is fairly general and it covers a large range of practical applications including, for example, per-antenna power constraint $|\mathcal{A}_n| = 1 \ \forall \ n$ as a special case.

The general system optimization objective is to maximize a function $f(\gamma_1, \ldots, \gamma_S)$ that depends on the individual SINR values $\gamma_s, s = 1, \ldots, S$. This can be formulated as

$$\begin{aligned} \text{max.} \ & f(\gamma_1, \ldots, \gamma_S) \\ \text{s. t.} \ & \Gamma_s \geq \gamma_s, \ \forall \ s \\ & \sum_{s=1}^{S} \left\| \mathbf{w}_s^{[n]} \right\|_2^2 \leq P_n, \ \forall \ n \end{aligned} \tag{5.12}$$

where the variables are $\mathbf{w}_s, \boldsymbol{v}_s \in \mathbb{C}^{N_{R_k}}$ and $\gamma_s, s = 1, \ldots, S$. The definition for Γ_s is in Equation (5.5).

The optimization criteria could be one of the following:

- Sum power minimization with fixed γ_s:
  $f(\gamma_1, \ldots, \gamma_S) = -\sum_n P_n$
- Maximization of minimum weighted SINR per data stream, that is, SINR balancing:
  $f(\gamma_1, \ldots, \gamma_S) = \min_{s=1,\ldots,S} \beta_s^{-1} \gamma_s$
- Weighted sum MSE minimization:
  $f(\gamma_1, \ldots, \gamma_S) = -\sum_{s=1}^{S} \frac{\beta_s}{(1+\gamma_s)}$
- Weighted sum rate maximization:
  $f(\gamma_1, \ldots, \gamma_S) = \sum_{s=1}^{S} \beta_s \log_2(1 + \gamma_s) = \log_2 \prod_{s=1}^{S} (1 + \gamma_s)^{\beta_s}$

- Maximization of weighted common user rate ($N_{R_k} > 1$):
  $f(\gamma_1, \ldots, \gamma_S) = \min_{k \in \mathcal{U}} \beta_k^{-1} \sum_{s \in \mathcal{P}_k} \log_2 (1 + \gamma_s), \mathcal{P}_k$ is a subset of data streams that correspond to UE k
- Maximization of the weighted harmonic mean of SINR:
  $f(\gamma_1, \ldots, \gamma_S) = \dfrac{1}{\sum_{s=1}^{S} \frac{\beta_s}{\gamma_s}}$

The stream or user-specific weights, $\beta_s > 0$, $\beta_k > 0$, can be selected arbitrarily and used to prioritize the data streams of different UEs. An extension to the frequency selective DL channel with additional quality of service constraints, such as minimum/maximum SINR or rate constraints per stream/UE can also be considered (Tölli et al. 2008b).

Sum Power Minimization Criterion

The minimum power beamforming design $f(\gamma_1, \ldots, \gamma_S) = -\sum_n P_n$, also known as sum power minimization under the fixed minimum SINR constraint per stream γ_s, was the first problem solved for the MU-MIMO downlink. The uplink-downlink SINR duality was utilized in Rashid-Farrokhi et al. (1998) and Visotsky and Madhow (1999) to solve this problem for single-antenna UEs in an iterative manner. It was shown that the minimum sum power required for satisfying the minimum SINR constraint per stream is equal both in the DL and in the dual UL. Schubert and Boche (2004) provided another solution for the minimum power beamformer design with faster convergence that accounts for the feasibility of the problem setting, as well as an optimal solution for the worst SINR per UE maximization problem under a sum power constraint. They utilized the property of the Perron-Frobenius theory (Horn and Johnson 1990) that states that the UL power assignment for maximizing the minimum SINR per UE is closely related to the Perron-Frobenius eigenvalue of the interference cross-coupling matrix between the UEs.

Bengtsson and Ottersten (Bengtsson and Ottersten 1999, 2001) presented a different solution for the minimum power beamforming problem based on the framework of convex optimization theory. They considered a general case with a statistical DL channel model, where the rank of the estimated channel covariance matrix per UE can have a value higher than one due to multipath propagation. In such a case, the beamforming problem with rank one beamformers is nonconvex. However, the rank constraint can be relaxed so that semidefinite programming (Vandenberghe and Boyd 1996) can be applied. Despite the nonconvexity of the problem, SDP relaxation achieves the global optimum of the problem (Bengtsson and Ottersten 1999, 2001). The relaxed problem can also be modified by introducing additional constraints, that is, to limit the dynamic range of the transmitted signal or to increase the robustness against channel estimation errors. The beamforming problem was later generalized to cope with additional (indefinite) quadratic side constraints in Hammarwall et al. (2006) and Huang and Palomar (2010). It was also noticed that the minimum power beamforming problem can be formulated as a second order cone problem (Lobo et al. 1998) for the special case of rank one channels, that is, for an instantaneous channel realization where the UEs are equipped with a single receive antenna or with fixed receiver beamformers. This can be achieved by reformulating the SINR constraint in (5.12) as a second-order cone constraint with fixed γ_s and v_s

$$\left\| \frac{\mathbf{W}^H \mathbf{H}_{k_s}^H v_s}{\sqrt{N_0}} \right\|_2 \leq \sqrt{1 + \frac{1}{\gamma_s}} v_s^H \mathbf{H}_{k_s} \mathbf{w}_s, \quad s = 1, \ldots S \qquad (5.13)$$

In addition to the min power beamformer design, the worst SINR maximization, $f(\gamma_1, \ldots, \gamma_S) = \min_{s=1,\ldots,S} \beta_s^{-1} \gamma_s$, subject to a sum power constraint was considered in (Wiesel et al. 2006). The

SINR maximization can be carried out via power optimization using one-dimensional bisection to search for the maximum SINR value that can be satisfied for all UEs/streams.

In Bengtsson (2002), Chang et al. (2002), Visotsky and Madhow (1999), the power minimization problem was extended to the situation where the UEs are also equipped with antenna arrays. However, the problems are no more convex and convergence of the algorithm to the global optimum cannot be guaranteed. A suboptimal iterative approach based on a coordinate ascent method, that is, optimization with respect to one set of variables ($\boldsymbol{v}_s$ or $\mathbf{w}_s$ $\forall$ s) at a time while keeping the other sets ($\mathbf{w}_s$ or $\boldsymbol{v}_s$ $\forall$ s) fixed, was proposed for providing a local minimum for the problem. Note that for a fixed $\mathbf{w}_s$ $\forall$ s, problem (5.12) has a unique solution given by the Linear Minimum Mean Square Error (LMMSE) receiver as defined in Equation (5.6).

Sum-MSE Minimization Criterion

A linear MU-MIMO transceiver design for solving the sum-MSE minimization criterion $f(\gamma_1, \ldots, \gamma_S) = -\sum_{s=1}^{S} \frac{\beta_s}{(1+\gamma_s)}$ with sum power constraint $|\mathcal{A}_n| = N_T$ was considered in Codreanu et al. (2007), Jorswieck and Boche (2003), Khachan et al. (2006), Mezghani et al. (2006), Serbetli and Yener (2004) and Shi and Schubert (2005), where iterative algorithms were proposed. Luo *et al.* (Luo et al. 2004) recognized that the optimal MMSE transceiver design problem for the UL can be reformulated as a Semidefinite Programming (SDP) that can be solved using highly efficient interior point methods (Boyd and Vandenberghe 2004). By using the UL-DL duality, the solution from Luo et al. (2004) was later extended to the DL case with a sum power constraint in (Codreanu et al. (2007) and Schubert et al. (2005). The sum-MSE minimization problem is convex only when the total number of UE antennas is lower than or equal to the number of transmit antennas. Otherwise, the nonconvex rank constraints become active. The rank constraints can be relaxed to obtain a lower bound for the sum-MSE (Codreanu et al. 2007; Serbetli and Yener 2004).

Sum-rate Maximization Criterion

Linear transceiver optimization for the sum-rate maximization criterion $f(\gamma_1, \ldots, \gamma_S) = \sum_{s=1}^{S} \beta_s \log_2(1 + \gamma_s)$ is a nonconvex problem that has so far eluded a simple solution. Suboptimal solutions were considered, for example in Boccardi et al. (2006) and Stojnic et al. (2006) for single-antenna UEs . In Codreanu et al. (2007), a general method for the joint design of linear transmit and receive beamformers subject to the sum power constraint was provided. The considered optimization criteria include the balancing of SINR values under the sum power constraint, the weighted sum rate maximization under the sum power constraint and the weighted sum MSE minimization under the sum power constraint. By exploiting the UL-DL SINR duality (Vishwanath et al. 2003; Viswanath and Tse 2003), the original optimization problems were decomposed into a series of remarkably simpler optimization subproblems that can be solved efficiently by using standard geometric program solvers (Boyd et al. 2007). Even though each subproblem is optimally solved, the global optimum cannot be guaranteed due to the nonconvexity of the original problems. The algorithms were shown to converge fast to a solution, which can be a local optimum, but remains efficient. In contrast to the previously proposed solutions, the method proposed in (Codreanu et al. 2007) can handle multiple antennas at the BS and at the UEs with an arbitrary number of data streams per scheduled UE. At each step of the iterative algorithm, the nonconvex objectives of the weighted sum rate maximization and the weighted sum MSE minimization problems were solved via signomial programming (Boyd et al. 2007), that is, solving a sequence of geometric programs which locally approximate the original problem. Similar approach was also proposed in Shi et al. (2008). Other optimization problems utilizing geometric programming can be found, for example, in Boyd et al. (2007), Chiang (2005) and

Chiang et al. (2007). In Christensen et al. (2008), an alternative method based on transmit/receive MMSE-designs was proposed for finding a local weighted sum-rate optimum. It was found out that the weighted sum rate problem can be also solved as a weighted MMSE problem with optimized MSE weights.

Common User Rate Maximization Criterion

For single-antenna receivers, the received SINR uniquely defines the UE rate, and hence, the minimum UE rate maximization is equal to minimum SINR per stream maximization, that is, rate balancing equals SINR balancing. With multiple receive antennas, the equal or weighted SINR per stream requirement is too restrictive, since it does not allow for the power to be freely allocated over the multiple streams that are assigned to a UE, unlike in the case of the rate maximization. UE rate balancing $f(\gamma_1, \ldots, \gamma_S) = \min_{k \in \mathcal{U}} \beta_k^{-1} \sum_{s \in \mathcal{P}_k} \log_2(1 + \gamma_s)$ with MMSE receivers was investigated in Shi et al. (2008) under a sum power constraint. Like Codreanu et al. (2007), they utilized the UL-DL duality, providing an iterative algorithm where the power allocations and linear MMSE receive filters were updated in an alternating manner in both virtual UL and DL channels. The UL power control step was formulated as a geometric program where the rate requirements per UE are handled via minimizing the product of MSEs per stream.

Maximization of Harmonic Mean of SINRs

A popular ad hoc alternative that allows decentralized implementation is the so-called Signal to Leakage and Noise Ratio (SLNR) criterion (Sadek et al. 2007), which is closely related to a design principle that can be traced back at least to Gerlach and Paulraj (1996) and Zetterberg and Ottersten (1995) and has appeared under many different names over the years. The basic idea is to design a transmit beamformer so that it maximizes the signal power at the desired receiver divided by the total interference contribution caused by the particular transmitter, at all the other receivers in the system. This cost function looks exactly like a receive beamforming problem, so one interpretation is that the optimal receive beamformer is used also for transmission. Another interpretation is that the harmonic mean of the SINRs is maximized, that is, that the sum of the inverse SINRs is minimized, see the derivation and discussion in Bengtsson and Ottersten (2001), which also highlights that there is no particular reason why the "noise" term in the SLNR should be set to the true noise level. A third interpretation is obtained from the previously mentioned UL-DL duality, which shows that the obtained solution is optimal for a certain set of weights of the interference terms (although these optimal weights typically are replaced by some ad hoc choice that is easier to obtain without coordination between the transmitting nodes).

Transmitter–Receiver Optimization with Generic Power Constraints

Any convex MIMO transceiver optimization problem can be extended with additional convex constraints such as limiting the dynamic range of the power amplifier at the transmitter (Bengtsson and Ottersten 1999, 2001; Palomar et al. 2003). In real systems, a maximum sum power for any subset of the transmit antennas $\mathcal{A}_n$ can be imposed in inequality (5.11). Such power constraints are useful in the case of distributed MIMO architectures, where a central BS is connected via a high speed backbone to multiple antenna heads, which are distributed over a large geographical area (Karakayali et al. 2006b; Tölli et al. 2008a). A per-antenna power constraint $|\mathcal{A}_n| = 1$ may also be required in practice, that is, in a case where each transmit antenna is equipped with a power amplifier and the linearity of the power amplifier is the limiting performance factor. Yu and Lan investigated the minimum-power beamformer design for DL MISO under per antenna power constraints in (Yu and Lan 2007), where the original DL problem was transformed into a dual UL minimax optimization problem with an uncertain noise covariance. The rate balancing for MU-MIMO with ZF transmission subject to per BS (or per antenna) power constraints was studied in (Karakayali et al. 2006a,b).

The general optimization problem (5.12) with generic power constraints is not known to be jointly convex in variables $\mathbf{w}_s$ and $\mathbf{v}_s$ for any of the aforementioned criteria, in general. Thus, finding optimal solutions is intractable. However, the nonconvex problem can be divided into subproblems where each convex subproblem can be optimally solved by using standard convex optimization tools. A general framework for joint design of the linear transmit and receive beamformers for several optimization criteria was proposed in Tölli et al. (2008a,b). The framework can accommodate a variety of supplementary constraints, for example the lower bounds for the SINR values of data streams or the minimum rate constraints per UE as well as per antenna group power constraints. In Tölli et al. (2008a,b) the optimization problems were decomposed into a series of remarkably simpler optimization subproblems that can be efficiently solved by using standard geometric program and/or second order cone program solvers. Even though each subproblem is optimally solved, the global optimum cannot be guaranteed in general due to the non-convexity of the optimization problems. However, the algorithms are shown to converge fast to a solution, which can be a local optimum, but remains very efficient.

Let us consider, for example, the SINR balancing problem $f(\gamma_1, \ldots, \gamma_S) = \max \min_{k=1,\ldots,S} \beta_s^{-1} \gamma_s$ with generic antenna group power constraints. At each iteration the objective is maximized with respect to one set of variables $\mathbf{v}_s$ (or $\mathbf{w}_s$) while considering the other set fixed. For fixed $\mathbf{v}_s$ the problem can be solved with any accuracy $\epsilon > 0$ by the bisection method (Boyd and Vandenberghe 2004). This leads to *Algorithm 5.1*.

Algorithm 5.1 Weighted SINR balancing under per antenna group power constraints

---

1. Initialize $\mathbf{w}_s^{(0)}$ such that antenna group power constraints are satisfied. Let $j = 1$.
2. Compute $\mathbf{v}_s^{(j)}$ given by Equation (5.6), where $\mathbf{w}_s = \mathbf{w}_s^{(j-1)}$, $s = 1, \ldots, S$.
3. Solve $f(\gamma_1, \ldots, \gamma_S) = \max \min_{s=1,\ldots,S} \beta_s^{-1} \gamma_s$ for the variables $\mathbf{w}_s$ by fixing $\mathbf{v}_s = \mathbf{v}_s^{(j)}$, $s = 1, \ldots, S$.
   (a) Initialize $\gamma_{\min} = \text{SINR}_{\min}$ and $\gamma_{\max} = \text{SINR}_{\max}$, where $\text{SINR}_{\min}$ and $\text{SINR}_{\max}$ define the range of relevant SINRs. Let $\epsilon > 0$ be the desired accuracy.
   (b) Set $\gamma_0 = (\gamma_{\max} + \gamma_{\min})/2$
   (c) Solve the following feasibility problem

$$\begin{aligned} \text{find} \quad & \mathbf{w}_s, \quad s = 1, \ldots, S \\ \text{subject to} \quad & \Gamma_s \geq \beta_s \gamma_0, \quad s = 1, \ldots, S \\ & \sum_{s=1}^{S} \left\| \mathbf{w}_s^{[n]} \right\|_2^2 \leq P_n, \quad n = 1, \ldots, M \end{aligned} \qquad (5.14)$$

   where the variables are $\mathbf{w}_s \in \mathbb{C}^{N_T}$, $s = 1, \ldots, S$. If the problem is feasible, then set $\gamma_{\min} = \gamma_0$. Otherwise, set $\gamma_{\max} = \gamma_0$.
   (d) If $(\gamma_{\max} - \gamma_{\min}) > \epsilon$ then go to Step 3(b). Otherwise, return $\mathbf{w}_s^{\star} = \mathbf{w}_s$, $s = 1, \ldots, S$, where $\mathbf{w}_s$ is the last feasible solution of problem (5.14) and STOP.
4. Denote the solution by $\mathbf{w}_s^{\star}$ and update the transmit beamformers $\mathbf{w}_s^{(j)} = \mathbf{w}_s^{\star}$, $s = 1, \ldots, S$. Test a stopping criterion. If it is not satisfied, let $j = j + 1$ and go to Step 2, otherwise STOP.

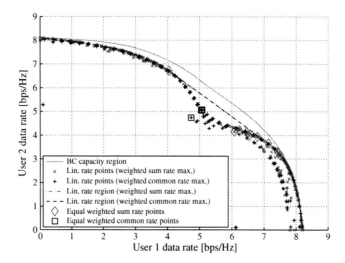

Figure 5.6 Broadcast capacity region and rate region with linear processing for $\{N_{\mathrm{T}}, N_{\mathrm{R}_k}, K, |\mathcal{A}_n|\} = \{4, 2, 2, 2\}$ system with per antenna group power constraint

Example 1

The impact of the beamformer initialization on the performance of two rate maximization algorithms, weighted common user rate $f(\gamma_1, \ldots, \gamma_S) = \min_{k \in \mathcal{U}} \beta_k^{-1} \sum_{s \in \mathcal{P}_k} \log_2 (1 + \gamma_s)$ and weighted sum rate maximization $f(\gamma_1, \ldots, \gamma_S) = \sum_{s=1}^{S} \beta_s \log_2(1 + \gamma_s)$, is studied in the following. Due to the nonconvexity of both rate maximization problems, different initial beamformer configurations $\{\mathbf{w}_1^{(0)}, \ldots, \mathbf{w}_S^{(0)}\}$ may end up in different locally optimal solutions. The behavior of both algorithms in a two UEs channel is compared at a 10 dB SNR in Figure 5.6. The rate pairs corresponding to different weight vectors and the resulting rate regions are plotted for a single random channel realization per UE and with a per BS power constraint. Rate pairs resulting from ten random beamformer initializations are plotted for each weight vector.

The example shown in Figure 5.6 demonstrates that the achievable rate region boundaries, which are plotted as convex hulls of all the achievable rate pairs, are identical for both rate maximization algorithms with linear processing. However, all the rate pairs with a weighted common rate constraint that deviate from the convex hull cannot be claimed as local optima, unlike in the weighted sum rate case. This is due to the different objectives of the two optimization criteria. Again, the linear part of the convex hull can be achieved only via time sharing. This disfavors the common rate maximization case especially in the region where the achievable points are inferior to the convex hull, since the UE rates are constrained to follow the ratio defined by the weights β_k. A somewhat higher common rate can still be achieved by allowing time sharing between two rate pairs in both ends of the time sharing region. For instance, a common rate about 0.4 bps/Hz higher could be achieved with time sharing in the example shown in Figure 5.6. However, finding the optimal pair of weight vectors and the corresponding beamformers for time sharing would require a full search on the rate region.

5.3.2 Transceiver Design with Interference Nulling

The linear multiuser transceiver design can be greatly simplified by imposing the constraint that all interuser interference terms must be zero, that is, the multiuser channel becomes diagonal. The simplest approach to achieve this is via channel inversion (Caire and Shamai 2003; Winters et al. 1994). However, when UEs are equipped with multiple antennas, channel inversion is sub-optimal since each UE is able to coordinate the processing of his own receiver outputs (Spencer et al. 2004a).

One solution is to use block channel inversion or block diagonalization (BD) (Spencer et al. 2004b), which essentially generalizes the channel inversion for a group of receive antennas rather than a single antenna. This is achieved by forcing the precoder for a given UE to lie in the null space of the other UEs' channels. Other solutions with some variations on the ZF constraint were proposed in (Choi and Murch 2004; Choi et al. 2004; Wong et al. 2003). The basic block-ZF method in (Spencer et al. 2004b) relies on the condition that the sum of the UE antennas is smaller than or equal to the number of transmit antennas. The method can be easily extended to operate with any number of UEs and receive antennas by coordinating the processing between the transmitter and the receivers. This can be achieved by estimating the receivers' linear combiners at the transmitter and applying the ZF solution (Spencer et al. 2004b) to a set of equivalent channels representing the transfer functions between the transmitter and the output of the receivers' linear combiners (Spencer and Swindlehurst 2004b). The design problem of jointly selecting the transmit and receive beamformers for providing interference free data streams is difficult to solve in closed form. This is due to the coupling between the optimal transmit and receive beamformers. Generalized iterative ZF solutions were proposed in Farhang-Boroujeny et al. (2003), Pan et al. (2004) and Spencer and Swindlehurst (2004b), where the BS and UE antenna weights are iteratively recomputed until they satisfy a convergence criterion. The iterative algorithms approach a solution where the UEs employ conventional Maximum Ratio Combining (MRC) receivers, which was suggested as a design criterion in Wong et al. (2003). While corresponding general closed-form solutions have not been presented, in Chae et al. (2008) it was derived for a two-UE case.

The ZF methods are generally power inefficient because beamforming weights are not matched to the UE channels. Inverting an ill-conditioned channel matrix results in a large power penalty that will dramatically reduce the SINR at the receivers. Allowing a limited amount of interference at each receiver generally provides a better solution than forcing the interference to be zero (Boccardi et al. 2006; Codreanu et al. 2007; Serbetli and Yener 2004; Stojnic et al. 2006). A simple noniterative approach utilizing this fact was proposed in Peel et al. (2005) for the case of single-antenna UEs. In Peel et al. (2005), the channel inversion is regularized with a term that depends on the SINR and, hence, the transmitter structure resembles an MMSE receiver.

Despite its suboptimality, the ZF transmission can still perform well when some multi-user diversity is available, that is, semiorthogonal UEs can be grouped together for ZF processing from a larger pool of UEs. The main benefit of the ZF method is that the solutions to different transceiver optimization problems can be simplified into simple power allocation problems. This is due to the fact that the ZF processing decouples the beamformer design and the power allocation. However, some performance penalty is caused by the ZF constraint. In Figure 5.4, the sum rate performance of the iterative multiuser ZF transmit-receive processing, in conjunction with greedy UE/beam selection and sum rate maximizing power allocation, is illustrated. As can be seen, multiuser diversity reduces the relative gap from ZF sum rate to the theoretical BC sum capacity.

As a result of the iterative ZF processing, the receiver can be replaced by a simple matched filter $v_s = H_{k_s} w_s$ and the inter-stream interference is eliminated, that is, $w_s^H H_{k_s}^H H_{k_s} w_i = 0$ for $i \neq s$. At the same time, $w_s^H H_{k_s}^H H_{k_s} w_s = \lambda_s p_s$, where λ_s represents the interference free channel gain of stream s and p_s the transmit power. Here, the precoder may be decomposed into the contributions of transmit power and transmit direction so that $w = \sqrt{p_s} v_s$, where $\|v_s\|_2 = 1$. For example, the weighted common rate maximization problem $f(\gamma_1, \ldots, \gamma_S) = \min_{k \in \mathcal{U}} \beta_k^{-1} \sum_{s \in \mathcal{P}_k} \log_2 (1 + \gamma_s)$ with ZF processing can be reformulated as the following convex optimization problem:

$$
\begin{aligned}
&\text{max. } r_o \\
&\text{s. t. } \sum_{s \in \mathcal{P}_k} \log_2 (1 + \lambda_s p_s / N_0) \geq \beta_k r_o, \ k \in \mathcal{U} \\
&\qquad \sum_{s=1}^{S} p_s \left\| v_s^{[n]} \right\|_2^2 \leq P_n \ n = 1, \ldots, M
\end{aligned}
\tag{5.15}
$$

where the variables are $r_0 \in \mathbb{R}_+$ and $p_s \in \mathbb{R}_+$. Again, $\mathbf{v}_s^{[n]}$ denotes the parts of $\mathbf{v}_s$ that contain the weights of the antennas belonging to $\mathcal{A}_n$. Note that this problem is reduced to the SINR balancing when $|\mathcal{P}_k| = 1 \; \forall \; k$.

In this section, full MIMO channel matrix knowledge of all UEs at the BS transmitter, obtained for example by perfect CSI sounding, was assumed. The following section discusses the methods for CSI feedback and channel sounding.

5.4   CSI Acquisition for Multiuser MIMO

5.4.1   Limited Feedback

As discussed in section 5.1.3, the radio resources can be assigned in space, time and frequency in order to efficiently exploit the available degrees of freedom, at the benefit of both the individual UEs and the system as a whole. This requires a certain level of channel knowledge. Ideally, this resource management should be done jointly in all the available dimensions but typically the task is divided into several subproblems, where the first scheduling is performed – which UE should use each particular time/frequency slot – and secondly the precoding and/or rate allocation is determined for the subset of UEs that were scheduled.

Unfortunately, channel knowledge at the transmitters does not come for free but requires feedback of CSI from the intended receivers to the transmitters. This can be done in several ways and it is not always an advantage to use feedback that tries to convey the full channel information. One famous example is random opportunistic beamforming (Sharif and Hassibi 2005; Viswanath et al. 2002), where the spatial processing is selected at random and the only feedback from each receiving node is the index of the preferred beam together with a CQI in the form of an estimated SINR, quantized using a few bits. This information is then used for scheduling and possibly for rate adaption. In practice, a fixed grid of beams can be used instead of the random beamformers (Svedman et al. 2009; Thiele et al. 2007). Such an approach is closely related to the use of codebook-based feedback, where each receiving node feeds back an index telling which precoder out of a predefined codebook of precoders should be used at the transmitter. Mostly, an additional CQI is needed, to be used for scheduling and possibly also for rate adaption.

In general, the feedback can typically be divided into a few bits for a CQI that primarily depends on the channel gain and a few bits of directional information, even though some attempts have been made to design a feedback scheme where all the feedback bits are used jointly for scheduling and choice of spatial precoding (Kim et al. 2007). One way to further reduce the feedback, is to use a two-stage approach where only a few bits of CQI are fed back from all the receivers and used to do a (preliminary) scheduling and then only the selected UEs are requested to provide a full feedback (Kountouris and Gesbert 2005; Kountouris et al. 2008; Zakhour and Gesbert 2007).

Another possibility for feedback reduction is to exploit the time correlation of the channels, and combine the feedback bits from several time frames, either using some kind of differential encoding or using hierarchical coding (Svedman et al. 2009; Trivellato et al. 2008).

Earlier, we claimed that channel knowledge does not come for free. However, there are possibilities to exploit duplex communication to obtain different levels of CSI without any explicit feedback. In TDD systems with well-calibrated radio chains, the reciprocity of the physical radio channel can be exploited. In FDD systems, the long-term CSI in the form of a channel covariance matrix can be estimated and used for the reverse link. It has been shown in Björnson et al. (2009) Hammarwall et al. (2008a,b) Kountouris et al. (2006) and Osseiran et al. (2007) that if such long-term covariance information is combined with a short-term CQI representing the Frobenius norm $\|\mathbf{H}\|_F^2$ of the MIMO channel matrix $\mathbf{H}$, it is possible to extract sufficient information to perform both scheduling and spatial precoding. The main idea is to obtain an estimate of the matrix $\mathbf{HH}^\mathbf{H}$ based on knowledge of the second-order statistics and the norm of the channel matrix, $(.)^\mathbf{H}$ is the Hermitian operator. This estimate can then be used instead of the true value of $\mathbf{HH}^\mathbf{H}$ in the precoding design and scheduling. More precisely, let $\mathbf{R}_{tx} = \mathrm{E}[\mathbf{HH}^\mathbf{H}]$ denote the transmit side channel covariance matrix with eigen-value decomposition $\mathbf{R}_{tx} = \mathbf{U\Lambda U}^\mathbf{H}$. Then, it is shown in Björnson et al. (2009) and Hammarwall et al. (2008a) that the MMSE estimate of $\mathbf{HH}^\mathbf{H}$ given $\|\mathbf{H}\|_F^2$ has the form $\mathbf{UDU}^\mathbf{H}$ for some diagonal matrix $\mathbf{D}$, which is a function of the channel eigenvalues $\mathbf{\Lambda}$ and $\|\mathbf{H}\|_F^2$. Intuitively, if the channel norm is large, then it is very likely that the channel is strongly aligned with the dominating eigenvector, whereas if the channel norm is lower, several of the eigenvectors will contribute. The corresponding MSE can also be calculated, as well as estimates of more general functions of $\mathbf{H}$. These MMSE esti-mates provide sufficient information to perform both scheduling and precoding with multiple cochan-nel UEs in the same cell (SDMA), as shown in (Björnson et al. 2009; Hammarwall et al. 2008b). In particular, they propose a generalization of ZF beamforming, where intercell interference is allowed in the dimensions corresponding to the smallest eigenvalues of the estimate of $\mathbf{HH}^\mathbf{H}$ but prevented in the subspace where the eigenvalues are larger. Numerical examples, showing the benefits of this scheme are provided in Björnson et al. (2009) and Hammarwall et al. (2008b).

This section has provided a brief overview of a selection of techniques for limited feedback. A more extensive coverage, as well as extensive reference lists can be found in the review papers Gesbert et al. (2007) and Love et al. (2008).

5.4.2 CSI Sounding

In networks employing TDD, unquantized instantaneous CSIT can be obtained at the transmitter. This allows more advanced or accurate multiuser transmitter and receiver design to be performed by the BS. The benefits of TDD are best available in local and metropolitan area deployments, where the cell sizes are relatively small and the channel changes slowly so that the coherence time is longer than a TDD frame. The assumption is that the uplink and downlink MIMO channels are reciprocal and that the transmit and receive RF chains in all transceivers are well calibrated. In order to facilitate MU-MIMO precoding in the downlink, mutually orthogonal antenna-specific uplink sounding pilot signals are needed. The number of UE antennas to be served in the same resource block defines the amount of orthogonal pilot resources needed. Thus the number of UEs that can be tracked simulta-neously is limited by the pilot overhead. Note that the uplink pilot is needed to keep the downlink MIMO channel open even when the UE has no data to transmit. If the traffic in uplink and downlink is relatively symmetric, the same subcarriers can be allocated to the same set of UEs in both directions. In this case the uplink data demodulation pilot can be reused as a reference for downlink transmit precoding as well. In addition to channel sounding, some form of scalar CQI feedback is needed to support rate allocation and adaptive modulation. This is due to the fact that the interference levels are not reciprocal, and the transmitter should know the SINR seen by the receiver.

Sounding Pilot Overhead Reduction

Antenna-specific uplink pilot streams cause an extensive overhead that restricts the size of the prac-tical UE group and the UE antenna setup that can be handled within the same time-frequency slot. Conventionally, the number of the required mutually orthogonal CSI sounding pilot streams corre-sponds to the aggregate number of UE antennas that are simultaneously active. That is, in a system with K UEs, each equipped with N_{R_k} antennas, $k = 1, \ldots, K$, the required number of mutually orthogonal – in time and/or frequency domain – pilot sequences becomes $\sum_k N_{R_k}$.

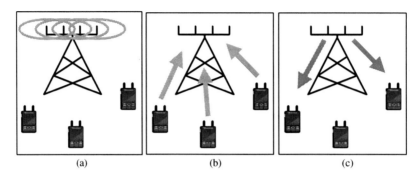

Figure 5.7   Reduced overhead sounding: (a) DL common pilot (b) UL CSI sounding (c) DL data transmission

The CSI sounding overhead may be reduced by letting the UEs transmit a reduced number of pilot signals (Hara and Oshima 2007; Komulainen et al. 2009). Instead of antenna-specific pilots, UE k forms $J_k < N_{R_k}$ uplink pilot beams by transmit precoding. As a result, the number of required pilot resources reduces to $\sum_k J_k$. Consequently, UE k appears as a J_k -antenna UE to the BS. The number J_k can be imposed either statically or dynamically by the BS. The CSI sounding beams are formed based on the knowledge of the UE-specific MIMO channels, obtained via a downlink common pilot signal. This way, part of the signalling overhead is moved to the downlink. The common pilot signal is resource efficient since only N_T orthogonal pilot sequences are needed, where N_T is the number of BS antennas.

The signaling stages are depicted in Figure 5.7. The UEs estimate their individual MIMO channels based on a transmit-antenna-specific downlink common pilot signal before performing CSI sounding. The best choice for the sounding beamformers is then based on the strongest spatial eigenmodes so that the precoding matrix contains the corresponding J_k left singular vectors of the estimated channel. As a result, the BS cannot determine the full channel matrices but only the J_k best eigenmodes per UE. From the system sum rate point of view, the optimal number of data streams to be allocated per UE is usually less than N_{R_k}, especially when either K or N_{R_k} is large. Therefore, the weak eigenmodes, neglected in the reduced overhead sounding concept, would rarely be utilized.

When taking the estimation noise into account the overhead reduction improves robustness and may even increase the average system capacity. This is due to the power efficiency of the CSI sounding concept: uplink sounding power is not wasted on the weak eigenmodes. Another benefit is that the physical layer specification of the communication system, such as LTE-A, does not need to explicitly support arbitrarily large or odd numbers of antennas at the UE. A UE can mimic to have some specification-compliant number of virtual antennas, for instance, a UE with five antenna elements may appear as a virtual two-antenna UE to the BS.

5.5   Future Directions of MIMO Techniques

A major challenge for wireless communication systems is how to allocate resources among UEs across the space (including different cells), frequency and time dimensions and jointly design all the transceivers with different system optimization objectives. This remains unresolved for a large variety of optimization criteria, especially when combined with practical modulation and coding schemes. The problem is a difficult non-convex combinatorial problem with integer constraints and finding jointly optimal solutions is most likely intractable. Therefore, efficient suboptimal solutions are required.

Another major challenge is to keep the CSI signaling overhead over the air at a minimum, calling for decentralized schemes that are primarily based on locally available information and are robust to heavily quantized channel information. In the case of sounding-based precoding, the CSI at the transmitter remains uncertain due to limited sounding power and estimation errors. Thus, the design of transmission methods that are robust to CSI uncertainty, and sounding strategies that provide a good tradeoff between performance and overhead, are needed.

In cellular systems, the evolution of multiantenna processing is heading towards CoMP techniques, where multiple BSs cooperate in order to provide spatial access to the UEs, as described in Chapter 6. Therefore, it is a requirement for the practical future MIMO methods to support also the CoMP paradigm, both from the signalling and the precoding point of view.

References

3GPP 2010 Further Advancements for E-UTRA Physical Layer Aspects. Technical Report 36.814 v9.0.0, 3rd Generation Partnership Project (3GPP), http://www.3gpp.org/ftp/Specs/2010-03/Rel-9/.

Akyildiz I, Gutierrez-Estevez D and Reyes E 2010 The evolution to 4G cellular systems: LTE-Advanced. *Physical Communication* **3**(4), 217–244.

Alamouti S 1998 A simple transmit diversity technique for wireless communications. *IEEE J. Select. Areas Commun.* **16**(8), 1451–1458.

Bengtsson M 2002 A Pragmatic Approach to Multi-user Spatial Multiplexing *IEEE Sensor Array and Multichannel Signal Processing Workshop*, pp. 130–134, Rosslyn, VA, USA.

Bengtsson M and Ottersten B 1999 Optimal downlink beamforming using semidefinite optimization *Proc. Annual Allerton Conf. Commun., Contr., Computing*, pp. 987–996, Monticello, IL.

Bengtsson M and Ottersten B 2001 Optimal and suboptimal transmit beamforming *Handbook of Antennas in Wireless Communications*, ed. L. C. Godara. CRC Press, Boca Raton, FL.

Björnson E, Hammarwall D and Ottersten B 2009 Exploiting quantized channel norm feedback through conditional statistics in arbitrarily correlated MIMO systems. *IEEE Transactions on Signal Processing* **57**(10), 4027–4041.

Bliss DW, Forsythe KW, Hero AO and Yegulalp AF 2002 Environmental issues for MIMO capacity. *IEEE Trans. Signal Processing* **50**, 2128–2142.

Boccardi F, Tosato F and Caire G 2006 Precoding schemes for the MIMO-GBC *Proc. Int. Zurich Seminar Broadband Commun.*, pp. 10–13, ETH Zurich, Switzerland.

Boyd S, Kim SJ, Vandenberghe L and Hassibi A 2007 A tutorial on geometric programming. *Optimization and Engineering* **8**(1), 67–127.

Boyd S and Vandenberghe L 2004 *Convex Optimization*. Cambridge University Press, Cambridge.

Cadambe V and Jafar S 2008 Interference alignment and the degrees of freedom of the k user interference channel. *IEEE Trans. Inform. Theory* **54**(8), 3425–3441.

Caire G and Shamai S 2003 On the achievable throughput of a multiantenna Gaussian broadcast channel. *IEEE Trans. Inform. Theory* **49**(7), 1691–1706.

Cepeda R, Fitton M and Nix A 2002 The performance of robust adaptive modulation over wireless channels with non reciprocal interference *Proc. VTC 2002 Spring-IEEE 55th Vehicular Technology Conf.*, vol. 3, pp. 1497–1501.

Chae CB, Mazzarese D, Jindal N and Heath Jr. RW 2008 Coordinated beamforming with limited feedback in the MIMO broadcast channel. *IEEE J. Select. Areas Commun.* **26**(8), 1505–1515.

Chan P and Cheng R 2007 Capacity maximization for zero-forcing MIMO-OFDMA downlink systems with multiuser diversity. *IEEE Trans. Wireless Commun.* **6**(3), 1880–1889.

Chang JH, Tassiulas L and Rashid-Farrokhi F 2002 Joint transmitter receiver diversity for efficient space division multiaccess. *IEEE Trans. Wireless Commun.* **1**(1), 16–27.

Cheng RS and Verdú S 1993 Gaussian multiaccess channels with ISI: Capacity region and multiuser water-filling. *IEEE Trans. Inform. Theory* **39**(3), 773–785.

Chiang M 2005 Geometric programming for communication systems. *Foundations and Trends in Communications and Information Theory* **2**(1–2), 1–154.

Chiang M, Tan CW, Palomar DP, O'Neill D and Julian D 2007 Power control by geometric programming. *IEEE Trans. Wireless Commun.* **6**(7), 2640–2651.

Choi LU and Murch RD 2004 A transmit preprocessing technique for multiuser MIMO systems using a decomposition approach. *IEEE Trans. Wireless Commun.* **3**(1), 20–24.

Choi RLU, Ivrlac MT, Murch RD and Utschick W 2004 On strategies of multiuser MIMO transmit signal processing. *IEEE Trans. Wireless Commun.* **3**(6), 1396–1941.

Christensen S, Agarwal R, Carvalho E and Cioffi J 2008 Weighted sum-rate maximization using weighted MMSE for MIMO-BC beamforming design. *IEEE Trans. Wireless Commun.* **7**(12), 4792–4799.

Codreanu M, Tölli A, Juntti M and Latva-aho M 2007 Joint design of Tx–Rx beamformers in MIMO downlink channel. *IEEE Trans. Signal Processing* **55**(9), 4639–4655.

Costa M 1983 Writing on dirty paper. *IEEE Trans. Inform. Theory* **29**(3), 439–441.

Cover T 1972 Broadcast channels. *IEEE Trans. Inform. Theory* **18**(1), 2–14.

Cover TM and Thomas JA 1991 *Elements of Information Theory*. John Wiley, New York, USA.

Dimic G and Sidiropoulos N 2005 On downlink beamforming with greedy user selection: Performance analysis and a simple new algorithm. *IEEE Trans. Signal Processing* **53**(10), 3857–3868.

Ding Y, Davidson TN, Luo ZQ and Wong KM 2003 Minimum BER block precoders for zero-forcing equalization. *IEEE Trans. Signal Processing* **51**(9), 2410–2423.

Farhang-Boroujeny B, Spencer QH and Swindlehurst AL 2003 Layering techniques for space-time communication in multi-user networks *Proc. VTC 2003 Fall–IEEE 58th Vehicular Technology Conf.*, vol. 2, pp. 1339–1343, Orlando, Florida.

Farrokhi FR, Foschini GJ, Lozano A and Valenzuela RA 2001 Link-optimal space-time processing with multiple transmit and receive antennas. *IEEE Commun. Lett.* pp. 85–87.

Foschini G 1996 Layered space–time architecture for wireless communication in a fading environment when using multi-element antennas. *Bell Labs Tech. J.* **1**(2), 41–59.

Foschini GJ and Gans MJ 1998 On limits of wireless communications in a fading environment when using multiple antennas. *Wireless Pers. Commun., Kluwer* **6**, 311–335.

Fuchs M, Galdo GD and Haardt M 2007 Low-complexity space-time-frequency scheduling for MIMO systems with SDMA. *IEEE Trans. Veh. Technol.* **56**(5), 2775–2784.

Gallager RG 1985 A perspective on multiple access channels. *IEEE Trans. Inform. Theory* **31**(2), 124–142.

Gerlach D and Paulraj A 1996 Base station transmitting antenna arrays for multipath environments. *Sig. Proc.* **54**(1), 59–73.

Gershman AB, Sidiropoulos ND, Shahbazpanahi S, Bengtsson M and Ottersten B 2010 Convex optimization-based beamforming: From receive to transmit and network designs. *IEEE Signal Processing Mag.* **27**, 62–75.

Gesbert D, Kountouris M, Heath Jr RW, Chae CB and Salzer T 2007 Shifting the MIMO paradigm. *IEEE Signal Processing Mag.* **24**(5), 36–46.

Godara LC 1997 Applications of antenna arrays to mobile communications. I. Performance improvement, feasibility, and system considerations. *Proc. IEEE* **85**(7), 1031–1060.

Goldsmith A, Jafar S, Jindal N and Vishwanath S 2003 Capacity limits of MIMO channels. *IEEE J. Select. Areas Commun.* **21**(5), 684–702.

Guo Y and Levy BC 2006 Worst-case MSE precoder design for imperfectly known MIMO communications channels. *IEEE Trans. Signal Processing* **54**(5), 1840–1852.

Hammarwall D, Bengtsson M and Ottersten B 2006 On downlink beamforming with indefinite shaping constraints. *IEEE Trans. Signal Processing* **54**(9), 3566–3580.

Hammarwall D, Bengtsson M and Ottersten B 2008a Acquiring partial CSI for spatially selective transmission by instantaneous channel norm feedback. *IEEE Trans. Signal Processing* **56**(3), 1188–1204.

Hammarwall D, Bengtsson M and Ottersten B 2008b Utilizing the spatial information provided by channel norm feedback in SDMA systems. *IEEE Trans. Signal Processing* **56**, 3278–3293.

Hara Y and Oshima K 2007 Spatial scheduling using partial CSI reporting in multiuser MIMO systems *Proc. VTC 2007 Spring – IEEE 65th Vehicular Technology Conf.*, pp. 1673–1677, Dublin, Ireland.

Heath RW and Love DJ 2005 Multimode antenna selection for spatial multiplexing systems with linear receivers. *IEEE Trans. Signal Processing* **53**(8), 3042–3056.

Heath RW and Paulraj AJ 2005 Switching between diversity and multiplexing in MIMO systems. *IEEE Trans. Commun.* **53**(6), 962–968.

Heath RW, Sandhu S and Paulraj AJ 2001 Antenna selection for spatial multiplexing systems with linear receivers. *IEEE Commun. Lett.* **5**(4), 142–144.

Horn R and Johnson C 1990 *Matrix Analysis*. Cambridge University Press, Cambridge.

Huang Y and Palomar DP 2010 Rank-constrained separable semidefinite program with applications to optimal beamforming. *IEEE Trans. Signal Processing* **58**(2), 664–678.

IEEE 802.16 2010 System Description Document (SDD). Document IEEE 802.16m-09/0034r3, IEEE 802.16 Broadband Wireless Access Working Group, Task Group m, http://www.ieee802.org/16/tgm/core.html.

Jindal N 2006 MIMO broadcast channels with finite-rate feedback. *IEEE Trans. Inform. Theory* **52**(11), 5045–5060.

Jindal N and Goldsmith A 2005 Dirty-paper coding versus TDMA for MIMO broadcast channels. *IEEE Trans. Inform. Theory* **51**(5), 1783–1794.

Jindal N, Vishwanath S and Goldsmith A 2004 On the duality of Gaussian multiple-access and broadcast channels. *IEEE Trans. Inform. Theory* **50**(5), 768–783.

Jongren G, Skoglund M and Ottersten B 2002 Combining beamforming and orthogonal space–time block coding. *IEEE Trans. Inform. Theory* **48**(3), 611–627.

Jorswieck E and Boche H 2003 Transmission strategies for the MIMO MAC with MMSE receiver: Average MSE optimization and achievable individual MSE region. *IEEE Trans. Signal Processing* **51**(11), 2872–2881.

Jorswieck EA and Boche H 2004 Channel capacity and capacity-range of beamforming in MIMO wireless systems under correlated fading with covariance feedback. *IEEE Trans. Wireless Commun.* **3**(5), 1543–1553.

Karakayali MK, Foschini GJ and Valenzuela RA 2006a Network coordination for spectrally efficient communications in cellular systems. *IEEE Wireless Commun. Mag.* **3**(14), 56–61.

Karakayali MK, Foschini GJ, Valenzuela RA and Yates RD 2006b On the maximum common rate achievable in a coordinated network *Proc. ICC 2006 – IEEE Int. Conf. Commun.*, vol. 9, pp. 4333–4338, Istanbul, Turkey.

Khachan A, Tenenbaum A and Adve R 2006 Linear processing for the downlink in multiuser MIMO systems with multiple data streams *Proc. ICC 2006 – IEEE Int. Conf. Commun.*, vol. 9, pp. 4113–4118, Istanbul, Turkey.

Kim TT, Bengtsson M and Skoglund M 2007 Quantized feedback design for MIMO broadcast channels *Proc. ICASSP 2007 - IEEE Int. Conf. Acoust. Speech and Signal Processing*.

Komulainen P, Tölli A, Latva-aho M and Juntti M 2009 Channel sounding pilot overhead reduction for TDD multiuser MIMO systems *IEEE Workshop on Broadband Wireless Access*, Honolulu, Hawaii.

Kountouris M and Gesbert D 2005 Robust multi-user opportunistic beamforming for sparse networks *Proc. SPAWC 2005 – Sig. Proc. Advances in Wireless Commun.*, pp. 975–979, New York, NY.

Kountouris M, Gesbert D and Pittman L 2006 Transmit correlation-aided opportunistic beamforming and scheduling *EUSIPCO 2006, 14th European Signal Processing Conference*, Florence Italy.

Kountouris M, Gesbert D and Sälzer T 2008 Enhanced multiuser random beamforming: Dealing with the not so large number of users case. *IEEE J. Select. Areas Commun.* **26**(8), 1536-1545.

Lebrun G, Gao J and Faulkner M 2005 MIMO transmission over a time-varying channel using SVD. *IEEE Trans. Wireless Commun.* **4**(2), 757–764.

Lee J and Jindal N 2006 Symmetric capacity of MIMO downlink channels *Proc. IEEE Int. Symp. Inform. Theory*, pp. 1031–1035, Seattle, USA.

Lee KH and Petersen D 1976 Optimal linear coding for vector channels. *IEEE Trans. Commun.* **24**(12), 1283–1290.

Lobo MS, Vandenberghe L, Boyd S and Lebret H 1998 Applications of second–order cone programming. *Linear Algebra and Applications* **284**, 193–228.

Love DJ and Heath RW 2005 Multimode precoding for MIMO wireless systems. *IEEE Trans. Signal Processing* **53**(10), 3674–3687.

Love DJ, Heath Jr RW, Lau VKN, Gesbert D, Rao BD and Andrews M 2008 An overview of limited feedback in wireless communication systems. *IEEE J. Select. Areas Commun.* **26**(8), 1341–1365.

Love DJ, Heath RW and Strohmer T 2003 Grassmannian beamforming for multiple-input multiple-output wireless systems. *IEEE Trans. Inform. Theory* **49**(10), 2735–2747.

Luo ZQ 2003 Applications of convex optimization in signal processing and digital communication. *Mathematical Programming* **97, Series B**, 177–207.

Luo ZQ, Davidson T, Giannakis G and Wong KM 2004 Transceiver optimization for block-based multiple access through ISI channels. *IEEE Trans. Signal Processing* **52**(4), 1037–1052.

Luo ZQ and Yu W 2006 An introduction to convex optimization for communications and signal processing. *IEEE J. Select. Areas Commun.* **24**(8), 1426–1438.

Maddah-Ali M, Motahari A and Khandani A 2008 Communication over MIMO X channels: interference alignment, decomposition, and performance analysis. *IEEE Trans. Inform. Theory* **54**(8), 3457–3470.

Mezghani A, Joham M, Hunger R and Utschick W 2006 Transceiver design for multi-user mimo systems *ITG/IEEE Workshop on Smart Antennas (WSA 2006)*, pp. 1–8, Ulm, Germany.

Mukkavilli KK, Sabharwal A, Aazhang B and Erkip E 2003 On beamforming with finite rate feedback in multiple-antenna systems. *IEEE Trans. Inform. Theory* **49**(10), 2562–2579.

Naguib AF, Tarokh V, Seshadri N and Calderbank AR 1998 A space–time coding modem for high-data-rate wireless communications. *IEEE J. Select. Areas Commun.* **16**(8), 1459–1478.

Osseiran A, Skillermark P and Olsson M 2007 Multi-Antenna SDMA in OFDM Radio Network Systems: Modeling and Evaluations *Personal, Indoor and Mobile Radio Communications, 2007. PIMRC 2007. IEEE 18th International Symposium on*, pp. 1–5.

Palomar DP 2003 A Unified Framework for Communications through MIMO Channels PhD thesis Department of Signal Theory and Communications, Technical University of Catalonia Barcelona, Spain.

Palomar DP 2005 Convex primal decomposition for multicarrier linear MIMO transceivers. *IEEE Trans. Signal Processing* **53**(12), 4661–4674.

Palomar DP and Barbarossa S 2005 Designing MIMO communication systems: Constellation choice and linear transceiver design. *IEEE Trans. Signal Processing* **53**(10), 3804–3818.

Palomar DP, Bengtsson M and Ottersten B 2005 Minimum BER linear transceivers for MIMO channels via primal decomposition. *IEEE Trans. Signal Processing* **53**(8), 2866–2882.

Palomar DP, Cioffi JM and Lagunas MA 2003 Joint Tx-Rx beamforming design for multicarrier MIMO channels: A unified framework for convex optimization. *IEEE Trans. Signal Processing* **51**(9), 2381–2401.

Palomar DP, Lagunas MA and Cioffi JM 2004 Optimum linear joint transmit-receive processing for MIMO channels with QoS constraints. *IEEE Trans. Signal Processing* **52**(5), 1179–1197.

Pan Z, Wong KK and Ng TS 2004 Generalized multiuser orthogonal space-division multiplexing. *IEEE Trans. Wireless Commun.* **3**(6), 1969–1973.

Park C, Wang Y, Jöngren G and Hammarwall D 2011 Evolution of Uplink MIMO for LTE-Advanced. *IEEE Communications Magazine* **49**(2), 76-83.

Pascual-Iserte A, Palomar DP, Perez-Neira AI and Lagunas MA 2006 A robust maximin approach for MIMO communications with imperfect channel state information based on convex optimization. *IEEE Trans. Signal Processing* **54**(1), 346–360.

Peel CB, Hochwald BM and Swindlehurst AL 2005 A vector-perturbation technique for near-capacity multiantenna multiuser communication part I: Channel inversion and regularization. *IEEE Trans. Commun.* **53**(1), 195–202.

Primolevo G, Simeone O and Spagnolini U 2005 Channel aware scheduling for broadcast MIMO systems with orthogonal linear precoding and fairness constraints *Proc. ICC 2005 - IEEE Int. Conf. Commun.*, vol. 4, pp. 2749–2753, Seoul, Korea.

Raleigh GG and Cioffi JM 1998 Spatio-temporal coding for wireless communication. *IEEE Trans. Commun.* **46**(3), 357–366.

Rashid-Farrokhi F, Liu K and Tassiulas L 1998 Transmit beamforming and power control for cellular wireless systems. *IEEE J. Select. Areas Commun.* **16**(8), 1437–1450.

Sadek M, Tarighat A and Sayed AH 2007 A leakage-based precoding scheme for downlink multi-user MIMO channels. *IEEE Trans. Wireless Commun.* **6**(5), 1711–1721.

Salz J 1985 Digital transmission over cross-coupled linear channels. *AT&T Technical Journal* **64**, 1147–1159.

Sampath H, Stoica P and Paulraj A 2001 Generalized linear precoder and decoder design for MIMO channels using the weighted MMSE criterion. *IEEE Trans. Commun.* **49**(12), 2198–2206.

Sartenaer T, Vandendorpe L and Louveaux J 2005 Balanced capacity of wireline multiuser channels. *IEEE Trans. Commun.* **53**(12), 2029–2042.

Sato H 1978 An outer bound to the capacity region of broadcast channels. *IEEE Trans. Inform. Theory* **24**(3), 374–377.

Scaglione A, Barbarossa S and Giannakis GB 1999 Filterbank transceivers optimizing information rate in block transmissions over dispersive channels. *IEEE Trans. Inform. Theory* **45**(3), 1019–1032.

Scaglione A, Stoica P, Barbarossa S, Giannakis G and Sampath H 2002 Optimal designs for space-time linear precoders and decoders. *IEEE Trans. Signal Processing* **50**(5), 1051–1064.

Schubert M and Boche H 2004 Solution of the multiuser downlink beamforming problem with individual SINR constraints. *IEEE Trans. Veh. Technol.* **53**(1), 18–28.

Schubert M, Shi S, Jorswieck E and Boche H 2005 Downlink sum-MSE transceiver optimization for linear multi-user MIMO systems *Proc. Annual Asilomar Conf. Signals, Syst., Comp.*, pp. 1424–1428, Pacific Grove, CA.

Serbetli S and Yener A 2004 Transceiver optimization for multiuser MIMO systems. *IEEE Trans. Signal Processing* **52**(1), 214–226.

Sesia S, Baker M and Toufik I 2011 *LTE – The UMTS Long Term Evolution: From Theory to Practice* 2nd edn. John Wiley & Sons, Ltd, Chichester.

Sharif M and Hassibi B 2005 On the capacity of MIMO broadcast channels with partial side information. *IEEE Trans. Inform. Theory* **51**(2), 506–522.

Sharif M and Hassibi B 2007 A comparison of time-sharing, DPC, and beamforming for MIMO broadcast channels with many users. *IEEE Trans. Commun.* **55**(1), 11–15.

Shen Z, Chen R, Andrews J, Heath R and Evans B 2006 Low complexity user selection algorithms for multiuser MIMO systems with block diagonalization. *IEEE Trans. Signal Processing* **54**(9), 3658–3663.

Shi S and Schubert M 2005 MMSE transmit optimization for multi-user multi-antenna systems *Proc. ICASSP 2005 – IEEE Int. Conf. Acoust. Speech and Signal Processing*, vol. 3, pp. 409–412, Philadelphia, PA.

Shi S, Schubert M and Boche H 2008 Rate optimization for multiuser mimo systems with linear processing. *IEEE Trans. Signal Processing* **56**(8), 4020 –4030.

Spencer QH and Swindlehurst AL 2004a Channel allocation in multi-user MIMO wireless communications systems *Proc. ICC 2004 – IEEE Int. Conf. Commun.*, vol. 5, pp. 3035–3039.

Spencer QH and Swindlehurst AL 2004b A hybrid approach to spatial multiplexing in multiuser MIMO downlinks. *EURASIP J. Wireless Comm. and Netw.* **2004**(2), 236–247.

Spencer QH, Peel CB, Swindlehurst AL and Haardt M 2004a An introduction to the multi-user MIMO downlink. *IEEE Commun. Mag.* **43**(10), 60–67.

Spencer QH, Swindlehurst AL and Haardt M 2004b Zero-forcing methods for downlink spatial multiplexing in multiuser MIMO channels. *IEEE Trans. Signal Processing* **52**(2), 461–471.

Stojnic M, Vikalo H and Hassibi B 2006 Rate maximization in multi-antenna broadcast channels with linear preprocessing. *IEEE Trans. Wireless Commun.* **5**(9), 2338–2342.

Svedman P, Jorswieck EA and Ottersten B 2009 Reduced feedback sdma based on subspace packings. *IEEE Trans. Wireless Commun.* **8**(3), 1329–1339.

Tarokh V, Jafarkhani H and Calderbank AR 1999 Space–time block codes from orthogonal designs. *IEEE Trans. Inform. Theory* **45**(5), 1456–1467.

Tarokh V, Seshadri N and Calderbank AR 1998 Space–time codes for high data rate wireless communication: Performance criterion and code construction. *IEEE Trans. Inform. Theory* **44**(2), 744–765.

Tejera P, Utschick W, Bauch G and Nossek JA 2006 Subchannel allocation in multiuser multiple-input–multiple-output systems. *IEEE Trans. Inform. Theory* **52**(10), 4721–4733.

Telatar E 1999 Capacity of multi-antenna Gaussian channels. *European Transactions on Telecommunications (ETT)* **10**(6), 585–595.

Telatar E 1995 Capacity of multi-antenna Gaussian channels. Technical report, Bell Laboratories. Internal Tech.Memo, pp. 1–28.

Thiele L, Jungnickel V, Schellmann M and Zirwas W 2007 Capacity scaling of multi-user MIMO with limited feedback in a multi-cell environment *Proc. 41st Asilomar Conf. Sig., Syst., Comput.*

Tölli A, Codreanu M and Juntti M 2007 Compensation of non-reciprocal interference in adaptive MIMO-OFDM cellular systems. *IEEE Trans. Wireless Commun.* **6**(2), 545–555.

Tölli A, Codreanu M and Juntti M 2008a Cooperative MIMO-OFDM cellular system with soft handover between distributed base station antennas. *IEEE Trans. Wireless Commun.* **7**(4), 1428–1440.

Tölli A, Codreanu M and Juntti M 2008b Linear multiuser MIMO transceiver design with quality of service and per antenna power constraints. *IEEE Trans. Signal Processing* **56**(7), 3049–3055.

Tölli A and Juntti M 2005 Scheduling for multiuser MIMO downlink with linear processing *Proc. PIMRC 2005 – IEEE 16th Int. Symp. on Pers., Indoor and Mobile Radio Commun.*, vol. 1, pp. 156–160, Berlin, Germany.

Trivellato M, Boccardi F and Huang H 2008 On transceiver design and channel quantization for downlink multiuser MIMO systems with limited feedback. *IEEE J. Select. Areas Commun.* **26**(8), 1491-1504.

Tse D and Viswanath P 2005 *Fundamentals of Wireless Communication*. Cambridge University Press.

Tu Z and Blum R 2003 Multiuser diversity for a dirty paper approach. *IEEE Commun. Lett.* **7**(8), 370–372.

Vandenberghe L and Boyd S 1996 Semidefinite programming. *SIAM Review* **38**(1), 49–95.

Vishwanath S, Jindal N and Goldsmith A 2003 Duality, achievable rates, and sum-rate capacity of Gaussian MIMO broadcast channels. *IEEE Trans. Inform. Theory* **49**(10), 2658–2668.

Visotsky E and Madhow U 1999 Optimum beamforming using transmit antenna arrays *Proc. IEEE Veh. Technol. Conf.*, vol. 1, pp. 851–856, Houston, TX.

Visotsky E and Madhow U 2001 Space–time transmit precoding with imperfect feedback. *IEEE Trans. Inform. Theory* **47**(6), 2632–2639.

Viswanath P and Tse D 2003 Sum capacity of the vector Gaussian broadcast channel and uplink-downlink duality. *IEEE Trans. Inform. Theory* **49**(8), 1912–1921.

Viswanath P, Tse D and Laroia R 2002 Opportunistic beamforming using dumb antennas. *IEEE Trans. Inform. Theory* **48**(6), 1277–1294.

Vorobyov SA, Gershman AB and Luo ZQ 2003 Robust adaptive beamforming using worst-case performance optimization: A solution to the signal mismatch problem. *IEEE Trans. Signal Processing* **51**(2), 313–324.

Weingarten H, Steinberg Y and Shamai S 2006 The capacity region of the Gaussian multiple-input multiple-output broadcast channel. *IEEE Trans. Inform. Theory* **52**(9), 3936–3964.

Wiesel A, Eldar YC and Shamai S 2006 Linear precoding via conic optimization for fixed MIMO receivers. *IEEE Trans. Signal Processing* **54**(1), 161–176.

Winters J 1984 Optimum combining in digital mobile radio with cochannel interference. *IEEE J. Select. Areas Commun.* **2**(4), 528–539.

Winters J 1987 On the capacity of radio communication systems with diversity in a Rayleigh fading environment. *IEEE J. Select. Areas Commun.* **5**(5), 871–878.

Winters JH, Salz J and Gitlin RD 1994 The impact of antenna diversity on the capacity of wireless communication systems. *IEEE Trans. Commun.* **42**(1/2/3), 1740–1751.

Wong KK, Murch RD and Letaief KB 2003 A joint-channel diagonalization for multiuser MIMO antenna systems. *IEEE Trans. Wireless Commun.* **2**(4), 773–786.

Yang J and Roy S 1994 On joint transmitter and receiver optimization for multiple-input-multiple-output (MIMO) transmission systems. *IEEE Trans. Commun.* **42**(12), 3221–3231.

Ye S and Blum RS 2003 Optimized signaling for MIMO interference systems with feedback. *IEEE Trans. Signal Processing* **51**, 2839–2847.

Yoo T and Goldsmith A 2006 On the optimality of multiantenna broadcast scheduling using zero-forcing beamforming. *IEEE J. Select. Areas Commun.* **24**(3), 528–541.

Yu W and Cioffi J 2004 Sum capacity of Gaussian vector broadcast channels. *IEEE Trans. Inform. Theory* **50**(9), 1875–1892.

Yu W and Lan T 2007 Transmitter optimization for the multi-antenna downlink with per-antenna power constraints. *IEEE Trans. Signal Processing* **55**(6, part 1), 2646–2660.

Yu W, Rhee W, Boyd S and Cioffi J 2004 Iterative water-filling for Gaussian vector multiple-access channels. *IEEE Trans. Inform. Theory* **50**(1), 145–152.

Zakhour R and Gesbert D 2007 A two-stage approach to feedback design in multi-user MIMO channels with limited channel state information *Proc. PIMRC 2007 – IEEE 18th Int. Symp. on Pers., Indoor and Mobile Radio Commun.*

Zetterberg P and Ottersten B 1995 The spectrum efficiency of a base station antenna array for spatially selective transmission. *IEEE Trans. Veh. Technol.* **44**, 651–660.

Zhang YJ and Letaief K 2005 An efficient resource-allocation scheme for spatial multiuser access in MIMO/OFDM systems. *IEEE Trans. Commun.* **53**(1), 107–116.

Zheng L and Tse DNC 2003 Diversity and multiplexing: A fundamental tradeoff in multiple-antenna channels. *IEEE Trans. Inform. Theory* **49**(5), 1073–1096.

Zhou S and Giannakis GB 2003 Optimal transmitter eigen-beamforming and space–time block coding based on channel correlation. *IEEE Trans. Inform. Theory* **49**(7), 1673–1690.

6

Coordinated MultiPoint (CoMP) Systems

Mauro Boldi, Antti Tölli, Magnus Olsson, Eric Hardouin, Tommy Svensson, Federico Boccardi, Lars Thiele and Volker Jungnickel

6.1 Overview of CoMP

Future cellular networks will need to provide high data rate services for a large number of User Equipments (UEs), which requires a high spectral efficiency over the entire cell area. Hence it is important that the radio interface is robust to interference. In particular Inter-Cell Interference (ICI) appears when the same radio resource is reused in different cells in an uncoordinated way. Naturally, ICI particularly degrades the performance of UEs located in the cell-edge areas, which creates a performance discrepancy between cell-edge and inner-cell UEs.

Over the years, several different methods have been studied in order to mitigate ICI. Interference averaging techniques (WINNER-II 2007a) aim at averaging the interference over all UEs, thereby reducing the interference experienced by some UEs. Frequency hopping (Carneheim 1994), which is used in GSM, is a well-known example of an interference averaging technique. Interference avoidance techniques (WINNER-II 2007b), on the other hand, aim at explicitly coordinating and avoiding interference, for example, by setting restrictions on how the radio resources are used. An example of this is Inter-Cell Interference Coordination (ICIC), which is available in the 3GPP Long Term Evolution (LTE) standard (see e.g. Dahlman et al. 2008).

Recently, even tighter interference coordination has gained significant interest under the name Coordinated Multipoint transmission or reception (CoMP). This refers to a system where the transmission and/or reception at multiple, geographically separated antenna sites is dynamically coordinated in order to improve system performance. The idea in CoMP is to proactively manage the interference to improve the performances of UEs, especially for those experiencing poor QoS.

Interest in the coordination among nodes either in transmission or reception has significantly grown in recent years. A comprehensive literature review of the studies that have led to the proposal of CoMP in future systems is given in section 6.1.3, and has been documented in several European Union (EU) research projects (ARTIST4G 2010; WINNER+ 2009a,b).

The adopted acronym to describe this coordinated approach (CoMP) directly drives the attention to the coordination that takes place among the involved entities, but it has to be reported that at the

Mobile and Wireless Communications for IMT-Advanced and Beyond, First Edition.
Edited by Afif Osseiran, Jose F. Monserrat and Werner Mohr.
© 2011 Afif Osseiran, Jose F. Monserrat and Werner Mohr. Published 2011 by John Wiley & Sons, Ltd.

moment many different definitions are given to systems generally considered under the framework of CoMP. In the literature it is also possible to find CoMP systems labeled as "Network MIMO" or "Multicellular MIMO" or "Multicellular cooperation" (Gesbert et al. 2010). The acronym CoMP will be used in this chapter due to its more generic meaning and its adoption in the 3GPP standard development organization (3GPP 2010).

In this section an introduction to CoMP types and the most feasible architectures to implement CoMP will be provided, together with the constraints that have to be taken into account. Closer view to specific CoMP techniques will be given in following sections. The chapter is completed by a description of a practical implementation of CoMP in a trial environment.

6.1.1 CoMP Types

The coordination among multiple entities in radio access networks can occur either as a transmission coordination, namely in the Downlink (DL) of the radio systems, or as a reception coordination, namely in the uplink. As a general view two main families of coordination methods have emerged:

Joint Processing means a coordination in which multiple entities are simultaneously transmitting/receiving to/from UEs located in the coordination area.

Coordinated scheduling or beamforming means a coordination where each UE is communicating with a single transmission/reception point, but the transmission/reception is made with an exchange of control information among the coordinated entities.

Downlink coordinated multipoint transmission implies dynamic coordination among multiple geographically separated transmission points, and the two main families described above can be further detailed as follows:

• In *Joint processing/transmission schemes*, data to a single UE is simultaneously transmitted from multiple transmission points in the coordination area, for example, to (coherently or noncoherently[1]) improve the received signal quality and/or cancel actively interference from transmissions intended for other UEs. This category of schemes put high requirements on the coordination links and the backhaul since the UE data need to be made available at the multiple coordinated transmission points. The amount of control data to be exchanged over the coordination links is also large, for example, channel knowledge and computed transmission weights. Joint processing/transmission is illustrated in the right panel of Figure 6.1. Fundamentally, multiple transmit antennas can be used to improve the DL performance in two ways:

1. the power of the desired signal received at each UE can be increased, that is, signals transmitted from multiple antennas are formed such that they add constructively at the desired UE;
2. the interference experienced by each UE can be suppressed, that is, signals transmitted to one UE from multiple antennas are formed such that they add destructively at the other UEs.

Joint processing/transmission aims to either accomplish one or both of the ways stated above. Note that the former inherently is a single-user transmission technique and can be referred to as

---

[1] The coherence in the transmission methods is a relevant issue in the CoMP techniques, depending on the level of synchronization among the coordinated entities.

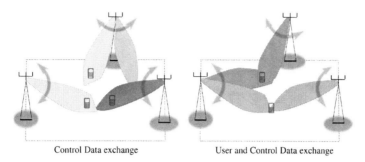

Control Data exchange   User and Control Data exchange

Figure 6.1   Coordinated beamforming/scheduling (on the left) and JP (on the right) exemplary schemes

single-user CoMP (SU-CoMP), while the latter is inherently a multiuser transmission technique and can be referred to as multiuser CoMP (MU-CoMP).

- In *coordinated scheduling and/or beamforming schemes*, data to a single UE is instantaneously transmitted from one of the transmission points in the CoMP set (the set of points/cells that are coordinated), and that scheduling decisions and/or generated beams are coordinated in order to control the interference created. An illustration is provided in the left panel of Figure 6.1. The main advantages of these schemes compared to schemes involving joint processing/transmission (see above) are that the requirements on the coordination links and on the backhaul network are much reduced, because typically
  - only information on scheduling decisions and/or generated beams (and information needed for their generation) need to be coordinated; and
  - the UE data do not need to be made available at the multiple coordinated transmission points, since there is only one serving transmission point for one particular UE.

Similarly, uplink coordinated multipoint reception can be categorized as follows:

- *Joint reception/processing* of signals at multiple, geographically separated points. The basic idea is to utilize multiple antennas at geographically separated sites to form a virtual receive array that is used to receive the signal transmitted by each UE, and then to combine and process the signals received at the different reception points. A simple variant of this is already used in 3GPP UTRAN systems and known as soft handover (Holma and Toskala 2004). The main benefits are that energy is collected at several reception points, the obtained macro-diversity gain, and also that advanced combining algorithms can be used in the receive processing in order to cancel out interference. However, as for DL joint processing/transmission described above, this approach puts high requirements on the coordination links since information based on the received signals needs to be exchanged.
- *Coordinated scheduling*, with reception of signals at one site only, scheduling decisions can be coordinated among cells to control interference. Again, the main advantage compared to schemes involving joint reception at several reception points (see above) is that the requirements on the coordination links are much reduced because only information on scheduling decisions need to be coordinated.

6.1.2   Architectures and Clustering

The possible introduction of CoMP implies a considerable impact on the architecture of the radio system and the feasibility of CoMP is extensively studied in the research community. Different

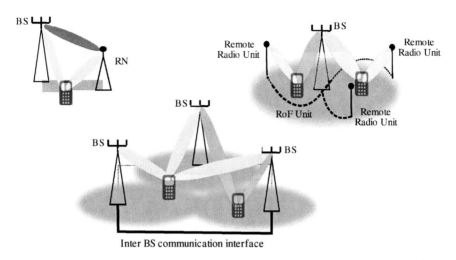

Figure 6.2   CoMP architectures

topologies are generally encompassed under the general definition of CoMP, each of them allowing a different degree of coordination among the involved nodes. In Figure 6.2 the architectures commonly considered when introducing CoMP are schematically represented. The central drawing sketches the most conventional architecture with a set of Base Stations (BSs) that coordinate their resources in order to improve the performances experienced by a set of UEs, in a defined area of cooperation. The fundamental issue in this simple architecture is the inter BS coordination interface. This interface may rely on existing logical interfaces, such as the X2 interface in 3GPP LTE systems, and the physical means adopted to implement this interface are important as well. The deployment of high capacity optical fiber connections in current networks paves the way for a growing adoption of optical fiber in the radio access network as well. Another feasible implementation making use of fiber links in CoMP is again in the Figure 6.2, upper right drawing, where fiber is now used in the front-haul link, following an approach often referred to as Radio over Fiber (RoF). In this scenario a BS Central Unit (CU) is connected by means of RoF links to a set of Radio Remote Units (RRUs) located closer to the UEs, enhances the overall capacity and coverage of a cell. The architecture described herein shows an application of CoMP in a so-called "intracell" or "intrasite" scenario, where the cooperation is performed among nodes belonging to the same cell. The straightforward advantage of an intrasite scenario is the reduction of information needed to be exchanged between the nodes. The case with Relay Node (RN) as cooperating entities is often included in the framework of CoMP architectures (see Figure 6.2, upper left). Depending on the type of RN different CoMP strategies can be applied (see Chapter 7 for more details on RN).

Different time scales are possible for coordination in all the CoMP architectures. The most efficient schemes require the information needed for scheduling to be available at each coordinated BS in the order of a millisecond, which calls for very low-latency information exchanges between cooperating nodes, or between the UE and all the cooperating nodes.

Two extreme approaches can be distinguished regarding how to make this information timely available at distant cooperating nodes: centralized and decentralized cooperation (see Figure 6.3). These approaches are described in the most general case where JP is performed across various nodes but they also apply in the case of coordinated scheduling/beamforming. *In a centralized approach the UE estimates the channel information from all the cooperating BSs and feeds it back to a central control unit, where scheduling operations are performed accordingly.* This central entity is a logical entity that can be accommodated at one of the collaborating BSs, by establishing a hierarchy

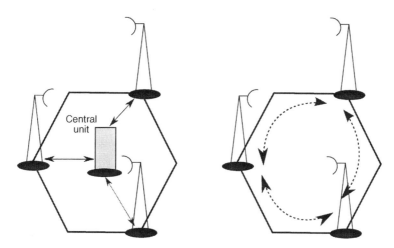

Figure 6.3 Centralized and decentralized architecture

between BSs. The need for a central entity entails the following changes upon the conventional architecture of cellular systems: collaborating BSs need to be interconnected via the central entity with low latency links in order to exchange local feedback information. Furthermore this information exchange needs to be coordinated with the use of additional communication protocols. *In a decentralized approach, instead, the UE feeds back the channel information to all the cooperating BSs. Therefore each BS gathers all the available feedback information, including those related to other BSs.* Under this decentralized framework, infrastructure cost and signaling protocol complexity can be minimized, because neither a central entity nor low latency links connecting it with the cooperating BS are required. The main obstacle associated with the decentralized collaborative framework is the handling of errors on the different feedback links; the impact of feedback errors on the system performance is discussed in Papadogiannis et al. (2009), and solutions for enhancing the robustness of decentralized multicell coordination were proposed in WINNER+ (2009a). Two types of solutions are usually introduced: some for reducing malfunction probability, and others for recovering from potential malfunctions, allowing a practical decentralized multicell coordination scheme to be designed. Other alternatives lying between these two extreme approaches are possible where some pieces of information are transmitted to a central entity and others are derived in a decentralized way. For instance, Song et al. (2007) describes a scheme with centralized scheduling but where the precoding weights are locally designed at each BS. Therefore, the amount of information to be exchanged between the collaborating nodes is reduced compared to the fully centralized approach.

In any case, whatever the chosen architecture, studies have been performed on how the distributions of UEs in a given ideal cellular layout could improve the performance with respect to an uncoordinated case; as the number of UEs and nodes increase, the signalling overhead required for the interbase information exchange and the amount of feedback needed from the UEs also increase. Therefore, cooperation should be restricted to a limited number of nodes. To achieve this goal, the network is thus divided into clusters of cooperative cells. Cluster selection is considered a key issue to cooperation algorithms performance and has been widely studied in the literature (description and definitions can be found in WINNER+ (2009b) and in the Third Generation Partnership Project (3GPP) Study Item on CoMP 3GPP (2010). Basically a cluster of cells for CoMP transmission can be formed in a UE-centric, network-centric or a hybrid fashion:

UE centric clustering: each UE chooses a small number of cells that are most suitable for cooperative transmission. In this case UE scheduling is challenging since the clusters for different UEs are chosen in a dynamic way and may overlap.

Network centric clustering: clusters are defined statically for all UEs of a given serving cell based on the neighborhood and try to combine the cells with strongest mutual interference. The performance for the UEs can be compromised at the cluster boundary.

Hybrid clustering: the network can predefine a set of clusters for a given area and the selection of the best cluster is assisted by the UE through feedback information.

Another relevant classification is related to dynamic or static formation of clusters for CoMP: if a cluster is statically formed, no fine tuning is allowed depending on the UE's actual channel and interference conditions, while a dynamic cluster formation can cope with them according to dedicated network algorithms aimed to the evolution of clusters. An ideal example of dynamic clusters is shown in Figure 6.4 where, as an example, three clusters of six cells are modified in the two CoMP scenarios presented. It is worth noting that the clusters could also be overlapped in some possible schemes; many works on dynamic or "semi-static" clustering approaches are now available in the literature; some examples can be found in Brueck et al. (2010), Papadogiannis et al. (2008b) and in Xiao et al. (2010) where thresholds are introduced to dynamically select the cells of the cluster.

6.1.3   Theoretical Performance Limits and Implementation Constraints

In this section the concept of CoMP is presented in the various steps from the introduction of simple relay-based approaches to recent JP schemes. This helps to explain the important enhancements that have been reached and at the same time indicates the main implementation constraints that have still to be overcome in this field.

Since the introduction of the relaying concept (Cover and Gamal 1979; Cover and Thomas 1991; van der Meulen 1977), network node cooperation and coordination has received significant attention in the context of improving the capacity or coverage of wireless communications (see Chapter 7 on Relaying for more details). In the classic relaying problem, UEs may cooperate in routing or relaying each others' data packets. A large number of papers have addressed the relaying problem with different assumptions on the transceiver capabilities. See, for example, (Cover and Gamal 1979;

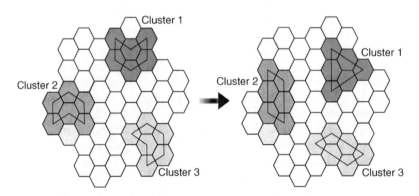

Figure 6.4   Example of dynamic clusters in a CoMP scenario

Cover and Thomas 1991; Gupta and Kumar 2000; Kramer et al. 2005; van der Meulen 1977) and the references therein.

The UE cooperation diversity builds upon the classical relay channel model (Cover and Gamal 1979; Cover and Thomas 1991; van der Meulen 1977), where the spatial diversity gains can be achieved both at the transmitters and the receivers using a collection of distributed antennas belonging to multiple nodes. Each node is responsible for transmitting not only its information, but also the information of the other nodes it receives and detects. Thus, spatial diversity is achieved through the joint use of the antennas of all the nodes. Sendonaris et al. (2003a,b) considered the beamforming approach where different nodes adjust their transmissions based on Channel State Information (CSI) knowledge so that the transmitted signals add up coherently at the destination. On the other hand, Laneman and Wornell (2003) and Laneman et al. (2004) assumed no CSI at the transmitters and proposed a variety of low-complexity, cooperative diversity protocols that enable wireless nodes to fully exploit spatial diversity in the channel. However, the cooperation is complicated due to the fact that the (wireless) channel between the nodes is noisy (Sendonaris et al. 2003a,b). Therefore, the UE cooperation often increases the interference level, the protocol overhead and the complexity of the transceivers (Bletsas and Lippman 2006; Sendonaris et al. 2003a).

In the ideal case with a noiseless link between the receiving or the transmitting nodes, the node cooperation is simplified to MIMO MAC or MIMO BC with per-node power constraints, respectively (Sendonaris et al. 2003a). In cellular networks, this can be achieved for example by a wired backbone connection between the distributed BS antenna heads, or by highly directional wireless microwave links (Zhang et al. 2005).

From Wyner Model to Joint Processing

Network infrastructure based coordinated processing across distributed BS antenna heads has received significant interest in the recent literature (Aktas et al. 2006; Bletsas and Lippman 2006; Choi and Andrews 2007; Jungnickel et al. 2009a; Karakayali et al. 2006b,c; Shamai and Zaidel 2001; Somekh et al. 2006; Wyner 1994; Zhang and Dai 2004a; Zhang et al. 2005). Wyner (1994) considered the Uplink (UL) of a cellular network with JP at the centralized controller, which has access to all the received signals at multiple receivers, and optimally decodes all the transmitted data in the entire network. In this simplistic model, each cell senses only the signal radiated from a limited number of neighboring cells, which yields closed-form expressions for the achievable rates and allows, to a certain extent, the analytical treatment of the distributed antenna systems with joint processing. It was shown in Wyner (1994) that a cellular network with such a JP receiver significantly outperforms a traditional network with individual processing per BS. A JP receiver for the UL was further considered in Aktas et al. (2006), Jafar et al. (2004), Liang et al. (2006) and Somekh et al. (2007).

Shamai and Zaidel (2001) were among the first to consider the DL sum rate and spectral efficiency optimization for coordinated Multiple-Input Multiple-Output (MIMO) systems with perfect data cooperation between BSs. They applied the ZF-DPC to the multicell JP in Wyner's scenario (Wyner 1994) with an average system power constraint and showed significant capacity enhancements from BS coordination. For a single-cell scenario, the power constraints are generally imposed on the total power radiated by all the elements of the array. Conversely, per antenna power constraints have to be enforced for distributed antenna systems, as each antenna head is provided with a separate power amplifier. In practice, each antenna element may also have a separate power amplifier.

Jafar and Goldsmith (2002) and Jafar et al. (2004) considered a multicell DL channel with perfect CSI at both ends, where an individual power constraint per BS is imposed. They proposed a heuristic but efficient suboptimal method based on iterative water-filling, which aimed at maximizing the sum-rate throughput of the coordinated system while meeting the individual BS power constraints. The sum capacity and the entire capacity region of the MIMO DL with per antenna or per BS power constraints were discovered in Yu (2006), Weingarten et al. (2006) and Yu and Lan (2007), respectively. These important findings can be utilized for finding the maximum achievable user rates from the rate region of a coordinated cellular MIMO system with practical peak power constraints per antenna or per BS .

Inspired by the pioneering work of Shamai, Jafar *et al.*, BS cooperation has been studied by several other authors (Karakayali et al. 2006b,c; Somekh et al. 2006, 2007; Zhang and Dai 2004a). Somekh et al. (2006, 2007) provided an information theory analysis of distributed antenna systems under the circular Wyner model (Wyner 1994), and derived bounds for the sum-rate capacity supported by the multicell DL under per BS power constraints. Karakayali et al. (2006b,c) studied the coordinated cellular DL using different ZF transmission schemes. In Karakayali et al. (2006b,c), the symmetric (or common) rate maximization in the coordinated cellular DL with Zero Forcing (ZF) transmission subject to per BS power constraints was formulated as a convex optimization problem, which can be efficiently solved.

Assuming linear transmitter processing, a coordinated antenna system with N antennas is ideally able to accommodate up to N streams without becoming interference limited. Both the inter-UE and intercell interference can be controlled or even completely eliminated by a proper precoder selection. This is especially true in the coherent multicell MIMO case, where UE data is conveyed from multiple antenna heads over a large virtual MIMO channel (Karakayali et al. 2006b).

CoMP Implementation Constraints

The coherent multiuser, multicell precoding techniques, however, have high requirements in terms of signaling and measurements as mentioned earlier. In addition to the complete channel knowledge of all jointly processed links, a tight synchronization across the transmitting nodes and centralized entities performing scheduling and computation of joint precoding weights is required in order to avoid carrier phase drifting at different transmit nodes. The theoretical studies referred above mostly assume perfect and complete CSI for all the transmitters, which is difficult to accomplish in practical cellular networks. In practical TDD MIMO-OFDM cellular systems, for example, UL transmissions from adjacent cells can be significantly more attenuated compared to the own cell UEs. Therefore, the joint channel estimation may be difficult, if not impossible, to implement in practice. In order to reduce the required information exchange between the network nodes, the JP of the transmitted signal from several MIMO BS antenna heads can be restricted to the cell edge region, where the available gains from the JP are the most beneficial (Tölli et al. 2008).

In order to perform joint transmission from all the distributed BS antenna heads, the baseband signals must have a common carrier phase reference. The Radio Frequency (RF) impairments and the impact of the propagation delay from each of the transmitters to the intended UE must be compensated for at the transmitter, for example, by using some feedback from UE. The BS antenna heads cannot fully synchronize the desired and the interfering signals received by different UEs due to different propagation times between the BS antenna heads and the UEs. Zhang et al. (2007) showed that significant performance degradation may follow if the asynchronous nature of the multiuser interference is not taken into account when designing the precoder for the coordinated DL. In MIMO-OFDM systems, however, this problem can be handled efficiently as long as the received signal paths are within the guard interval. Specifically, the increase of the delay spread is not necessarily so dramatic if the coordinated processing is limited to the cell edge region.

Table 6.1 CoMP implementation constraints

Constraint	Main issue
Backhauling	The amount of data to be transmitted over the backhaul depends on the chosen CoMP strategy, that is, whether the cooperation is on control plane only (e.g. in case of coordinated scheduling for interference avoidance) or control plane and user plane (for joint transmission from different sites).
Pilot design	The difficulty in obtaining the precise DL CSI imposes requirements on the pilot design to enable the channel estimation with sufficient quality and to be able to separate the pilots from different cells. In addition, the UE in some approaches needs to be able to decode a control channel of the neighboring cells. In order to attain the full CSI between all UEs and BS antennas in the cellular network, the UE channels should be jointly estimated at each antenna head.
Feedback design	Time Division Duplex (TDD) systems can exploit channel reciprocity to attain most of the required CSI at the transmitters. In FDD systems, the design of suitable feedback channels is an important issue. The information to be exchanged over the feedback link can include, for example, short-term or long-term CSI, preferred precoding matrix indices, received power from all the nodes, long term fading, etc.
Synchronization	A tight frequency synchronization, carrier phase synchronization across the transmitting nodes and complete CSI of all jointly processed links is required in coherent approaches. Noncoherent approaches are less demanding, but they could still need centralized resource management mechanisms. Time synchronization is required for all CoMP techniques, with looser requirements for coordinated beamforming than for JP.

The general constraints that have to be taken into account when considering a possible and feasible introduction of CoMP in a radio access system are summarized in Table 6.1, see also (WINNER+ 2009c, p. 28).

6.2   CoMP in the Standardization Bodies

CoMP has been studied widely in 3GPP and IEEE during the feasibility studies of the LTE-Advanced and 802.16m WiMAX, respectively. In this section the focus will be on the standardization activity performed in the framework of LTE-Advanced.

CoMP was one of the main techniques investigated during the 3GPP LTE-Advanced feasibility study (called Study Item in 3GPP) between May 2009 and March 2010. In the end, no consensus was achieved on the maturity of the CoMP technology for being standardized in 3GPP Release 10 (Rel-10). Nevertheless, several decisions were made on the design of a CoMP functionality for LTE that will be able to serve as a basis for future specification works. Indeed, CoMP will continue to be studied in 3GPP for potential inclusion in future releases. It is worth noting that even though Rel-10 does not provide any specific support for CoMP, some CoMP schemes can be implemented in a proprietary manner in LTE Rel-10 networks.

This section is organized in two main parts: section 6.2.1 gives an overview of the work carried out on CoMP in 3GPP for LTE-Advanced, and section 6.2.2 describes the main decisions that have been taken on the design of the CoMP functionality.

6.2.1   Overview of CoMP Studies

Both UL and DL CoMP techniques were studied in 3GPP. Most of the attention was devoted to DL techniques because UL techniques were expected to have only a limited impact on the physical layer specifications. The main reason for this situation is that LTE Release 8 (Rel-8) already allows the UE transmission to be demodulated at several eNodeBs (eNBs) (i.e. BSs). How the eNBs exchange the information to perform the coordination is up to the upper layers (typically the Medium Access Control (MAC) and Radio Resource Control (RRC) layers), which were not part of the initial studies on CoMP.

Nevertheless, some CoMP techniques were proposed for the UL and two families of techniques were identified: joint reception of the transmitted signal at multiple reception points and/or coordinated scheduling decisions among cells to control interference (3GPP 2010). In addition, UL coordinated link adaptation was discussed (Deutsche Telekom AG and Vodafone 2009).

In the DL, coordinated beamforming/scheduling and JP were both studied. For coordinated beamforming, the best companion approach (see also section 6.5.3) and ZF-based precoding approaches received particular attention. Within JP, two subcategories were identified: joint transmission, where the data to a single UE is simultaneously transmitted from multiple transmission points, and dynamic cell selection, where a single transmission point is active at a time within a set of cooperating points (3GPP 2010). For joint transmission, various techniques were considered, assuming different levels of coordination. Many of the proposed DL techniques were assumed to work in a MU-MIMO fashion, that is, where several UEs could be served on the same resources across the same cell or a set of cooperating cells.

As already stated from a general standpoint in section 6.1, for both UL and DL two types of coordination are considered in 3GPP for LTE-Advanced: inter-eNB and intra-eNB coordination. Inter-eNB CoMP in a multivendor environment is more complicated to standardize because standardized interfaces have to be used to exchange the necessary information between the coordinated eNBs. In addition, depending on the used backhaul technology, a non-negligible delay can affect the information availability, and inter-eNB CoMP techniques need to be robust to this. The delay in the information exchange was considered in the 3GPP work on CoMP. At the end of the LTE-Advanced Study Item, it was decided to restrict potential CoMP solutions for Rel-10 to intra-eNB CoMP, in order to reduce the standardization effort. It is worth noting that the absence of a standardized interface would not mean that inter-eNB CoMP is not possible; however, proprietary solutions would then have to be used, which would only be applicable in areas where the BSs are supplied by the same equipment vendor. It is useful to note that the UE does not know whether cells belong to the same or different eNBs, so it cannot distinguish between intra- and inter-eNB CoMP operation.

Even though standard support was not agreed for CoMP in Rel-10, the discussions in 3GPP have led to structuring decisions on the design of a CoMP functionality, which are described in section 6.2.2. Both coordinated beamforming/scheduling and JP are supported in the agreed framework. The main remaining areas to be addressed to complete the definition of the CoMP concept are the measurements needed at the UE and the feedback design principles.

By the end of the work on CoMP in 3GPP, an evaluation campaign evaluated the benefits of CoMP against Multi-User (MU)-MIMO. A majority of earlier performance evaluations had shown that either MU-MIMO or CoMP were necessary to meet the International Mobile Telecommunications Advanced (IMT-Advanced) spectral efficiency requirements (3GPP 2010). The results were not consistent enough between the companies to make conclusions. In fact some companies showed gains, whereas others found no gain but loss. There were also disagreements on the realism of the simulations depending on the overhead assumptions. However, it should be noted that the MU-MIMO versus CoMP comparison was undertaken with the full buffer traffic model only, most companies simulating a fairly high number of UEs per cell. In reality, the traffic is generally rather bursty, so that pairing UEs to serve them on the same resource is more difficult.

Overall, the CoMP technology was not recognized as mature enough to be standardized. Therefore, it was agreed that there would be no standard support for CoMP in Rel-10. One exception to this rule was the design of Reference Signal (RS) targeting CSI estimation (called CSI-RS), which takes into account CoMP needs for future proof (see section 6.2.2). Despite the lack of specific standard support, CoMP techniques may be feasible by proprietary solutions, for example, by taking advantage of the (long-term for Frequency Division Duplex (FDD)) UL-DL channel reciprocity property (Ericsson and ST-Ericsson 2010) to acquire the necessary multicell CSI, in particular when limited to intra-eNB coordination. Studies on CoMP continue within 3GPP in 2011, through a dedicated CoMP Study Item.

6.2.2   Design Choices for a CoMP Functionality

As indicated earlier, UL CoMP was expected to have only a limited impact on the physical layer specifications, and was therefore allocated only a limited attention. As a consequence, all the decisions specific to the CoMP design addressed the DL. The decisions addressed the design areas of *reference signals, control signalling design and feedback*. These decisions are explained in the following.

Reference Signals

Reference signals (RSs) are a key element of the system design, because they determine the performance of the channel estimation. Channel knowledge is key to allow coherent demodulation at the receiver, and CSI allows advanced MIMO transmitter processing when known at the BS. Compared to single-cell processing, DL CoMP requires channel estimates to be obtained from several cells, which constrains the ability to receive the RS from multiple cells.

LTE Rel-10 has defined a new framework for RSs. Even though most of the related design was not motivated by CoMP, some of the RSs' characteristics are useful for CoMP operation, so it is useful to briefly recall them here. Two types of reference signals are introduced in Rel-10: CSI-RSs are cell-specific and target CSI estimation (to be subsequently fed back to the transmitter), whereas DM RSs are UE-specific and target the demodulation of the scheduled transmissions. The split of the RS types according to the RS functionality was motivated by two objectives: one is to limit the overhead for a high number of transmit antennas (up to eight in Rel-10). Therefore, the CSI-RSs are transmitted sparsely in time and frequency to allow CSI to be acquired on the whole bandwidth with affordable overhead. On the other hand, the DM-RSs are more dense and only transmitted on the resources assigned to the UE. This allows to fine tune the tradeoff between overheads and CSI estimation accuracy compared to a fixed RS design fulfilling both demodulation and CSI estimation constraints. The other objective was to allow powerful transmitter processing at the BS. To this end, the DM-RSs are precoded with the same precoding as the data. Precoded RSs suppress the need to signal the precoding weights (necessarily selected from a predefined set) to the UE, which gives total freedom to the transmitter to apply any type of precoding. Indeed, the UE directly estimates the composite channel formed by the precoding operation and the actual propagation channel. One example of powerful transmitter processing enabled by this approach is ZF precoding. It is therefore useful to remark that the specifications will not mandate a single particular precoding method, but rather will allow a family of methods.

One immediate consequence of the DM-RSs adoption on the CoMP functionality design is that the UE does not need to know what transmission points or cells are involved in the transmission: provided the coordinated cells use the same DM-RS sequence to serve a given UE, the UE will only see a unique channel, formed by the sum of the channels to the cells participating in the transmission. This enables the multicell transmission as well as transmission points to be dynamically switched on/off without any signaling to inform the UE.

In contrast, the CSI-RSs are cell-specific and not precoded in order to allow the UE to estimate and report the CSI related to various neighboring cells. To ensure proper CoMP operation, the CSI-RS need to allow accurate estimation of multiple cells' channels. Even though CoMP is not supported in

Rel-10, the CSI-RS design needs to account for this need in Rel-10, because such a design is difficult to change in the future without creating backward compatibility problems. It was therefore agreed that the CSI-RS design in Rel-10 will take into account the CoMP needs, and should in particular allow accurate multi-cell measurements. As a result, Rel-10 provisions the possibility to mute data resource elements (i.e. one subcarrier from one OFDM symbol) at one cell to protect the CSI-RSs transmission on the same resources at a neighboring cell.

Control Signalling

The UE will receive its control channel (called PDCCH in LTE), which carries in particular the scheduling information, from a single cell. This cell is called the serving cell, and is the cell the UE is served by in the case of a single-cell transmission. This choice limits the UE complexity by avoiding the need to monitor the control channels from several cells. In addition, this enforces the principle that the UE is only aware of its serving cell, the other cells potentially participating in the transmission remaining transparent.

CSI Feedback

Feedback is the main remaining open issue for CoMP standardization. Although the complete design was not agreed, a number of decisions were taken. The feedback was an important design issue in Rel-10, since new feedback mechanisms were discussed for SU- and MU-MIMO in addition to CoMP. Therefore, the discussion on feedback for CoMP was embedded into a global feedback framework, and can not be considered isolated.

One design principle for feedback in Rel-10 is scalability, so that the same feedback allows the network to select dynamically between SU-MIMO and MU-MIMO. More specifically to CoMP, the UE feedback supporting CoMP transmission should be such that it also enables the network to switch dynamically to single-point transmission. To facilitate this switch, it was agreed that the UE reception and demodulation of CoMP transmissions is the same as for single-cell SU/MU-MIMO. Furthermore, if a new form of feedback was needed for CoMP, scalable feedback for different CoMP categories (e.g. coordinated beamforming and JP) should be aimed at, which means that feedback in support of CoMP JP should be a superset of the feedback in support of CoMP coordinated beamforming/scheduling.

Two main types of feedback have been identified: the explicit feedback involves reporting the raw channel as observed by the receiver (together with interference information), without assuming any transmission or receiver processing. Examples of explicit feedback include reporting the channel impulse response, or the channel covariance matrix. In contrast, the implicit feedback involves reporting channel-related characteristics assuming hypotheses on a particular transmission and/or reception processing; one typical example of implicit feedback is the Channel Quality Indicator (CQI)/PMI/RI feedback used in Rel-8. In addition, UE transmission of UL Sounding Reference Signal (SRS) was identified as being usable for CSI estimation at multiple cells, in particular via the exploitation of the channel reciprocity property (both for FDD and TDD). More details on the different feedback alternatives within each type can be found in (3GPP 2010). Note that the framework finally adopted for single cell MIMO was the implicit one. In principle, this does not preclude explicit feedback to be adopted for CoMP in the future, but the constraint of easy switch between single-cell and multicell operation tends to render implicit feedback easier to adopt also for CoMP.

Several decisions have been made regarding how to report the CoMP-specific information. When an eNB-to-eNB communication interface (called X2 for LTE) is available and is adequate for CoMP operation in terms of latency and capacity, the starting-point solution is that UE CoMP feedback targets the serving cell. In this case, the reception of UE reports at cells other than the serving cell is possible but is left to implementation. For cases where X2 interface is not available or not adequate (with respect to latency and capacity), the feedback reporting needs further consideration. It was also agreed that the starting point for schemes that need feedback is individual per-cell feedback (i.e. no differential CSI information between two cells). Nevertheless, it was noted that complementary intercell feedback might be needed. In addition, the starting point is that the feedback related to each individual cooperating cell is transmitted only to the serving cell. For the purpose of measurements, the *CoMP measurement set* was defined as the set of cells about which the UE performs CSI measurement; the cells effectively reported are possibly a subset of the CoMP measurement set.

The feedback design for CoMP is still an open discussion topic in 3GPP. The choice of a feedback strategy, is in particular between explicit and implicit feedback, or a combination of both, in order to take a tradeoff between the UL overhead and the DL performance. Nevertheless, although it was not completed in Rel-10, the standardization of CoMP has progressed substantially. For an overview of the latest CoMP views within 3GPP, the reader can be referred to 3GPP (2010).

6.3   Generic System Model for Downlink CoMP

In order to have a common framework for the CoMP schemes a system model is briefly outlined in this section. A cellular Orthogonal Frequency Division Multiplexing (OFDM) DL is considered where a center site is surrounded by multiple tiers of sites. Each site is partitioned into three $120°$ sectors, that is, a set $\mathcal{B} = \{1, \ldots, L\}$ consisting of L sectors (also called a cell or a BS) in total. Consider that each BS has N_T transmit antennas and the UE m is equipped with N_{R_m} antennas. A set $\mathcal{U}$ with size $M = |\mathcal{U}|$ includes all UEs active at the given time instant, while a subset $\mathcal{U}_l \subseteq \mathcal{U}$ includes the UEs allocated to BS l. Each UE m can be served by K_m BSs which define the cluster set $\mathcal{B}_m$ for the UE m, and $\mathcal{B}_m \subseteq \mathcal{B}$. The signal $\mathbf{y}_m \in \mathbb{C}^{N_{R_m}}$ received by the UE m consists of the desired signal, intracluster and intercluster interference, and can be expressed as

$$\mathbf{y}_m = \sum_{l \in \mathcal{B}_m} \mathbf{H}_{l,m}\mathbf{d}_{l,m} + \sum_{l \in \mathcal{B}_m} \mathbf{H}_{l,m} \sum_{i \in \mathcal{U}_l \setminus m} \mathbf{d}_{l,i}$$
$$+ \underbrace{\sum_{l \in \mathcal{B} \setminus \mathcal{B}_m} \mathbf{H}_{l,m} \sum_{i \in \mathcal{U}_l} \mathbf{d}_{l,i} + \mathbf{n}_m}_{\mathbf{z}_m} \quad (6.1)$$

where the vector $\mathbf{d}_{l,m} \in \mathbb{C}^{N_T}$ is the transmitted signal from the l'th BS to UE m, $\mathbf{n}_m \sim \mathcal{CN}(0, N_0\mathbf{I}_{N_{R_m}})$ represents the additive noise sample vector with noise power density N_0, $\mathbf{z}_m$ is the inter-cluster interference plus noise, and $\mathbf{H}_{l,m} \in \mathbb{C}^{N_{R_m} \times N_T}$ is the channel matrix from BS l to UE m. The transmitted vector for UE m is generated at BS l as

$$\mathbf{d}_{l,m} = \mathbf{W}_{l,m}\mathbf{x}_m \quad (6.2)$$

where $\mathbf{W}_{l,m} \in \mathbb{C}^{N_T \times N_{S_m}}$ is the pre-coding matrix, $\mathbf{x}_m = \left[x_{m,1}, \ldots, x_{m,N_{S_m}}\right]^T$ is the vector of normalized complex data symbols, and $N_{S_m} \leq \min(N_T K_m, N_{R_m})$ denotes the number of active data streams within the cluster set $\mathcal{B}_m$. Here, $x_{m,s}$ is the data symbol on the s'th data stream designated to UE m. It should be noted that the first term in Equation 6.1 includes interstream interference.

6.3.1   SINR for Linear Transmissions

We focus on linear transmission schemes, where the L BS transmitters send altogether $S = \sum_{m=1}^{M} N_{S_m}$ independent data streams. Per data stream processing is considered, where for each data stream $s = 1, \ldots, S$ the scheduler unit associates an intended UE m_s, with the channel matrices $\mathbf{H}_{l,m_s}, l \in \mathcal{B}_{m_s}$.

Let $\mathbf{w}_{l,s}$ (a column of $\mathbf{W}_{l,m}$) and $\mathbf{v}_s \in \mathbb{C}^{N_{R_m}}$ be arbitrary transmit and receive beamformers for the stream s. Hence the achievable data rate of the stream s intended to the UE m_s is given by $\log(1 + \Gamma_s)$, where Γ_s is the DL Signal-to-Interference-plus-Noise Ratio Signal to Interference plus Noise Ratio (SINR) of the data stream s and can be expressed as

$$\Gamma_s = \frac{\left| \sum_{l \in \mathcal{B}_{m_s}} \mathbf{v}_s^{\mathrm{H}} \mathbf{H}_{l,m_s} \mathbf{w}_{l,s} \right|^2}{N_0 \|\mathbf{v}_s\|_2^2 + \sum_{i=1, i \neq s}^{S} \left| \sum_{l \in \mathcal{B}_{m_i}} \mathbf{v}_s^{\mathrm{H}} \mathbf{H}_{l,m_s} \mathbf{w}_{l,i} \right|^2} \tag{6.3}$$

Note that in case of nonlinear receivers, such as the Successive Interference Cancellation receiver, Equation 6.3 is not valid.

6.3.2 Compact Matricial Model

For simplicity, let us designate K the cluster size and assume it is not UE specific. Hence, BSs are grouped in subsets of $N = \lceil \frac{L}{K} \rceil$ clusters. As as an example, Figure 6.4 depicts the case of $N = 3$ clusters where each of them contains $K = 6$ sectors. Further, let N_R and N_S be the number of receive antennas and maximum number of data streams per UE, respectively. Let $M_C = |\mathcal{U}_C|$ be the number of active UEs in a cluster.

Let $\mathbf{W}_m = [\mathbf{W}_{1,m}^{\mathrm{T}}, \ldots, \mathbf{W}_{K,m}^{\mathrm{T}}]^{\mathrm{T}}$ be the $K N_T \times N_S$ precoding matrix of a defined cluster for UE m. Moreover, let $\mathbf{H}_m = [\mathbf{H}_{1,m}^{\mathrm{T}}, \ldots, \mathbf{H}_{K,m}^{\mathrm{T}}]^{\mathrm{T}}$ be the $N_R \times K N_T$ channel matrix between all the transmit antennas of a cluster and the N_R receive antennas of UE m. Further, define the transmit data vector, transmit precoding matrix, channel matrix, inter-cluster interference plus noise vector, and receive data vector, respectively as follows:

$$\mathbf{x} = [\mathbf{x}_1^{\mathrm{T}}, \mathbf{x}_2^{\mathrm{T}}, \ldots, \mathbf{x}_{M_C}^{\mathrm{T}}]^{\mathrm{T}} \tag{6.4}$$

$$\mathbf{W} = [\mathbf{W}_1, \mathbf{W}_2, \ldots, \mathbf{W}_{M_C}] \tag{6.5}$$

$$\mathbf{H} = [\mathbf{H}_1^{\mathrm{T}}, \mathbf{H}_2^{\mathrm{T}}, \ldots, \mathbf{H}_{M_C}^{\mathrm{T}}]^{\mathrm{T}} \tag{6.6}$$

$$\mathbf{z} = [\mathbf{z}_1^{\mathrm{T}}, \mathbf{z}_2^{\mathrm{T}}, \ldots, \mathbf{z}_{M_C}^{\mathrm{T}}]^{\mathrm{T}} \tag{6.7}$$

$$\mathbf{y} = [\mathbf{y}_1^{\mathrm{T}}, \mathbf{y}_2^{\mathrm{T}}, \ldots, \mathbf{y}_{M_C}^{\mathrm{T}}]^{\mathrm{T}} \tag{6.8}$$

Finally, the receive data vector, on a per subcarrier basis for a cluster, can be written in the following compact matrix form:

$$\mathbf{y} = \mathbf{H}\mathbf{W}\mathbf{x} + \mathbf{z} \tag{6.9}$$

In this model, the precoding matrix $\mathbf{W}$ inherently carries the transmit powers of each data stream. Another common practice is to decompose the precoding matrix into a product of two matrices, where the first one includes the spatial transmit directions as unit-norm column vectors, and the second one is a diagonal matrix that indicates the power weighting of each stream.

6.4 Joint Processing Techniques

Motivated by the predicted theoretical gains, joint transmission or JP between BSs has been identified as one of the key techniques for mitigating intercell interference in future broadband communication systems (Karakayali et al. 2006a). As discussed in section 6.2 JP techniques are included in the more general framework of CoMP transmission schemes within the 3GPP standardization activities concerning LTE-A (3GPP 2010).

As discussed earlier, from a practical point of view one of the major drawbacks related to JP is its high complexity, that is, large backhaul and signaling overhead. To reduce the complexity, *clustering* solutions that restrict JP techniques to a limited number of BSs have been proposed. In these approaches, the network is statically or dynamically divided into clusters of cells (Boccardi and Huang 2007a; Papadogiannis et al. 2008c; Thiele et al. 2009). Moreover, the cluster formation may be performed and optimized by a central entity (*network-centric*), or in a per-UE way (*user-centric*). In the following, JP is only allowed between BSs belonging to the same cluster, whereas BSs belonging to different clusters are not coordinated. The clusters are considered disjoint, that is, a given BS cannot belong to more than one cluster.

In this section, some of the main recent achievements on JP for CoMP are presented. Section 6.4.1 complements the state-of-the art overview of JP CoMP in the downlink. In section 6.4.2 results from a system level performance evaluation of coherent JP CoMP are shown and compared with a baseline LTE Rel-8 system configuration. Section 6.4.3 investigates possible dynamic JP approaches for user-centric, and BS-centric based clustering. Finally, in section 6.4.4, an insight about uplink JP CoMP is given.

6.4.1   State of the Art

A comprehensive description of the state of the art of CoMP has been presented in section 6.1.3. In this section some further references are given for some JP-specific aspect of CoMP.

An interesting topic extensively investigated in the literature is the case of JP with BSs equipped with antenna arrays. Motivated by current antenna array designs where a separate amplifier is provided for each antenna, Boccardi and Huang (2006) and Wei and Lan (2007) consider the transmitter optimization problem subject to per-antenna power constraints. These results can be extended to the case when a single antenna array is partitioned into antenna groups and a per-group power constraint is imposed, which is equivalent to a fully coordinated multi-cell system with per-BS power constraints. The power minimization problem for CoMP systems was addressed in Botella et al. (2008) and Dahrouj and Wei (2008). The main contribution of (Dahrouj and Wei 2008) is a practical algorithm that is capable of finding the jointly optimal beamformers for all BSs globally and efficiently.

Recent works about CoMP systems can be divided into two main areas of research: on one hand, some authors give practical insights for the implementation of coordinated multicell systems (Hongyuan et al. 2008; Papadopoulos and Sundberg 2008; Zarikoff and Cavers 2008) and, on the other hand, others addressed the increased complexity of implementing multiuser MIMO schemes and the large signaling overhead associated with these systems (Boccardi and Huang 2007a; Papadogiannis et al. 2008a,c; Skjevling et al. 2008; Venkatesan 2007).

In a CoMP system, the signal arriving at the UE experiences different propagation delays from each BS, that is, it is fundamentally asynchronous. Most previous studies simplified the system model by assuming synchronous reception at the UE. In Hongyuan et al. (2008), a mathematical model for this synchronicity is proposed, whereas in Botella et al. (2008) a spatial equalizer is assumed at the receiver of each UE. On the other hand, perfect CSI at the BSs is one of the requirements for implementing transmit beamforming from the set of all coordinated antennas. In Zarikoff and Cavers (2008), the estimation of the carrier frequency offset is addressed using training sequences taking into account the possible loss of orthogonality. Finally, asynchrony-resilient space-time codes are proposed in Papadopoulos and Sundberg (2008) for coordinated multicell systems.

Another relevant topic in this field, as stated in section 6.1, is the adoption of clusters, formed either in a static, dynamic or hybrid way. Some recent achievements concerning clustering are reported in section 6.1.2. The concept of clustering is not actually restricted to JP CoMP schemes only, and an important implementation of clustering in a system encompassing both joint processing in an *inner* cluster and coordinated processing in an *outer* cluster can be found in Björnson et al. (in press).

6.4.2 Potential of Joint Processing

A system-level performance evaluation of JP CoMP is presented here in order to estimate the possible expected benefits. Coherent joint transmission CoMP based on traditional ZF precoding (Caire and Shamai 2003) is compared with a baseline LTE Rel-8 system configuration. Then the goal is to show the potential of JP CoMP.

The overall assumptions (WINNER+ 2009a) follow to a large extent the deployment scenario and system parameters as specified for the 3GPP case 1 scenario, which is an urban macro scenario defined in 3GPP 2006. Each sector is equipped with an antenna array comprising four elements separated by ten wavelengths. CoMP transmission is carried out over a number of traditional cell sites, which are controlled by one eNB. The underlying assumption is that the coordinated cell sites are connected to the eNB via high-speed connections in order to allow for the fast coordination. Clusters of nine and 21 cells are taken into consideration.

The UEs have two antenna elements each separated by half a wavelength and employ ideal Minimum Mean Square Error (MMSE) receivers with additional Successive Interference Cancellation (SIC) functionality between the two streams of own UE. The DL transmission scheme is 4×2 MIMO on all links, and as reference case LTE Rel-8 with fast codebook switching (Dahlman et al. 2008) is simulated. The simulations assume perfect channel and interference estimation at the UE. Furthermore, all UEs are assumed to have full buffers, and the average number of UEs per cell is varied from 0.1 to five. The UEs are uniformly distributed across the simulated area, and each UE moves at the speed of 3 km/h and is assumed to be indoors (modeled with a 20 dB additional path loss).

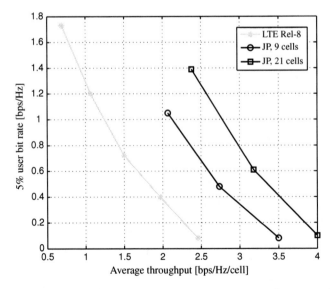

Figure 6.5   Simulation results for ZF based JP CoMP over 9 and 21 cells compared to LTE Rel-8

Figure 6.5 presents the average throughput versus 5th percentile UE bit rate for coordination over nine cells and 21 cells, compared to LTE Rel-8, with the UE density varied along the curves. Overall it can be seen that the JP based on ZF provides a significant gain both in terms of average system throughput as well as in cell-edge UE throughput (5th percentile UE bit rate). It can be seen that the performance is clearly improved when the coordination is carried out over 21 cells compared to nine cells (also for less significant impact of edge effects on the borders). This is explained by the fact that in the case with 21 cells a larger effective antenna array is considered, and hence higher degrees of freedom can be exploited.

6.4.3 Dynamic Joint Processing

Motivated by the need to reduce the complexity of JP with full cooperation, in this section JP CoMP within a cluster area with a dynamic utilization of the involved BSs is taken into consideration.

Partial Joint Processing: a User-Centric Clustering

In the Partial Joint Processing (Partial Joint Processing (PJP)) scheme presented here, the UE receives its data from a dynamically selected subset of the K BSs or from an active set (Botella et al. 2008). From the system point of view, three benefits are provided adopting PJP with respect to more conventional JP methods: reduced feedback from the UEs, lower interbase information exchange and transmit power saving. However, the PJP scheme introduces multiuser interference in the system, because less CSI is available at the CU to design the linear precoding matrix $\mathbf{W}$.

In order to define the subset of BSs transmitting to a given UE, assume that the UE is assigned to a master BS, which is the one with the highest channel gain. The UE estimates the average channel gain from the remaining BSs in the cluster, $K - 1$, and compares it to the channel gain from the master BS. Base stations are included in the active set only if their channel gains are above a relative threshold with respect to the master BS. By doing so, BSs related to poor quality channels do not transmit to the UE and the cluster becomes partially coordinated. The threshold value is specified by the cluster management, and different degrees of JP can be obtained by modifying its value. Therefore, the PJP scheme includes as a particular case full cooperation as defined in Equation 6.9, here denoted as Centralized Joint Processing (CJP).

Both the CJP and the PJP approaches imply the need of a CU to design the precoding scheme. For comparison purposes, a Decentralized Joint Processing (DJP) scheme is considered, where only local CSI is available at each BS and the power allocation and the precoding are locally implemented at each BS. However, the UE may receive its data from several BSs, depending on its given channel conditions. Hence, the cardinality of the set of spatially separated UEs that can be served by each BS in the cluster is reduced to N_T, and a multibase scheduling algorithm is required in order to assign UEs to BSs. Here, the multibase scheduling problem is solved allowing each BS in the cluster to serve its set of $\mathcal{U}_k$ UEs. As shown in Skjevling et al. (2008), with this solution, each of the M UEs can be served by a number of BSs that ranges from zero to K. Hence, the DJP scheme implies that a certain number of UEs in the cluster may remain without service.

To evaluate the PJP scheme, single-antenna UEs ($N_R = 1$) are considered, whereas a ZF precoder is jointly designed by the BSs, $\tilde{\mathbf{W}} = \mathbf{H}^H (\mathbf{H}\mathbf{H}^H)^{-1}$. The maximum available transmit power at each BS is restricted to a P_{max} value, and hence, under an equal received UE power allocation assumption, the final precoder matrix can be written as $\mathbf{W} = \sqrt{p} \cdot \tilde{\mathbf{W}}$, where $p = \{\min_{k=1,\dots,K}(P_{max}/||\tilde{\mathbf{W}}^{(k)}||_F^2)\}$, in which $\tilde{\mathbf{W}}^{(k)}$ are the rows of matrix $\tilde{\mathbf{W}}$ related to the kth BS (Zhang and Dai 2004b).

The Quality of Service (QoS) experienced by a UE should preferably not be location dependent - that is, the JP scheme should provide a uniform performance over the cluster area. For this reason, we

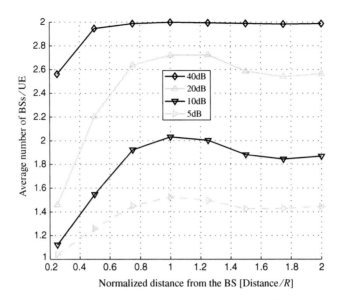

Figure 6.6   Average number of BSs assuming PJP transmitting to each UE, $\bar{N}_{BS}$, for different threshold values versus normalized distance. $M = 6$ UE and an edge-of-cell SNR of 15 dB (Thiele et al. 2010a)

characterize and compare the performance of the CJP, PJP and DJP schemes over the whole cluster area. A cluster is considered of $K = 3$ BSs, each one equipped with an array of $N_T = 3$ antennas, and $M = 6$ single-antenna UEs. The cluster radius is $R = 500$ m. The channel vector between the mth UE and the kth BS is modeled as $\mathbf{h}_{mk} = \mathbf{h}'_{mk}\sqrt{\gamma_s \gamma_p}$, where the shadow fading is a random variable described by a log-normal distribution, $\gamma_s \sim \mathcal{N}(0, 8\,\text{dB})$, the pathloss follows the 3GPP Long Term Evolution (LTE) model, $\gamma_p(dB) = 148.1 + 37.6\log_{10}(r_{mk})$, and $\mathbf{h}'_{mk}$ includes the small-scale fading coefficients, which are i.i.d. complex Gaussian values according to $\mathcal{CN}(0, 1)$ (WINNER+ 2009b).

In this case, $M = 6$ UEs are placed in each position. The PJP plots stand for active set threshold values of 10, 20 or 40 dB, respectively. For low mobility UEs, the backhaul overhead related to exchanging the UE data between the BSs is higher than that required for exchanging the channel coefficients. Then, the combined amount of backhaul exchange and feedback from the UEs can be roughly estimated by means of the average number of BSs that are transmitting to a UE, $\bar{N}_{BS}$. In the CJP and DJP schemes, this parameter remains fixed regardless of the location of the UEs in the cluster area. However, for the PJP scheme, $\bar{N}_{BS}$ depends both on the active set threshold value and the UE position over the cluster area. This is shown in Figure 6.6, where the average number of BSs used along a line from one BS towards the far most point in the cluster is plotted. The median Throughput (TP) gain of CJP, DJP and PJP is shown in Table 6.2. As it can be seen from the table and from Figure 6.6 the active set threshold value is a suitable parameter to define the PJP scheme, in order to trade-off the cooperation gain versus the backhaul load to locations in the cluster area where the gain is substantial. Additional results are included in Appendix C.1.1.

Table 6.2   Gains with respect to 1BS case and at $Distance/R = 1$

	1BS	DJP	2BS	PJP–10 dB	PJP–20 dB	PJP–40 dB	CJP
Median TP	1.0	1.6	2.0	2.1	3.4	4.7	4.7

Distributed Joint Processing: a User-Centric Clustering with Multi-antenna Receivers

This section presents a distributed JP scheme, where K BSs perform the DL CoMP transmission to M UEs each equipped with $N_R = 2$ receive antennas. In order to use the same ZF beamformer, appropriate virtual receive antennas out of the $M \times N_R$ antennas have to be selected. At maximum, the cluster can provide $K N_T$ coherently transmitted data streams.

A total number of $K N_T$ Multiple-Input Single-Output (MISO) channels, selected from a sufficiently large set of UEs, are composed to form a compound MIMO channel matrix of size $N_R \times K N_T$. The basic idea is to enable each UE to generate and provide CSI feedback by selecting a preferred receive strategy v_m, which can differ from the equalizer v_m used in (6.3). Therefore, each UE can choose its desired receive strategy according to its own computational capabilities and knowledge about CSI at the Receiver (CSIR) including interference, independently from other UEs.

Each UE is assumed to use linear receive filters to transform the MIMO channel into an effective MISO channel (Thiele et al. 2010b, 2009) $\mathbf{h}^{U E_m} = v_m^H \mathbf{H}_m$. We limit the evaluation to a Multiuser Eigenmode Transmission Multi-user Eigenmode Transmission (MET)-based (Boccardi and Huang 2007b) approach: each UE decomposes its channel $\mathbf{H}_m$ by a Singular Value Decomposition (SVD) into orthogonal eigenspaces, that is, $\mathbf{H}_m = \mathbf{U}\Sigma\mathbf{V}^H$. Further, each UE is applying for a single data stream only. Thus, it is favorable to select the dominant mode, that is, the singular vector corresponding to the highest singular value. The effective channel after decomposition using the dominant left singular vector, that is, $v_m = \mathbf{u}_1$ is given by $\mathbf{h}^{U E_m} = \mathbf{u}_1^H \mathbf{U}\Sigma\mathbf{V}^H = \Sigma_1 \mathbf{v}_1^H$. The scheme maximizes the signal power transferred from a cluster to the UE. UEs should preferably be grouped such that their modes show highest orthogonality. This keeps the costs in received power reduction due to ZF precoding as small as possible. The approach has two major advantages: first, the multiple receive antennas are efficiently used for suppression of interference at the UE side. Second, by reducing the number of data streams per UE, the system is enabled to serve a larger set of active UEs instantaneously and thus benefiting from multiuser diversity.

The scheme has been evaluated with a system simulator using $L = 57$ multiantenna BS sectors and a wraparound technique using the 3GPP Spatial Channel Model Extended (SCME) channel model (Baum et al. 2005). The system performance is determined by assuming a dynamic and UE-driven clustering method. However, the system does not utilize any additional gains from multiuser scheduling, that is, active set of UEs is selected according to the following metric. A set of active multiantenna UEs is uniformly distributed in a cluster of the cellular environment. The UE selection for each cell is done by a round-robin scheduling policy, yielding a set of UEs $\mathcal{U}_k$ of size N_T. Note that the UEs in $\mathcal{U}_k$ experience highest channel gain to the k-th BS, that is, all UEs are connected to a master BS. In addition, all UE sets $\mathcal{U}_k$ are disjoint for different BSs $k \in K$. Further, we emulate a cluster selection which is UE-centric and dynamic over frequency: the K strongest channel gains of the UEs in $\mathcal{U}_k$ are the ones of the K BSs within the cluster.

Results are provided for different cluster sizes of $K \in \{1, 2, 3, 4, 5, 10\}$. All results in Figure 6.7 are based on an equal per-beam power constraint with a Per-Antenna Power Constraint (PAPC) (Zhang and Dai 2004a), which is aligned with the assumptions made in 3GPP LTE. For reference purpose, the performance results for interference-limited Single Input Single Output (SISO) as well as a MIMO 2×2 transmission are included. For $N_T = 2$, two active fixed beams are sent to $M = 2$ different UEs in a round-robin manner.

As a next step, CSI feedback is introduced from each UE to its serving BS. In particular, each UE is assumed to decompose its MIMO channel matrix into its eigenspaces, where only the dominant one is used as feedback (Boccardi and Huang 2007b; Noda et al. 2007). Based on this feedback, the

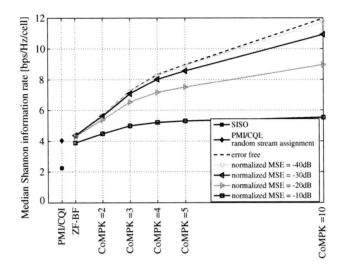

Figure 6.7 Performance of distributed DL CoMP results as a function of the cluster size K. The normalized MSE is given per subchannel, that is, in case of $K = 10$ sectors in the cluster, the UE estimates $N_T = 2$ times $10 = 20$ subchannels in total with an i.i.d. Gaussian normalized MSE. Reproduced by permission of © 2009 IEEE

BS can serve its UEs using ZF beamforming. From Figure 6.7 rather small gains are observed from channel adaptive precoding in a single sector, that is, $K = 1$. This is mainly caused by the following fact: a simplified PAPC is assumed, which leads to a suboptimal power allocation where only one antenna transmits with full power and all others are scaled accordingly (Zhang and Dai 2004a). In contrast, in the case of fixed PMI-based precoding all BS antennas transmit with full power.

Now, the cluster size is increased from $K = 1$ to $K = 10$. Figure 6.7 depicts the achievable Shannon information rate per sector as a function of the cluster size K and accuracy of CSI feedback, that is, in case of error free and erroneous CSI. For erroneous feedback, an additive Gaussian i.i.d. normalized Mean Square Error (MSE) is considered per subchannel – that is, per antenna port. The error variance μ is normalized with respect to channel power. Throughput results are provided for different values of μ, where the precoding weights are obtained by the erroneous CSI estimates. From this figure it is obvious that an MSE of $\mu = -10$ dB would restrict the useful cluster size to $K = 3$. In essence, the CoMP gains as function of the cluster size show less saturation behavior for improved multicell channel knowledge.

In conclusion, it can be observed that the median sector spectral efficiencies are increased by 220%, 300% and 430% for coordinating 3, 5 and 10 cells for error-free CSI feedback, respectively, compared to a noncoordinated SISO setup. These numbers are reduced to 190%, 230% and 300% in case of erroneous feedback with an MSE of $\mu = -20$ dB.

Table 6.3 Performance in terms of gains with respect to the non-CoMP case. The cell-edge performance is measured at the 5%-tile of the UE throughput CDF

	non-CoMP	static CoMP	proposed algorithm	full CoMP
Median(cell-edge) throughput	1.0 (1.0)	0.9 (1.1)	1.3 (2.0)	1.6 (2.8)

Dynamic Base Station Clustering

The previous two methods presented here were JP user-centric based clustering. Dynamic BS clustering as its name indicates is a dynamic JP BS-centric clustering method. In fact, in the PJP method a threshold was chosen to limit the cluster size in order to keep the overhead rate reasonable. As a further extension the cluster can be created in a dynamic way in order to maximize a given objective function, for example, $\sum_{n=1}^{N} R(\mathcal{C}, \mathcal{U}, \mathcal{W}, \mathcal{P})$ which is function of $\mathcal{C}$ the BS clusters, $\mathcal{U}$ the UEs scheduled in each cluster, $\mathcal{W}$ the beamforming coefficients, and $\mathcal{P}$ the power allocation. The algorithm is described in details in Appendix C.1.2.

In Table 6.3 the performance of this algorithm is summarized in terms of average rate per cell and cell-edge rate (5% tile) respectively for a cluster size of 10 BSs, which corresponds to a 50% reduction in the number of BSs sharing the data of the UEs scheduled for transmission in a given frame. Four different techniques are compared: noncooperating BSs, static coordination, that is, clusters of cooperating BSs are kept fixed during all the simulation and in each cluster the UEs are selected for transmission using a Proportional Fair (PF) scheduler, dynamic coordination and full coordination, that is, all the K BSs cooperate together and up to M UEs are scheduled for transmission in each frame with a PF approach. As seen in Table 6.3, a substantial fraction of the gains with full cooperation can be achieved with this technique. Note that the simulation assumption are shown in Appendix C.1.2.

6.4.4 Uplink Joint Processing

As already stated in section 6.1 the concept of CoMP encompasses both DL and uplink of a radio access network. We have so far put an emphasis in this chapter on Downlink techniques, but JP CoMP has the same potential in the uplink. JP CoMP in the uplink is more straightforward to implement, with respect to backward compatibility, since it can be made transparent for the UEs. Intrasite CoMP can also be efficiently implemented, and there are large gains in average spectral efficiency and cell-edge performance to obtain, 22% and 26% respectively, as shown in an LTE-like evaluation in Frank et al. (2010).

Intersite CoMP can provide additional gains, but the backhaul traffic between BSs is a challenge. Limited central JP in the form of joint detection is performed today by coordinated BSs in Third Generation (3G) Wideband Code Division Multiple Access (WCDMA) networks, denoted soft handover for multiple participating sites and softer handover for cells/sectors of the same site. Decentralized implementations and quantization (Marsch and Fettweis 2007) are suggested as the main approaches for achieving lower backhaul traffic for JP. Decentralized implementations include distributed decoding (Aktas et al. 2008), with local interaction between neighboring BSs, and distributed iterative detection Khattak et al. (2008), where each BS performs single UE detection and exchanges data in an iterative manner. In Marsch and Fettweis (2007), quantized data is exchanged among BSs. More basic analysis and approaches can be found in Yifan and Goldsmith (2006) and Yifan et al. (2006).

To illustrate the potential gains of intersite JP CoMP in an LTE uplink, the system level simulation results in Hoymann et al. (2009) and Frank et al. (2010) are highlighted. In Hoymann et al. (2009), a straightforward algorithm is considered, where the received complex baseband signals from cooperating BSs are distributed in a decentralized manner over the backhaul links to the serving BS, provided the received signal strength difference is within a certain range. The results are reproduced in Table 6.4, cf. Hoymann et al. (2009) for further simulation assumptions.

Similar gains for intersite cooperation are reported in (Frank et al. 2010). In addition, in this work an intersite joint detection scheme with reduced backhaul load is proposed and evaluated within an LTE system. The idea is to restrict the cooperation to a subset of the available subcarriers per Physical

Resource Block (PRB), and the scheme is combined with a threshold on received signal strength difference as in Hoymann et al. (2009). The results show that performance close to full cooperation can be reached if only eight out of 12 subcarriers per PRB are considered for cooperation, with a corresponding reduction of 33% in the backhaul.

6.5 Coordinated Beamforming and Scheduling Techniques

As shown in section 6.4, JP is a very effective technique to enhance both cell-average and cell-edge throughput. Nevertheless, from section 6.4 it appears clear that many problems remain to be solved to apply the JP paradigm to realistic systems. In this section a different class of techniques is considered, where each UE may connect to a single BS at the same time.

> Coordination is either used to jointly schedule UEs belonging to BSs in the same coordinated cluster (coordinated scheduling) or to calculate beamforming coefficients and power allocation, jointly in the set of coordinated cells (coordinated beamforming).

Due to the reduced constraints in terms of amount of exchanged data, coordinated scheduling/beamforming is likely to be a candidate for the first implementations in realistic systems. In this section, after a summary of the state of the art, two methods are described in more details, the first one based on coordinated beamforming (section 6.5.2), the second on coordinated scheduling (section 6.5.3).

6.5.1 State of the Art

The idea of coordinated beamforming comes from the mid-1990s, mainly targeting a so-called *SINR leveling* problem, that is, the power levels and the beamforming coefficients are calculated to achieve some common SINRs in the system or to maximize the minimum SINR. Usually in these first implementations there is no scheduling involved, as the set of UEs is fixed and the SINRs are optimized for this fixed set of UEs.

In Rashid-Farrokhi et al. (1998) an UL system is considered where power and receive beamforming coefficients are calculated in order to achieve a minimum target SINR for each UE and minimize the sum of the powers. Two algorithms can be used to calculate the powers and beamforming coefficients: a centralized one, and a distributed iterative one that updates powers and coefficients based on the last interference measurement.

Table 6.4 Average cell throughput and 5%-percentile UE throughput for different cooperation parameters (Hoymann et al. 2009)

	Average cell throughput		5% percentile UE throughput	
	[bps/Hz]	[%]	[bps/Hz]	[%]
No cooperation	2.35		0.134	
3 supporting cells, 3 dB range	2.52	+7	0.150	+12
3 supporting cells, 10 dB range	2.87	+22	0.174	+30
3 supporting cells, 20 dB range	3.29	+40	0.217	+62
1 supporting cell, 10 dB range	2.67	+14	0.161	+20
5 supporting cells, 10 dB range	2.91	+24	0.175	+31

In Bengtsson (2001), an algorithm to determine the jointly optimal DL beamformers and assignment of UEs to BSs for a set of cochannel UEs is presented. The algorithm is based on global knowledge about the channels from all BSs to all UEs.

Referring to coordinated beamforming methods, an alternative method where the optimal minimum power beamformers can be obtained locally relying on some coupled parameters exchanged between adjacent BSs is shown in Tölli et al. (2011). More details on this method are reported in section 6.5.2.

Coordinated scheduling is a relatively new idea and the theoretical limits of this approach have been explored only in the last past years. In the following, some of the main works in this field are recalled.

In Gesbert and Kountouris (2011) and Choi and Andrews (2008) the problem of joint multicell power control and scheduling to maximize the sum rate is considered. In Gesbert and Kountouris (2011) it is shown that, when the number of UEs goes to infinity, multicell interference (no matter how strong) does not affect the asymptotic scaling of the network, if a rate-optimal scheduling is applied. Moreover, a simple scheduler based on each cell measuring a worst-case SINR and not requiring any exchange of information between the cells results in a quasi-optimal behavior. For example, when UEs have the same average SINR the capacity has the same scaling law as in a single-cell broadcast channel with random beamforming. In Choi and Andrews (2008), the effect of intercell scheduling on multiuser diversity is studied. A full-reuse system with intercell scheduling is compared with a system with frequency reuse. It is proved that the former system has a higher multiuser diversity order due to the effect of intercell scheduling.

From a more practical perspective, different works have studied the possibility of exploiting information about the intercell interference in the scheduling process. In the following, two of the major contributions in this area are recalled. In Das et al. (2003) the entire network is divided into clusters and centralized scheduling is assumed for cells belonging to the same cluster such that only the long term average channel conditions are assumed available to the central scheduling entity. Scheduling involves determining which BSs in the cluster should transmit in each time slot and to which UEs. In Alcatel-Lucent (2008) a technique is considered where the UE sends a feedback to the serving BS which includes also information about the worst interfering beams belonging to nonserving BSs. A distributed scheduler is used that prioritizes the beam/UE selection across BSs belonging to the same coordination cluster.

It is worth noting that coordinated beamforming/scheduling techniques present another important advantage with reference to JP methods: the enhanced robustness to UEs mobility that could be achieved in case of slower update change of coordination data. In particular, studies reported in (WINNER+ 2009b) have shown that coordinated beamforming is competitive compared to non-CoMP and JP for the International Telecommunication Union (ITU) Urban Macro model scenario.

6.5.2 *Decentralized Coordinated Beamforming*

In this section, a decentralized coordinated beamforming method first introduced in Tölli et al. (2009, 2011) is outlined. In this method the optimal minimum power beamformers are considered and they are obtained locally at the BSs relying on some coupled parameters exchanged between adjacent BSs. The problem of finding such optimal beamformers is reformulated such that the BSs are coupled by real-valued intercell interference terms, that are handled by taking local copies of the terms at each BS and enforcing consistency between them. Thus, the coupling in the interference terms is transferred to a coupling in the consistency constraints, which can then be decoupled by a standard dual decomposition approach allowing a distributed algorithm. The consistency is

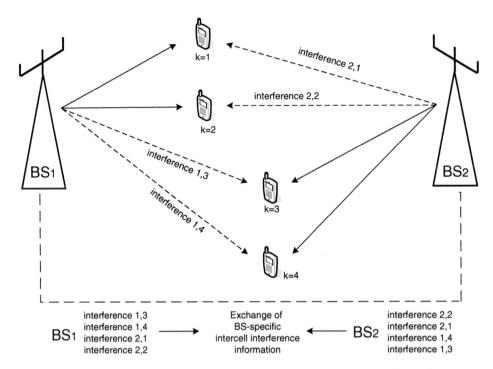

Figure 6.8 Distributed implementation of interference aware beamformer design Tölli et al. (2011). Reproduced by permission of © 2011 IEEE

enforced by exchanging the local intercell interference terms between adjacent BSs as depicted in Figure 6.8.

The proposed approach is able to guarantee feasible solutions for intermediate number of iterations. The UE specific SINR targets can be guaranteed even if the update rate of the coupled interference terms between BSs is relatively low, at the possible cost of increased sum power. In addition, the dual decomposition approach allows for a number of special cases, where the backhaul information exchange is further reduced at the cost of somewhat suboptimal performance.

The general system optimization objective considered here is to minimize the total transmitted power subject to fixed UE-specific SINR constraints $\gamma_m \ \forall \ m$. This can be formulated as

$$\text{minimize} \qquad \sum_{l=1}^{L} P_l \qquad\qquad (6.10)$$

$$\text{subject to } \Gamma_m \geq \gamma_m, \forall \ m$$

where L is number of BSs, P_l is the total power transmitted by the l-th BS and Γ_m is the m-th UE's SINR.

A decentralized method was proposed in Tölli et al. (2011) for the minimum power beamforming problem in (6.10). The problem can be solved by using a dual decomposition approach, which is appropriate when the optimization problem has a coupling constraint such that the problem decouples into several subproblems after relaxing the constraint. Thus, the original one-level optimization problem can be separated into two-level optimization, that is, subproblems for fixed dual variables (prices) and a master dual problem in charge of updating the dual variables (Boyd et al. 2007). In other words, the master problem sets the price for the resources in each subproblem, which in turn

Table 6.5 Main performance results of decentralized coordinated beamforming

Transmission scheme	Average sum power [dB]: $k = 0$ dB	Average sum power [dB]: $k = 10$ dB
Coherent JP-CoMP	−1.45	9.71
CB-CoMP: UE-specific ICI constraint	3.90	19.44
ZF for ICI	8.92	24.81
ZF for intra- and intercell interference	15.21	25.21

has to decide the amount of used resources depending on the price. By using this approach a decentralized solution of (6.10) can be found where the beamformer vectors are obtained locally relying on coupled real-valued intercell interference parameters exchanged between adjacent BSs. A more detailed description and the mathematical derivations can be found in Appendix C.2.1.

A detailed performance evaluation of the decentralized coordinated beamforming with a wide range of numerical results, for example in time-correlated fading scenario, is provided in Tölli et al. (2011). Some of the numerical examples from (Tölli et al. 2011) are provided in the following. Table 2 presents the average sum power of a system with $M = 2$ UEs, $L = 2$ BSs and N_T transmit antennas. The fixed SINR target per UE are 0 dB and 10 dB. Different coordinated beamforming cases and two ZF (interference nulling) approaches are compared with coherent transmission at the cell edge, where each UE has similar large-scale fading properties. As expected, coherent CoMP greatly outperforms the coordinated beamforming cases at the cell edge. The coordinated beamforming case required about 5-6 dB more power than the JP CoMP case in order to meet the 0 dB SINR target. There is a large gain from the optimal intracell beamformer design (ZF for ICI) as compared to the channel inversion (ZF for both intra- and inter-cell interference). It shall be noted that without coordination, the required power is infinite (i.e. SINR constraints can not be guaranteed).

6.5.3 Coordinated Scheduling via Worst Companion Reporting

In Alcatel-Lucent (2008), an approach to realize coordinated scheduling is proposed for LTE. The basic idea behind this proposal, nicknamed as *worst companion* reporting, is to allow the UE to provide feedback information about how the interference level could be reduced by forbidding the interfering eNBs to use one or more beams, belonging to a codebook of possible transmit beams.

The scheme can be summarized as follows. Each UE measures the channel of the serving eNB and of neighboring eNBs; the feedback signaling from UE to the serving eNB is designed such that in addition to the index of the preferred precoding codeword and the associated CQI (as for example in LTE), the UE also reports the index of the most interfering eNBs. Based on this additional information, the eNBs can coordinate to serve the UEs using appropriated precoding weights in order to minimize the inter-cell interference. The principle of the scheme is summarized in Figure 6.9.

Coordinated scheduling can be realized either with a centralized or with a distributed processing approach. An effective distributed scheduling technique, nicknamed as *cyclically prioritized scheduling*, has been proposed in Alcatel-Lucent (2010). The principle is that a scheduling priority is assigned to each eNB, and scheduling between the eNBs of the same coordination cluster is realized starting from the eNBs with high priority and ending to eNBs with low priority. More in details, an eNB with a given priority schedules a set of UEs based on

- the PMI report and the associated CQI;
- the worst companion report and the associated Δ-CQI, which is the CQI when the worst companion is not used in the neighbor cell;
- the constraints on the precoding codewords coming from eNBs with a higher priority.

For example an eNB with a priority i could decide, based on the worst companion report, to forbid an eNB with priority $i - 1$ to use the precoding codeword corresponding to the worst companion report.

The scheduling constraints (or forbidden codewords) must be passed to eNBs with a lower priority (for example through the X2 interface in LTE). Besides this, no other information exchange is assumed between the cooperating BSs. The required bandwidth on the backhaul is therefore small. From the latency point of view, the backhaul and the schedulers must be designed taking into account that overall latency is given by the time to realize the entire coordination cycle from the eNBs with the highest priority to the eNBs with the lowest priority.

The performance is assessed considering jointly cell edge and cell average performance, as shown in Figure 6.10. As the main goal of the worst companion scheme described herein is to improve the cell-edge performance, a possible way to assess the achieved gain is by fixing the cell-average

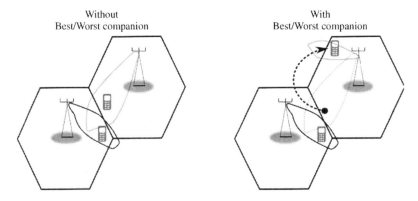

Figure 6.9   Principle of the interference reduction via worst companion signaling

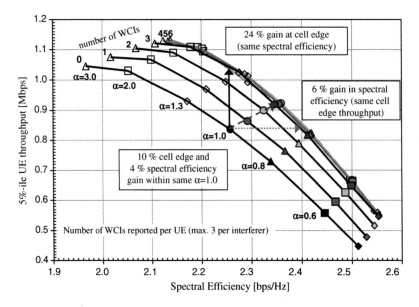

Figure 6.10   Performance of the worst companion reporting, in terms of cell average and cell-edge performance

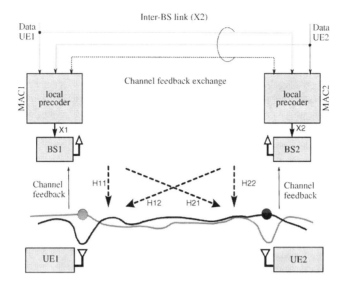

Figure 6.11 Distributed DL CoMP (Jungnickel et al. 2009c). © 2009 IEEE

throughput and measure the gain in cell-edge performance. For instance, by choosing an appropriate PF scheduling α factor, we can achieve a gain of about 24%. Further results and the simulation assumptions of this coordinated scheduling method are shown in Appendix C.2.2.

6.6 Practical Implementation of CoMP in a Trial Environment

In this section an initial real-time implementation and testing of distributed JP CoMP in the Downlink of an LTE-Advanced trial network, taking place in Berlin in the framework of the LTE-Advanced testbed, is described. Enabling features such as distributed synchronization (Jungnickel et al. 2008), cell- and UE-specific pilots have been implemented and tested in real-time; a fast backbone network has been realized to transport data for the JP coordination. The trial layout shown in Figure 6.11 has been implemented; it is based on two BSs and two UEs having two antennas each at 2.68 and 2.53 GHz, respectively. The duplex is based on FDD mode, using 20 MHz bandwidth in the DL and 10 MHz bandwidth for each UE in the UL. For more implementation details, please refer to Jungnickel et al. (2010, 2009c).

Transmission experiments, limited by interference, have been conducted using three multicell transmission modes:

- mode with no coordination between the cells, considered as a lower bound, applying per-antenna rate control with interference-aware equalization at the UE;
- mode with distributed DL CoMP, where the mutual interference between the cells is canceled;
- mode with isolated cells, considered as an upper bound where the interference from other cells is switched off.

Trials over the air are described here for indoor and outdoor-to-indoor scenarios covering both intrasite and intersite CoMP. For an outdoor-to-outdoor scenario refer to Jungnickel et al. (2010). Quite impressive observations regarding the reduced outage probability at the cell edge are described and demonstrate the huge overall performance improvements when using CoMP JP instead of interference-limited transmission. Clearly these gains are obtained only in the absence of interference

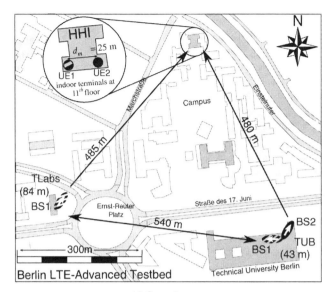

(a) Scenario map

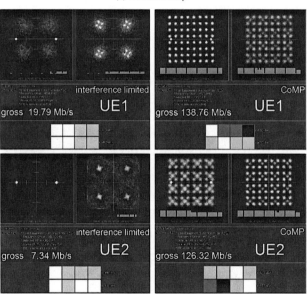

(b) Test equipment

Figure 6.12 Scenario for CoMP trials in Berlin LTE-Advanced testbed (Thiele et al. 2010b). Reproduced by permission of © 2010 IEICE

from non-coordinated cells. The performance in presence of residual Co-Channel Interference (CCI) were shown in section 6.4. Nonetheless, the following experiments demonstrate that the implementation challenges of DL CoMP can be overcome and that similarly high gains as predicted by the theory can be realized in practice.

6.6.1 Setup and Scenarios

Scenarios comprise indoor and outdoor-to-indoor configurations, refer to Figure 6.12(a) for the geography of the trial setup. In the indoor test, both BSs are located in the same lab. In the outdoor-to-indoor scenarios, two BS sites transmit and two sectors are selected either at the same site or at two different sites in order to realize intra- and inter-site cooperation, respectively. Sites are located on the Deutsche Telekom Laboratories (TLabs) building at Ernst-Reuter-Platz (84 m antenna height) and on the Technical University of Berlin (TUB) main building, Straße des 17 Juni (43 m, see Figure 6.12(a)) in Berlin. The estimated height of the buildings in the area is between 25 and 35 m. For more insights, refer to Jungnickel et al. (2009b). Sites are interconnected by optical fibers deployed in the campus with a length of 4.5 km. The X2 signaling over the fiber is based on 1 Gbps Ethernet.

For indoor and outdoor-to-indoor scenarios, both UEs are located on the 11th floor at the Heinrich Hertz Institute (HHI). Both UEs are placed at the south front of the building with the windows facing towards both BSs either in the same lab or in two different labs which are 25 m separated. UE2 is at a fixed location. In order to capture the local fading statistics, UE1 moves at low speed with approximately 3 cm/s. In our implementation, UE1 is always assigned to BS1 and UE2 to BS2, that is, handover is not performed. For performance evaluation, the overall statistics from UEs placed in the same lab and different labs are considered (see 6.12(b) for an example of results in the interference limited and CoMP case for both UEs).

6.6.2 Measurement Results

Indoor and outdoor-to-indoor results are plotted in Figure 6.13. In the setup implemented in practice, a single UE is assigned to BS1 and a single UE to BS2. In the case in which both UEs are at the cell border between BS1 and BS2, both interference scenarios are very similar and similar throughput statistics are then obtained. In Figure 6.13 the throughput is shown together with reference bounds. As explained above, some different transmission modes are taken into consideration in Figure 6.13:

- as a lower bound, interference-limited transmission is introduced using an identity matrix as an independent precoder for BS1 and BS2, in combination with Optimum Combining at the receiver side; in this case, a single UE is assigned to each BS and, due to limited hardware, the trial system cannot utilize multiuser diversity gains in the frequency selective scheduling, as discussed in section 6.4.3, which implies higher outage probability;
- CoMP JP transmission is used from both sites with a fixed number of four streams on the air;
- as an upper bound, there is the case of isolated cells where the interference from the other cell is not present, that is, either BS1 or BS2 is switched off.

In the indoor scenario (i.e. Scenario 1 in Figure 6.13), both BSs are received with the same average power in a rich scattering environment. The average SINR is around 0 dB in both cells simultaneously. Due to multiple reflections in the room, however, both the signal and the interference experience fading. Statistically independent fading of both signals creates a crucial throughput situation for a UE: when moving the UE by a few cm only, we can realize situations where either the signal channel is strong while the interference is in a fade, and correspondingly the serving BS assigns data transmission in a certain part of the whole frequency band, as well as the reverse situation where the interference channel is strong while the signal is in a fade so that no more data is usually transmitted. Note that the BS assignment is always kept fixed, that is, BS1 serves UE1 and BS2 serves UE2. As a consequence, the UE suffers from bad SINR conditions and thus the outage probability amounts to 50%. Thus, the data traffic is not continuous but frequently interrupted when moving through the lab

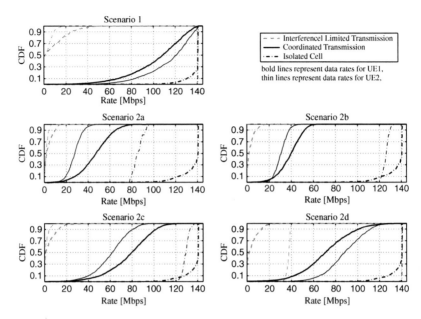

Figure 6.13 Performance results of CoMP trials in Berlin LTE-Advanced testbed (Jungnickel et al. 2010). Reproduced by permission of © 2010 IEEE

and the UE experience may be quite poor. If CoMP is enabled in such a bad interference scenario, significant improvements are observed for the data throughput. Despite the critical interference situation and although the data rate still varies, CoMP removes outage completely. Using CoMP in such indoor deployments provides 18 times higher data rates for both UEs in both cells simultaneously with respect to the interference limited setup. The CoMP setup realizes approximately 78% of the rate achievable in the isolated cell scenario.

Next consider the intrasite scenarios where the intrasite interference is canceled (see Scenarios 2a and 2b in Figure 6.13). It is typical in the distributed multicell network that path losses are not equal for different pairs of BSs and UEs. Nonetheless, the basic observations remain similar. In the interference-limited case, again there is significant outage. Using CoMP, in contrast, both UEs can realize 30% of the peak data rate on average. Note that in this scenario both UEs are placed in the same building, while both serving BSs are situated at the same site, that is, the same antenna pole. In this case, signals transmitted from both sectors show high correlation (Jaeckel et al. 2009).

Finally consider the intersite scenarios, called Scenarios 2c and 2d in Figure 6.13, where the intersite interference from the other site is suppressed. These scenarios provide superior performance compared to the intrasite scenarios, despite distributed synchronization of both BSs. Studies of the underlying channel correlations suggest that the higher data rates observed with CoMP in the intersite scenarios may also be attributed to the lower transmit antenna correlation (Jaeckel et al. 2009) if eNB antennas as well UEs are sufficiently separated in the deployment. However, in (Jungnickel et al. 2010) the difference between intrasite and intersite CoMP throughput compared to isolated cell transmission is rather small when UEs are placed at different buildings or streets in the outdoor-to-outdoor scenario. The mobility of the UE in the field was rather limited due to the time variance of the channel. The inevitable feedback delay causes a mismatch between the precoder and the actual channel as soon as the UE is moved. Compensation is possible by using channel prediction to make the interference suppression more robust.

6.7 Future Directions

The CoMP systems represent a topic of wide interest in the research community and are widely studied towards their introduction in practical Radio Access Networks. Nevertheless CoMP is still in a phase of uncertainty due to the complexity of its implementation and some unresolved issues, such as the CSI estimation accuracy, and at the moment the fundamental step from theoretical research to practical industrial exploitation has yet to be done. The information and the encouraging results shown in this chapter, especially regarding the trial, have been aimed at illustrating CoMP issues.

As explained earlier, CoMP adoption would have implications for the networks, both from a technical and also from a financial standpoint. Additional costs have to be expected for the significant exchange of information that some practical implementations of CoMP could require, leading even to a complete update of the backhaul technologies adopted for this information sharing. In this sense the choice of centralized or decentralized solutions for CoMP, as well as of the level of coordination, are not a minor issue, together with a more pervasive deployment of fiber back-haul and front-haul connections to support higher capacity data exchange among cooperating nodes.

Currently the CoMP studies are mostly related to two main areas, and standardization activities have followed this classification as well: coordinated beamforming/scheduling and JP schemes. Considering DL, these are the most suitable practical implementations of CoMP. It has been pointed out how JP performances are significantly better than coordinated scheduling/beamforming, at the expense of greater, and often excessive, complexity. As for uplink almost the same applies, even if it has to be added that the somewhat lower complexity could allow a possible quicker adoption of uplink CoMP, currently foreseen as a proprietary implementation in cooperative nodes.

It is expected that the future directions will focus on several aspects of CoMP: architectural, theoretical and practical.

There are many architectural aspects to be tackled, like the backhaul design and the adopted simplifications to be sought: decentralized approaches, distributing complexity among nodes, and clustering. In particular, dynamic clustering is considered a viable solution for future CoMP methods. Otherwise, at least in the first implementations, cooperation among sectors of the same cell could be attempted. The introduction of Heterogeneous Network (HetNet), then, could be an option to exploit coordination also among different cells in a multilayer environment.

From a theoretical point of view, the feedback design and robust channel estimation algorithms are of crucial importance. In fact, the question to be solved, regardless of the adopted coordination solution, is the achievable tradeoff between the amount of information exchanged among the nodes and the expected performance gain. Another related important tradeoff, with achievable performance, is between the amount of feedback and performance gain, discussed in Section 6.2.

From a practical perspective, future research trends should consider reference signal design, CSI impairments, and hardware limitations such as synchronization. In fact, the performance of any CoMP system must take into consideration the imperfect CSI, addressing scenarios less ideal than those analyzed so far in the first stages of CoMP studies. Upper bounds about achievable capacity with realistic assumptions on CSI are being searched. A possible further direction of research about CoMP could be also the performance estimation in case of different mobility scenarios; especially for JP the high mobility scenario could be very challenging for CoMP. Finally, synchronization (and the cyclic prefix design) of the cooperative entities in CoMP schemes is another research topic that shall deserve attention in the future, especially for the schemes requiring higher coherence. There are already functionalities aiming to distribute clock signals among the coordinated nodes or relying on GPS, but further efforts in this field are foreseen in the coming years.

References

3GPP 2006 Physical Layer Aspects for Evolved Universal Terrestial Radio Access (UTRA) (Release 7). Technical Report 25.814 v7.1.0, 3rd Generation Partnership Project (3GPP), http://www.3gpp.org/ftp/Specs/2006-09/Rel-7/.

3GPP 2010 Further Advancements for E-UTRA Physical Layer Aspects. Technical Report 36.814 v9.0.0, 3rd Generation Partnership Project (3GPP), http://www.3gpp.org/ftp/Specs/2010-03/Rel-9/.

Aktas D, Bacha M, Evans J and Hanly S 2006 Scaling results on the sum capacity of cellular networks with MIMO links. *IEEE Trans. Inform. Theory* **52**(7), 3264–3274.

Aktas E, Evans J and Hanly S 2008 Distributed decoding in a cellular multiple-access channel. *IEEE Trans. Wireless Commun.* **7**(1), 241–250.

Alcatel-Lucent 2008 UE PMI feedback signalling for user pairing/coordination. Technical Report R1-083759, 3GPP Third Generation Partnership Project, Working Group RAN1, meeting 54, Prague, Czech Republic.

Alcatel-Lucent 2010 Performance of coordinated beamforming with multiple PMI feedback. Technical Report R1-100944, 3GPP Third Generation Partnership Project, Working Group RAN1, meeting 60, San Francisco, USA.

ARTIST4G 2010 *Definitions and Architecture Requirements for Supporting Interference Avoidance Techniques* (ed. D'Amico V. and Gresset N.). Public Deliverable D1.1, Advanced Radio InTerface TechnologIes for 4G SysTems - ARTIST4G, August 2010.

Baum DS, Salo J, Milojevic M, Kyösti P and Hansen J 2005 MATLAB implementation of the interim channel model for beyond-3G systems (SCME), http://radio.tkk.fi/en/research/rf_applications_in_mobile_communication/radio_channel/scm.html.

Bengtsson M 2001 Jointly optimal downlink beamforming and base station assignment *in Proc. IEEE Int. Conf. Acoust., Speech, Signal Processing*, Salt Lake City, Utah, USA.

Björnson E, Jaldén N, Bengtsson M and Ottersten B In press Optimality properties, distributed strategies, and measurement-based evaluation of coordinated multicell OFDMA transmission.*IEEE Trans. Signal Processing*.

Bletsas A and Lippman A 2006 Implementing cooperative diversity antenna arrays with commodity hardware. *IEEE Commun. Mag.* **44**(12), 33–40.

Boccardi F and Huang H 2006 Zero-forcing precoding for the mimo broadcast channel under per-antenna power constraints *Signal Processing Advances in Wireless Communications, 2006. SPAWC '06. IEEE 7th Workshop on*, pp. 1–5.

Boccardi F and Huang H 2007a Limited downlink network coordination in cellular networks. *IEEE International Symposium on Personal, Indoor and Mobile Radio Communications*.

Boccardi F and Huang H 2007b A near-optimum technique using linear precoding for the MIMO broadcast channel. *IEEE International Conference on Acoustics, Speech and Signal Processing*.

Botella C, Piñero G, González A and de Diego M 2008 Coordination in a multi-cell multi-antenna multi-user W-CDMA system: a beamforming approach. *IEEE Trans. Wireless Commun.* **7**(11), 4479–4485.

Boyd S, Xiao L, Mutapcic A and Mattingley J 2007 Notes on decomposition methods. Course reader for convex optimization II, Stanford, http://www.stanford.edu/class/ee364b/.

Brueck S, Zhao L, Giese J and Awais Amin M 2010 Centralized scheduling for joint transmission coordinated multi-point in LTE-Advanced. International ITG Workshop on Smart Antennas (WSA), pp. 177–184.

Caire G and Shamai S 2003 On the achievable throughput of a multiantenna gaussian broadcast channel. *IEEE Trans. Inform. Theory* **49**(7), 1691–1706.

Carneheim 1994 FH-GSM Frequency Hopping GSM *Proceedings of the 44th IEEE Vehicular Technology Conference*.

Choi W and Andrews J 2007 Downlink performance and capacity of distributed antenna systems in a multicell environment. *IEEE Trans. Wireless Commun.* **6**(1), 69–73.

Choi W and Andrews J 2008 The capacity gain from intercell scheduling in multi-antenna systems. *IEEE Trans. Wireless Commun.* **7**(2), 714–725.

Cover TM and Gamal AAE 1979 Capacity theorems for the relay channel. *IEEE Trans. Inform. Theory* **25**(5), 572–584.

Cover TM and Thomas JA 1991 *Elements of Information Theory*. John Wiley & Sons, Inc., New York.

Dahlman E, Parkvall S, Sköld J and Beming P 2008 *3G Evolution - HSPA and LTE for Mobile Broadband*. Academic Press, Amsterdam.

Dahrouj H and Wei Y 2008 Coordinated beamforming for the multi-cell multi-antenna wireless system *Information Sciences and Systems, 2008. CISS 2008. 42nd Annual Conference on*, pp. 429–434.

Das S, Viswanathan H and Rittenhouse G 2003 Dynamic Load Balancing Through Coordinated Scheduling in Packet Data Systems *IEEE Infocom*, San Francisco.

Deutsche Telekom AG and Vodafone 2009 Coordinated link adaptation based on multi-cell channel estimation in the LTE-A uplink. Technical Report R1-095067, 3GPP Third Generation Partnership Project, Working Group RAN1, meeting 59, Jeju, South Korea.

Ericsson and ST-Ericsson 2010 Channel reciprocity in FDD systems including systems with large duplex distance. Technical Report R1-100853, 3GPP Third Generation Partnership Project, Working Group RAN1, meeting 60, San Francisco, USA.

Frank P, Müller A and Speidel J 2010 Inter-site joint detection with reduced backhaul capacity requirements for the 3GPP LTE uplink *Proc. IEEE VTC-Fall 2010*, pp. 1–5.

Gesbert D, Hanly S, Huang H, Shamai Shitz S, Simeone O and Yu W 2010 Multi-Cell MIMO Cooperative Networks: A New Look at Interference. *IEEE J. Selected Areas in Commun.* **28**(9), 1380–1408.

Gesbert D and Kountouris M 2011 Rate Scaling Laws in Multicell Networks Under Distributed Power Control and User Scheduling. *IEEE Trans. Inform. Theor.* **57**(1), 234–244.

Gupta P and Kumar PR 2000 The capacity of wireless networks. *IEEE Trans. Inform. Theory* **46**(3), 388–404.

Holma H and Toskala A 2004 *WCDMA for UMTS: Radio Access for Third Generation Mobile Communications* 3rdf edn. John Wiley & Sons, Ltd, Chichester.

Hongyuan Z, Mehta NB, Molisch AF, Jin Z and Huaiyu D 2008 Asynchronous interference mitigation in cooperative base station systems. *IEEE Trans. Wireless Commun.* **7**(1), 155–165.

Hoymann C, Falconetti L and Gupta R 2009 Distributed uplink signal processing of cooperating base stations based on IQ sample exchange *Proc. IEEE ICC 2009*, pp. 1–5.

Jaeckel S, Jiang L, Jungnickel V, Thiele L, Jandura C, Sommerkorn G and Schneider C 2009 Correlation properties of large and small-scale parameters from multicell channel measurements *European Conference on Antennas and Propagation (EuCAP 2009)*, Berlin, Germany.

Jafar SA and Goldsmith AJ 2002 Transmitter optimization for multiple antenna cellular systems *Proc. IEEE ISIT 2002*, vol. 1, p. 50, Lausanne, Switzerland.

Jafar SA, Foschini GJ and Goldsmith AJ 2004 PhantomNet: Exploring optimal multicellular multiple antenna systems. *EURASIP J. Applied Signal Processing* **5**, 591–604.

Jungnickel V, Forck A, Jaeckel S, Bauermeister F, Schiffermller S, Schubert S, Wahls S, Thiele L, Haustein T, Kreher W, Müller J, Droste H and Kadel G 2010 Field trials using coordinated multi-point transmission in the downlink *Proc. 3rd International Workshop on Wireless Distributed Networks (WDN), held in conjunction with IEEE PIMRC 2010*, Istanbul, Turkey.

Jungnickel V, Jaeckel S, Thiele L, Jiang L, Krger U, Brylka A and Helmolt C 2009a Capacity measurements in a cooperative multicell mimo network. *IEEE Trans. Veh. Technol.* **58**, 2392–2405.

Jungnickel V, Schellmann M, Thiele L, Wirth T, Haustein T, Koch O, Zirwas W and Schulz E 2009b Interference-aware scheduling in the multiuser MIMO-OFDM downlink. *IEEE Communicaions Magazine* **47**(6), 56–66.

Jungnickel V, Thiele L, Wirth T, Haustein T, Schiffermüller S, Forck A, Wahls S, Jaeckel S, Schubert S, Juchems C, Luhn F, Zavrtak R, Droste H, Kadel G, Kreher W, Mueller J, Stoermer W and Wannemacher G 2009c Coordinated multipoint trials in the downlink *Proc. 5th IEEE Broadband Wireless Access Workshop (BWAWS), Co-located with IEEE GLOBECOM 2009*, Honolulu, Hawai.

Jungnickel V, Wirth T, Schellmann M, Haustein T and Zirwas W 2008 Synchronization of cooperative base stations. *IEEE International Symposium on Wireless Communication Systems 2008 (ISWCS2008)*.

Karakayali M, Foschini G and Valenzuela R 2006a Network coordination for spectrally efficient communications in cellular systems. *IEEE Trans. Wireless Commun.* **13**(4), 56–61.

Karakayali MK, Foschini GJ and Valenzuela RA 2006b Network coordination for spectrally efficient communications in cellular systems. *IEEE Trans. Wireless Commun. Mag.* **3**(14), 56–61.

Karakayali M, Foschini GJ, Valenzuela RA and Yates RD 2006c On the maximum common rate achievable in a coordinated network *Proc. IEEE ICC 2006*, vol. 9, pp. 4333–4338, Istanbul, Turkey.

Khattak S, Rave W and Fettweis G 2008 Distributed iterative multiuser detection through base station cooperation. *EURASIP Journal on Wireless Communications and Networking*. Article ID 390489.

Kramer G, Gastpar M and Gupta P 2005 Cooperative strategies and capacity theorems for relay networks. *IEEE Trans. Inform. Theory* **51**(9), 3037–3063.

Laneman JN, Tse DNC and Wornell GW 2004 Cooperative diversity in wireless networks: Efficient protocols and outage behavior. *IEEE Trans. Inform. Theory* **50**(12), 3062–3080.

Laneman JN and Wornell GW 2003 Distributed space-time-coded protocols for exploiting cooperative diversity in wireless networks. *IEEE Trans. Inform. Theory* **49**(10), 2415–2425.

Liang Y, Yoo T and Goldsmith A 2006 Coverage spectral efficiency of cellular systems with cooperative base stations *Proc. IEEE GLOBECOM 2006*, pp. 1–5, San Francisco, CA, USA.

Marsch P and Fettweis G 2007 A framework for optimizing the downlink performance of distributed antenna systems under a constrained backhaul *13th European Wireless Conference*.

Noda M, Muraguchi M, Khanh TG, Sakaguchi K and Araki K 2007 Eigenmode tomlinson-harashima precoding for multi-antenna multi-user MIMO broadcast channel *Information, Communications Signal Processing, 2007 6th International Conference on*, pp. 1–5.

Papadogiannis A, Bang HJ, Gesbert D and Hardouin E 2008a Downlink overhead reduction for multi-cell cooperative processing enabled wireless networks *Personal, Indoor and Mobile Radio Communications, 2008. PIMRC 2008. IEEE 19th International Symposium on*, pp. 1–5.

Papadogiannis A, Gesbert D and Hardouin E 2008b A dynamic clustering approach in wireless networks with multi-cell cooperative processing *Proc. IEEE ICC 2008*, pp. 4033–4037.

Papadogiannis A, Gesbert D and Hardouin E 2008c A dynamic clustering approach in wireless networks with multi-cell cooperative processing. *IEEE International Conference on Communications*.

Papadogiannis A, Hardouin E and Gesbert D 2009 Decentralising multi-cell cooperative processing on the downlink: a novel robust framework. *EURASIP J. Wireless Commun. Networking*, Special issue on broadband wireless access.

Papadopoulos H and Sundberg CEW 2008 Space-time codes for mimo systems with non-collocated transmit antennas. *IEEE J. Selected Areas in Commun.* **26**(6), 927–937.

Rashid-Farrokhi F, Tassiulas L and Liu K 1998 Joint optimal power control and beamforming in wireless networks using antenna arrays. *IEEE Trans. Commun.* **46**(10), 1313–1324.

Sendonaris A, Erkip E and Aazhang B 2003a User cooperation diversity–part I: System description. *IEEE Trans. Commun.* **51**(11), 1927–1938.

Sendonaris A, Erkip E and Aazhang B 2003b User cooperation diversity–part II: Implementation aspects and performance analysis. *IEEE Trans. Commun.* **51**(11), 1939–1948.

Shamai S and Zaidel B 2001 Enhancing the cellular downlink capacity via co-processing at the transmitting end *Proc. IEEE VTC Spring 2001*, pp. 1745–1749, Rhodes, Greece.

Skjevling H, Gesbert D and Hjørungnes A 2008 Low-complexity distributed multibase transmission and scheduling. *EURASIP Journal on Advances in Signal Processing*.

Somekh O, Simeone O, Bar-Ness Y and Haimovich M 2006 Distributed multi-cell zero-forcing beamforming in cellular downlink channels *Proc. IEEE GLOBECOM 2006*, pp. 1–5, San Francisco, CA, USA.

Somekh O, Simeone O, Bar-Ness Y, Haimovich A, Spagnolini U and Shamai S 2007 *An Information Theoretic View of Distributed Antenna Processing in Cellular Systems, in Distributed Antenna Systems: Open Architecture for Future Wireless Communications* first edn. Auerbach Publications, CRC Press.

Song Y, Cai L, Wu K and Yang H 2007 Collaborative mimo based on multiple base station coordination. *Contribution to IEEE 802.16m*.

Thiele L, Boccardi F, Botella C, Svensson T and Boldi M 2010a Scheduling-Assisted Joint Processing for CoMP in the Framework of the WINNER+ Project *Proc. Future Network & Mobile Summit 2010*, Florence, Italy.

Thiele L, Jungnickel V and Haustein T 2010b Interference management for future cellular OFDMA systems using Coordinated Multi-Point transmission. *IEICE Transactions on Wireless Distributed Networks*.

Thiele L, Wirth T, Haustein T, Jungnickel V, Schulz E and Zirwas W 2009 A unified feedback scheme for distributed interference management in cellular systems: Benefits and challenges for real-time implementation. *17th Euopean Signal Processing Conference (EUSIPCO2009)*.

Tölli A, Codreanu M and Juntti M 2008 Cooperative MIMO-OFDM cellular system with soft handover between distributed base station antennas. *IEEE Trans. Wireless Commun.* **7**(4), 1428–1440.

Tölli A, Pennanen H and Komulainen P 2009 Distributed coordinated multi-cell transmission based on dual decomposition *Proc. IEEE GLOBECOM 2009*, pp. 1–6, Honolulu, Hawai'i, USA.

Tölli A, Pennananen H and Komulainen P 2011 Decentralized minimum power multi-cell beamforming with limited backhaul signaling. *IEEE Trans. Wireless Commun.* **10**(2), 570–580.

van der Meulen EC 1977 A survey of multiway channels in information theory: 1961–1976. *IEEE Trans. Inform. Theory* **23**(1), 1–37.

Venkatesan S 2007 Coordinating base stations for greater uplink spectral efficiency in a cellular network. *IEEE International Symposium on Personal, Indoor and Mobile Radio Communications*.

Wei Y and Lan T 2007 Transmitter optimization for the multi-antenna downlink with per-antenna power constraints. *IEEE Trans. Signal Processing* **55**(6), 2646–2660.

Weingarten H, Steinberg Y and Shamai S 2006 The capacity region of the Gaussian multiple-input multiple-output broadcast channel. *IEEE Trans. Inform. Theory* **52**(9), 3936–3964.

WINNER+ 2009a Celtic Project CP5-026 *Initial Report on Advanced Multiple Antenna Systems* (ed. Komulainen P. and Boldi M.). Public Deliverable D1.4, Wireless World Initiative New Radio - WINNER+, January 2009, http://projects.celtic-initiative.org/winner+/index.html (accessed June 2011).

WINNER+ 2009b Celtic Project CP5-026 *Intermediate Report on CoMP (Coordinated Multi-Point) and Relaying in the Framework of CoMP* (ed. Mayrargue S.). Public Deliverable D1.8, Wireless World Initiative New Radio - WINNER+, November 2009, http://projects.celtic-initiative.org/winner+/index.html (accessed June 2011).

WINNER+ 2009c Celtic Project CP5-026 *Preliminary WINNER+ System Concept* (ed. Osseiran A. and Gouraud A.). Public Deliverable D2.1, Wireless World Initiative New Radio – WINNER+, May 2009, http://projects.celtic-initiative.org/winner+/index.html (accessed June 2011).

WINNER-II 2007a IST-4-027756 *Interference averaging concepts* (ed. Bublin M.). Public Deliverable D4.7.1, Wireless World Initiative New Radio – WINNER, 2007, http://projects.celtic-initiative.org/winner+/index.html (accessed June 2011).

WINNER-II 2007b IST-4-027756 *Interference avoidance concepts* (ed. Mange G.). Public Deliverable D4.7.2, Wireless World Initiative New Radio – WINNER, 2007, http://projects.celtic-initiative.org/winner+/index.html (accessed June 2011).

Wyner AD 1994 Shannon–theoretic approach to a Gaussian cellular multiple-access channel. *IEEE Trans. Inform. Theory* **40**(6), 1713–1727.

Xiao S, Zhang Z and Shi Z 2010 Clustered multi-point coordinating transmission systems with the reduced overhead. *IEEE Trans. Commun.* **5**(6), 493–500.

Yifan L and Goldsmith A 2006 Symmetric rate capacity of cellular systems with cooperative base stations *Global Telecommunications Conference, 2006. GLOBECOM '06. IEEE*, pp. 1–5.

Yifan L, Taesang Y and Goldsmith A 2006 Coverage spectral efficiency of cellular systems with cooperative base stations *Global Telecommunications Conference, 2006. GLOBECOM '06. IEEE*, pp. 1–5.

Yu W 2006 Uplink-downlink duality via minimax duality. *IEEE Trans. Inform. Theory* **52**(2), 361–374.

Yu W and Lan T 2007 Transmitter optimization for the multi-antenna downlink with per-antenna power constraints. *IEEE Trans. Signal Processing* **55**(6, part 1), 2646–2660.

Zarikoff B and Cavers J 2008 Multiple frequency offset estimation for the downlink of coordinated mimo systems. *Selected Areas in Communications, IEEE Journal on* **26**(6), 901–912.

Zhang H and Dai H 2004a Cochannel interference mitigation and cooperative processing in downlink multicell multiuser MIMO networks. *EURASIP J. Wireless Comm. and Netw.* **2004**(2), 222–235.

Zhang H and Dai H 2004b Cochannel interference mitigation and cooperative processing in downlink multicell multiuser MIMO networks. *EURASIP J. Wireless Comm. and Netw.* **2004**(2), 222–235.

Zhang H, Mehta NB, Molisch AF, Zhang J and Dai H 2007 On the fundamentally asynchronous nature of interference in cooperative base station systems *Proc. IEEE ICC 2007*, vol. 4, pp. 6073–6078, Glasgow, Scotland, UK.

Zhang P, Tao X, Zhang J, Wang Y, Li L and Wang Y 2005 A vision from the future: beyond 3G TDD. *IEEE Commun. Mag.* **43**(1), 38–44.

7

Relaying for IMT-Advanced

Afif Osseiran, Ahmed Saadani, Peter Rost and Alexandre Gouraud

This chapter overviews relaying techniques in general, and describes them within the International Mobile Telecommunications Advanced framework in particular. The most common relaying deployment scenarios are analyzed and the efficient combination of Coordinated Multipoint transmission or reception and relaying for an International Mobile Telecommunications Advanced cellular networks is evaluated. Relaying for indoors scenario, is compared to femtocells. Finally an overview of cooperative relaying is given.

7.1 An Overview of Relaying

The main driving force in the development of wireless communication networks and systems is to provide, among other aspects, increased coverage and support for higher data rates. At the same time, the cost of building and maintaining the system is of great importance and is expected to become even more so in the future. Moreover, higher demand on data rates and on longer battery life are requested from the end-user perspective. Until recently the main topology of wireless communication systems has been fairly unchanged, for the first three generations of cellular networks. The topology of existing wireless communication systems has been hitherto characterized by the cellular architecture, which consists of fixed radio BS and UEs as the only transmitting and receiving entities in the network. Alternatively a radio device node called a Relay Node (RN) can be deployed in order to help wirelessly conveying the information between the BS and the UE. These RNs may consist mainly of passive repeaters that forward analog signals in scenarios where a wired backhaul is not feasible or is very costly. Alternatively active and smart Relay Nodes (RNs) can be deployed. In that case the RNs may communicate with other network elements (e.g. Base Station (BS), another RN or a User Equipment (UE)).

Although some cellular operators have been deploying RNs in form of repeaters in order to extend the coverage of the networks, it was only from 2005 that RNs were specified in major cellular standards. IEEE was quick to adopt it in 2007, in its specification (IEEE 2007), later in 2010, 3GGP followed suit (3GPP 2010).

Mobile and Wireless Communications for IMT-Advanced and Beyond, First Edition.
Edited by Afif Osseiran, Jose F. Monserrat and Werner Mohr.
© 2011 Afif Osseiran, Jose F. Monserrat and Werner Mohr. Published 2011 by John Wiley & Sons, Ltd.

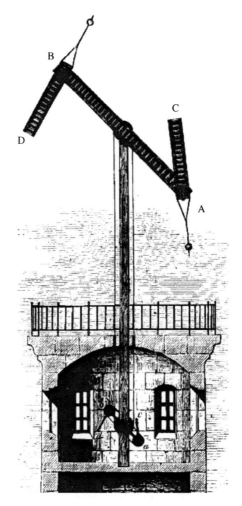

Figure 7.1   Example of Chappe brothers' RN (Figuier 1868, p.52)

7.1.1   Relay Evolution

Relaying was a common technique used to convey messages over large distances, in ancient empires such as Egypt, Babylon, China, Greece, Persia, Rome and the Omeyyad. The messages were in various form such as beacon fire relayed by tower or mountain peak. A more common method was sending messengers on horseback between RNs until the final destination is reached. With the advent of science, communication techniques improved. In 1793, the Chappe brothers proposed a telegraph system relying on RNs equipped with telescopes and lighted by lamps. An example of the Chappe brothers' RN is shown in Figure 7.1.

In modern time, RNs were initially simple devices, which amplify a signal and forward it immediately, and were mainly intended to extend the coverage of the wireless system. These were low-cost devices, compared to BSs, and didn't include any base band processing and hence no network protocol operation was possible. The backhaul connection was usually implemented with microwave links in own frequency bands to avoid interference with the access link, or in-band with appropriate (receive and transmit) antenna isolation avoiding Larsen-like effects. However, this flexible approach revealed significant drawbacks after its deployment, such as difficult monitoring of a RN's operation as a physical site visit was required, which significantly increased the operational costs. Operators were therefore reluctant to deploy relay-based solutions and used them on a case-by-case basis when no other alternative was suitable.

With the growth of data traffic and the emergence of new services, network densification is one solution to increase the network capacity. Increasing the density of deployed BSs implies additional backhaul that corresponds to increased deployment costs. One solution to this problem is to use RNs that are smarter than a simple repeater. This encourages the definition of new RN families that are monitored by the network and could have different levels of intelligence. The simplest RN case is called "Layer 1" RN where the RN is able to receive, Amplify-and-Forward (AF) the received signal and also to exchange some control information with the network. "Layer 1" RN can be seen as a repeater with control signalings. "Layer 2" RNs are able to Decode-and-Forward (DF) the success-fully decoded signal. It also has its own Layer 2 protocols such as Medium Access Control (MAC) and Radio Link Control (RLC) layers. Finally, "Layer 3" RNs are comparable to BSs with wireless backhaul (i.e. the three layers are present in that type of RN). The "Layer 3" RN has been standard-ized and will be described in the following. It should be noted that a Layer-2 relaying solution (IEEE 2007; Pabst et al. 2004; Döttling et al. 2009, Chapter 8) was developed on how to integrate into the LTE standard, by the Wireless World Initiative New Radio (WINNER) project.

7.1.2 Relaying Deployment Scenarios

Relay nodes, by construction, are flexible devices in terms of deployment positions and transmit power. Those inherent qualities make them very useful in several deployment scenarios. In particular, three main scenarios, which are depicted in Figure 7.2, can be identified:

- **Network densification** where the aim is to increase the network capacity with the introduction of new BS sites. However in urban and suburban zones it is difficult and very time-consuming to negotiate contracts for new sites. In addition, the installation of backhaul is quite costly, which makes RNs a cost-efficient solution as they are exploiting wireless backhaul. Further, the RNs' positioning is more flexible. In fact, they can be deployed on street lamps, building walls, or even buses. Hence, RNs represent an opportunity to ubiquitously provide high data rates.

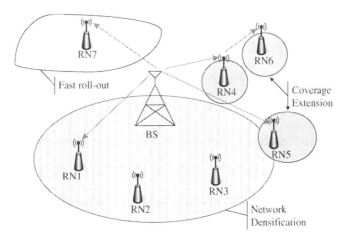

Figure 7.2   Relay deployment scenarios

- **Coverage extension** relates to dead zones not covered by the network in urban and suburban environments. Using RNs will extend the coverage to these zones without the need for a wired backhaul. On the other hand, the lack of a dense network can cause bad coverage in indoor environments. For instance, rural buildings without a broadband connection can suffer from a lack of multimedia services. The use of indoor RNs is one solution to cover them with the required data rates.
- **Fast roll-out** where the fast penetration with new wireless services requires a very dynamic and fast change of an operator's infrastructure. Relay nodes can be deployed much faster and support operators to be the first to offer new services.

7.1.3   Relaying Protocol Strategies

This section gives a brief overview of Layer 1 and Layer 2 relaying. A more detailed overview is given in Kramer et al. (2005) where DF- and CF-based protocols are analyzed.

Layer 1 Relaying

Layer 1 relaying, also called Amplify-and-Forward (AF), forwards a linear function $X_r = f_r(Y_r)$. More specifically, the transmitted RN signal X_1 is an amplified version of its own received signal, i.e. $X_r = \beta \cdot Y_r$. Amplify-and-Forward is a very simple and low-complexity protocol, which requires no digital processing. However, it also does not separate the useful source signal X_s and noise signal at the RN. Hence, it also forwards an amplified version of the receiver noise at the RN. An overview on AF is given in Hammerström et al. (2004), where AF is applied to a block-fading multiple-relay scenario. In particular, each RN forwards a phase-rotated version of its received signal, which introduces additional time diversity due to the time-variance of the effective channel. However, in Laneman et al. (2004) it was shown that the achievable Diversity-Multiplexing-Tradeoff (DMT) of digital relaying (for instance incremental relaying) always dominates AF.

Decode-and-Forward

By contrast to AF relaying, DF and CF apply nonlinear functions on the received signal at the RN. Figure 7.3(b) illustrates the DF strategy for a single communication pair, where the RN r decodes the source message X_s submitted in block t using its own receive signal Y_r. Based on the decoded source message a RN message X_r is selected and transmitted in the next block $t + 1$, which does not necessarily have the same codebook (i.e. the set of all possible codewords) size as the source message. More specifically, the RN can assign multiple source messages to the same RN message. The destination node then decodes first the RN message X_r, which helps to reduce the number of possible source messages X_s to the set of messages, which is assigned to the decoded RN message. Afterwards, the destination may decode the source message X_s using this prior knowledge. In the case of cellular systems, the main benefit of relaying is the significantly reduced path loss. Hence, the destination solely exploits the RN signals, which is called in the following *noncooperative relaying*. By contrast, if the destination experiences similar channel gains towards source and relay, it might be beneficial to exploit the diversity provided by combining both paths, which is referred to as *cooperative relaying*.

The achievable rates of the DF strategy are given in Cover and Gamal (1979, Theorem 1). It is relevant to mention that after decoding at the RN only correctly decoded message should be re-encoded as otherwise the system diversity is decreased (Laneman et al. 2004). A practical implementation of DF is to determine parity bits at the RN and forward those to the destination. Finally, we distinguish

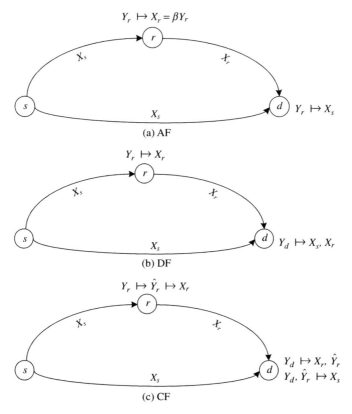

$$Y_r \mapsto X_r = \beta Y_r$$

(a) AF

$$Y_r \mapsto X_r$$

(b) DF

$$Y_r \mapsto \hat{Y}_r \mapsto X_r$$

(c) CF

Figure 7.3 *Logical* information exchange of the protocols AF, DF, and CF. Arc labels show which information is exchanged between nodes and node labels show the decoding order where $Y \mapsto X$ symbolizes that in this particular decoder stage, Y is mapped to X. s is the source, r the RN and d the destination

DF with different codebook sizes at source and relay, which is also referred to as *irregular coding* while using the same codebook size at each terminal is referred to as *regular coding* (Xie and Kumar 2004, 2005).

Compress-and-Forward

While DF provides transmit diversity, CF provides receive diversity and is illustrated in Figure 7.3(c). Instead of decoding the source message, the RN quantizes its received signal using $\hat{Y}_r$, which then represents a digital version of the analog received signal Y_r. Using the RN message X_r, the quantized signal is communicated to the destination. At the destination, this quantized signal $\hat{Y}_r$ is restored using the destination's own receive signal Y_d as side information. Once the quantized RN signal $\hat{Y}_r$ is known at the destination, the destination can use both Y_d and $\hat{Y}_r$ to decode the source message. The achievable rates of the CF strategy are given in Cover and Gamal (1979, Theorem 6).

The major benefit of CF over DF is the ability to provide a second (quantized) received signal to the destination. This is of particular interest if the RN is placed close to the destination node and experiences similar channel conditions. In this case, DF cannot improve the data rates as the link between source and RN is the bottleneck and provides the same data rates as a direct transmission from source to RN. In Rost and Fettweis (in press), a regular encoding approach for CF has been presented, which is able to improve data rates in networks with multiple RNs.

It should be mentioned that another relaying decoding strategy, called Demodulate-and-Forward (DmF) relaying, or uncoded DF (Chen and Laneman 2006), exists. In DmF, the RN simply demodulates the received signals and retransmits it. For instance, assume a Binary Phase Shift Keying (BPSK) modulation is used, then applying DmF is equivalent to use CF with one bit of quantization.

7.1.4 Half Duplex and Full Duplex Relaying

Relay Nodes can be further differentiated into those operating in half-duplex and those operating in full-duplex. Half-duplex RNs are subject to an orthogonality constraint, which implies that they either transmit or receive on a time-frequency resource and therefore must operate in a half-duplex mode. This requires that the RN coordinates its resource allocation with UEs in the uplink and the assigned BS in the downlink. Such a coordination can be done using a predefined RN-zone resulting in a static centralized solution or using more dynamic solutions, where the RN is allowed to assign resources (distributed coordination). In addition, half-duplex RNs have less stringent requirements on the deployed hardware compared to full-duplex RNs.

In the case of full duplex RNs, they receive the signal on one carrier frequency, process it and transmit it on the same frequency with a small delay compared to the received frame duration. However, RNs continue simultaneously receiving from the source. This assumes that there is good isolation between the receiver antennas and the transmit antennas of RNs. There is only a minor rate loss caused by the delay. There is hence a tradeoff between performance and RN cost (due to the antenna isolation).

7.1.5 Numerical Example

In order to illustrate the operating regions of the discussed protocols (in Section 7.1.3), Figure 7.5 shows the achievable end-to-end rates for the line network illustrated in Figure 7.4 and for the case of full- and half-duplex RNs. For these results, we use the same transmit power at source and RN ($P_s = P_r$) as well as AWGN with the same noise power at RN and destination ($N_r = N_d$). Furthermore, the Signal to Noise Ratio (SNR) $\gamma_{s,d}$ of the source-destination channel is given by $P_s/N_d = 10$ dB and we use the path loss exponent $\theta = 4$. Figure 7.5(a) shows the achievable rates for full-duplex relaying and for the previously discussed protocols as well as an upper bound on the capacity and a combined approach Cover and Gamal (1979, Theorem 6), which integrates both DF and CF. The shown upper bound uses the max-flow min-cut theorem and is achievable in very specific cases such as the degraded RN channel. In addition, Figure 7.5(a) shows the performance of direct communication with power normalization ($P_s/N_d = 13$ dB) and without power normalization ($P_s/N_d = 10$ dB).

We can observe in Figure 7.5(a) that full-duplex relaying has the potential to significantly increase the achievable rate particularly for $x \approx D/2$. Some interesting points can be singled out such as when $x = 0$. In that case CF provides the same performance as direct communication with power normalization. This shows that CF implements a distributed form of receive diversity. At $x = D$, DF

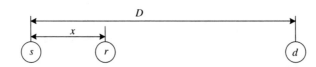

Figure 7.4   Setup for the analysis of the Gaussian single-relay channel

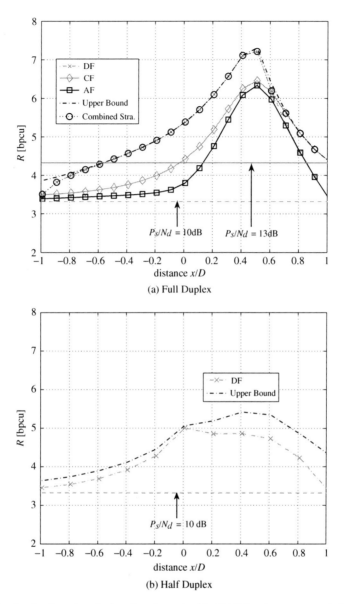

Figure 7.5 Results for the Gaussian single-relay channel with AWGN, coherent transmission, path loss exponent $\theta = 4$, and $\gamma_{s,d} = 10$ dB

provides the same performance as direct communication without power normalization as both RN and destination experience the same channel.

Figure 7.5(b) shows DF and a max-flow min-cut upperbound for the same scenario but using half-duplex RN nodes. It draws a similar picture as for full-duplex relays, that is, instead of a maximum performance of about 100% provided by DF and full-duplex relays, half-duplex relaying only improves the maximum by about 50%. While full-duplex DF achieves capacity for a wide range of parameters, half-duplex DF only achieves performance close to capacity at $x = 0$ when the time slot for communicating the data to the RN is very short. A detailed discussion of achievable rates in half-duplex multiple-relay channels can be found in Rost and Fettweis (in press).

In fading environments the picture changes as the RN introduces a diversity gain (Laneman et al. 2004). Finally, we can see in Figure 7.5(a) that digital relaying always outperforms analog relaying, which shows that the lower complexity of analog RNs comes at the expense of performance. We compare the relaying protocols with power-normalized direct communication, which is not always possible in mobile communication systems as the maximum transmission power is limited by the hardware employed such as the power amplifier. In a relaying system the power is more homogeneously distributed in the network and power constraints do not need to be changed.

7.2 Relaying in the Standard Bodies

3GPP started to specify relaying functionalities in LTE-Advanced (LTE-A) Release 10 (Rel-10) (3GPP 2010), while in IEEE 802.16m the relaying definition started in 2009. In 3GPP, a donor cell is defined as a cell via which the RN is wirelessly connected to the radio access network. One donor cell can support more than one RN and each RN can be transparent or nontransparent towards the UE: the difference is whether the UE is aware of the RN (nontransparent case) or not (transparent case) when it communicates with the network via the RN. The backhaul wireless connection could be done in the system frequency band or in another available band. However, in the latter case, the operator needs to acquire additional bands hence increasing the total deployment cost. The use of free bands does not allow operators to control the backhaul quality. For this reason 3GPP focuses in its specification process on the International Mobile Telecommunications Advanced (IMT-Advanced) bands. One can distinguish two ways for relaying: inband and outband RNs, which are further detailed in the following.

7.2.1 Relay Types in LTE-Advanced Rel-10

Two categories of RNs are defined in LTE-A Rel-10. In the first category, Type 1, the RN has its own Identity (ID) and own reference signalling. In the second category, Type 2, the RN is transparent to the UE and does not have its own reference signalling.

In case the BS-RN link shares the same carrier frequency with the RN-UE links the RN is said to be "inband". In fact, in order to be backward compatible with Long Term Evolution (LTE) Release 8 (Rel-8) where the UEs should be able to be connected to the donor cell, the backhaul and UEs share the same resources.

For "outband" RNs, the BS-RN link does not operate in the same carrier frequency as the RN-UE links. In this case, Rel-8 UEs should also be able to be connected to the donor cell.

In the following, the main features of the RN types in Rel-10 are described and summarized in Table 7.1.

Table 7.1 Relay classification in 3GPP Release 10

Class	Cell ID	Inband/Outband
Type 1	Yes	Inband half duplex
Type 1.a	Yes	Outband full duplex
Type 1.b	Yes	Inband full duplex
Type 2	No	Inband full duplex

Type 1 Relay Nodes are inband RNs controlling their own cells and using their own cell ID. Thus, they have specific synchronization channels and reference signals. User Equipment will receive the scheduling information and the Hybrid Automatic Repeat-reQuest (HARQ) feedback directly from the RN.

Type 1.a Relay Nodes are outband RNs and have the same properties as Type 1 RNs . They can transmit and receive at the same time. Their performance are expected to be similar to RN Type 1.

Type 1.b Relay Nodes are inband RNs and have an adequate antenna isolation between the antenna connected with the UE and the antenna connected with the donor cell. The isolation could be done with a signal processing mechanism that cancels self-interference or by spatial isolation. This isolation will have an impact on the RN cost, however, their performance are expected to be similar to the performance provided by femtocells as they are full duplex.

Type 2 Relay Nodes are low-cost devices, operating inband and as part of the donor cell. They do not have a physical cell ID and are transparent to Rel-8 UEs. Furthermore, the radio resource management is partly controlled by the BS.

Resource Partitioning for Inband RNs

In an LTE-A system the backhaul link (Un) and access link (Uu) are multiplexed in time on a single carrier frequency (3GPP 2010).

- For Frequency Division Duplex (FDD) mode, in the Un link, the BS to RN transmissions are done in the Downlink (DL) frequency band, whereas the RN to BS transmissions are done in the Uplink (UL) frequency band.
- For the Time Division Duplex (TDD) mode, the BS to RN transmissions are done in the DL sub-frames of the BS and RN. The RN to UE transmissions are done in the UL subframes of the BS and RN. This resource partitioning does not impact the LTE Rel-8 UEs.

Backward Compatibility Issues

When the RN is inband it should operate in a half-duplex mode unless its antennas are well separated and isolated. Consequently, the RN can either transmit or receive. In order to make this "gap in time" unnoticeable to previous LTE releases and in particular Rel-8, the MBSFN feature is exploited (3GPP 2010) as shown in Figure 7.6. In fact the MBSFN subframes are configured as frames where the RN transmits to (resp. receives from) BS and no UE is allowed to transmit (resp. receive) to the BS. In LTE, it is possible to allocate up to six MBSFN subframes per frame for the backhaul link. The allocation is done in a semistatic way.

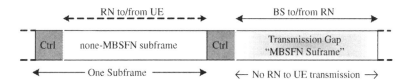

Figure 7.6 Relaying exploiting MBSFN feature in LTE Rel-8

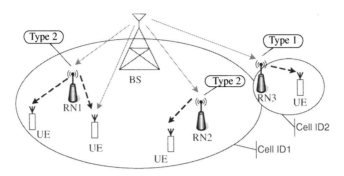

Figure 7.7 Type 1 (resp. nontransparent) and Type 2 (resp. transparent) RN in LTE Rel-10 (resp. IEEE 802.16m)

7.2.2 Relay Nodes in IEEE 802.16m

As in LTE-A Rel-10, two categories of RNs are defined in IEEE 802.16m (IEEE 2009, 2010), namely transparent and nontransparent. The nontransparent and transparent categories correspond to RN Type 1 and Type 2 in LTE-A Rel-10, respectively. A nontransparent RN can operate in both centralized and distributed scheduling mode, while a transparent RN can only operate in centralized scheduling mode. An illustration of the RNs types used in the standard bodies is shown in Figure 7.7. As it can be seen, RN3 is of Type 1 (or non-transparent) since it has its own cell ID. On the other hand, RN1 and RN2 are of Type 2 (or transparent) and both have the same cell ID as the donor BS.

7.3 Comparison of Relaying and CoMP

Although Coordinated Multipoint transmission or reception (CoMP), introduced in the previous chapter, offers remarkable improvements. It still faces the problem of shadowed areas where not necessarily intercell interference but strong path loss impairs the cell throughput. In those situations relaying is the means of choice to improve the quality of service. Hence, IMT-Advanced cellular networks will not face questions about whether *either* CoMP *or* relaying is used but rather about how both can be efficiently combined.

Figure 7.8 illustrates different interference situations in cellular networks, where the following interference regions can be defined:

A : Between two physically separated sites, the *intersite* interference region is located, where UEs experience uncorrelated channel conditions to both sites.
B : By contrast, the *intersector* interference region is located between two sectors hosted by the same site. In this case, UEs might experience correlated channel conditions to both sectors.
C : In the case of dominant *intrasector* interference, RNs deployed in the same cell are interfering with each other.
D : Finally, in this region, UEs neither experience significant interference from other BSs nor other RNs but only receive signals either from a dominating BS or RN.

In the following, we will explore the possibilities and challenges of integrating CoMP and relaying in a cellular network.

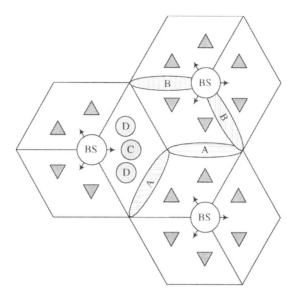

Figure 7.8 Interference regions in a RN communication system. RNs are indicated by triangles. Each BS covers one site and is divided into three sectors. Arrows indicate the main lobe direction

7.3.1 Protocols and Resource Management

The integration of CoMP and relaying can be well described by the relay-assisted interference channel (Rost et al. 2009), which is illustrated in Figure 7.9. It models the case where two communication pairs are supported by two RNs and both transmitters are connected by a finite capacity backhaul link. A link-level analysis of this channel has been conducted in Rost et al. (2009) and in Rost et al. (in press) system-level results have been presented. Furthermore, in Somekh et al. (2010) analytical results for a similar scenario based on a linear Wyner model and using DF RNs have been presented.

CoMP Transmission and Detection

As was mentioned in Chapter 6, CoMP transmission exploits the possibility of cooperatively transmitting from multiple BSs and therefore improving the Signal to Interference plus Noise Ratio (SINR) particularly of interference-limited UEs. One prominent approach is to apply the Wiener transmit

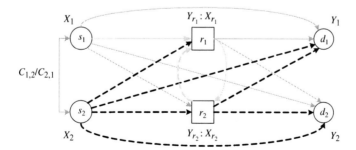

Figure 7.9 Relay-assisted interference channel with two sources $S = \{s_1, s_2\}$, two RNs $R = \{r_1, r_2\}$, and two destinations $D = \{d_1, d_2\}$. Both source nodes are connected, with capacities $C_{1,2}$ and $C_{2,1}$, respectively. Channel inputs are denoted by X and channel outputs are denoted by Y

filter (Joham et al. 2005) based on the knowledge of the compound channel and the transmit data of the cooperating BSs. This implies strong requirements on the backhaul as a significant amount of Channel State Information (CSI) and data must be exchanged between the individual BSs. In addition, particularly nonlinear approaches such as Dirty Paper Coding (DPC) (Costa 1983) are very sensitive to imperfect CSI caused by low-rate feedback and imperfect channel state estimation. Hence, it is more likely to apply CoMP at colocated BSs, which share the same cabinet at one site. This allows for an efficient exchange of data between the individual BSs and counteracts intersector interference, which poses one of the main challenges in cellular networks. Like CoMP transmission, CoMP detection can exploit the availability of compound CSI and data knowledge in order to improve the array gain (centralized approach) and to improve the multiplexing gain (Marsch and Fettweis 2007).

Coordinated Relaying with CoMP feeder link

One possibility for combining the advantages of CoMP with the benefits of relaying is to use CoMP on the link between BSs and RNs. RNs most likely experience more static channels than UEs as RNs will be fixed in their position (at least for a long period of time).

Hence, the required CSI updates are less frequent and the overhead, which needs to be exchanged between cooperating BSs, can be reduced. This allows for more efficient backhaul usage and much improved performance on the link between BSs and RNs, which usually represents the bottleneck in a relay-assisted cellular network. In addition, the signaling overhead can be reduced as less RNs than UEs must be served and therefore the scheduling and the Radio Resource Management (RRM) signaling require less overhead.

On the link between RNs and UEs we need to apply an interference mitigation scheme (Han and Kobayashi 1981), which does not require common CSI and data knowledge at the transmitters. One such approach is Han-Kobayashi (HK) coding where both transmitters divide their messages into a common and private part. The common part is decoded by both receivers in order to reduce the interference for the private part, which is only decoded by the respective receiver. Although the common part is decoded by both receivers, it is only exploited by the actual destination while the other receiver only uses it for interference reduction. However, this approach requires a very complex and resource-intensive optimization over all possible power allocations, which makes the approach less suitable for practical implementations. Etkin *et al.* introduced, in Etkin et al. (2008), an approach that always achieves rates within 1 bit per channel use (bpcu) of capacity[1] and deterministically assigns the individual power levels. Furthermore, empirical observations in Rost (2009) showed that it is sufficient to apply the described scheme based on the long-term SINR and only using two possible power assignments: one where only a private message and one where only a common message is transmitted. In this way, the receiver either ignores interference (only private message) or jointly decodes interference and useful signal (common message).

Integrated Approach

User Equipments that are located in the main lobe direction of a BS or have a Line of Sight (LoS), are likely to experience very high SINR towards the BS. In those situations it is suitable not to serve UEs using relaying but using either a conventional Multiple-Input Multiple-Output (MIMO) transmission

---

[1] Bits per channel use (i.e. bits per transmission) is a generic unit for measuring channel capacity.

Table 7.2 The individual protocols classified by their ability to cooperate

		Interpath	
		Coordination	*Cooperation*
Intersubsystem	Coordination	• Noncooperative relaying • HK on each hop • Single-Hop transmission	• CoMP on BSs→RNs or BSs→UEs • HK or Time Division Multiple Access (TDMA) on RNs→UEs
	Cooperation	• Cooperative relaying • BSs and RNs form one Virtual Antenna Array (VAA)	• All nodes form one VAA

or CoMP transmission in the case that the UE experiences strong interference. On the other hand, if UEs are located in shadowed areas or have a very weak non-LoS connection, it is preferable to use relaying. It is difficult to predict in real time the optimal choice with regard to whether relaying or CoMP should be used. Alternatively, it is much simpler to assign UEs to the radio access points with the smallest effective path loss towards the UE, that is, considering the different transmission power at RNs and BSs. In the following analysis, a snapshot-based simulation has been employed and UEs are assigned to the RN or BS, which has the smallest effective path loss towards the UE. Such an approach is also used in systems with femtocells, which face similar problems, and it provides remarkable performance gains (Rost 2009; Rost et al. in press).

The channel as illustrated in Figure 7.9 as well as the protocols applied to this channel can be categorized as shown in Table 7.2. In order to categorize the protocols, we group them in terms of their ability to cooperate or to coordinate across different communication pairs (multiple paths) and across different hops (multiple subsystems, that is, one between BSs and RNs, and one between RNs and UEs). In the case that neither BSs nor RNs share common CSI among each other, we apply intercell TDMA or noncooperative relaying with full reuse of resources. If we allow for interpath cooperation we obtain the previously discussed integrated approach where BSs jointly transmit and receive while RNs apply TDMA between both communication pairs or HK coding. However, we do not consider the case of joint RN-transmission and detection as it is unlikely that RNs obtain the required common CSI and data knowledge in order to coherently transmit. We can further consider protocols where the individual subsystems are cooperating. An example is cooperative relaying, which is detailed later in this chapter.

7.3.2 Simulation Results

Finally, we examine results for a system-level simulation using the channel model defined by the European research project WINNER (WINNER-II 2006). In particular, we apply Orthogonal Frequency Division Multiplexing (OFDM) with 100 MHz bandwidth and 2048 subcarriers. The applied simulator uses the channel fading model defined by WINNER for the Urban Macro scenario. Furthermore, in each sector two RNs are deployed at a radius of about 333 m, which corresponds to one-third of the inter-site distance of 1000 m. User Equipment is randomly and uniformly distributed such that on average about 29 UEs are assigned to each cell. We apply a fair scheduler for the assignment of UE resources based on history acquired for each UE. The results presented are obtained for one central

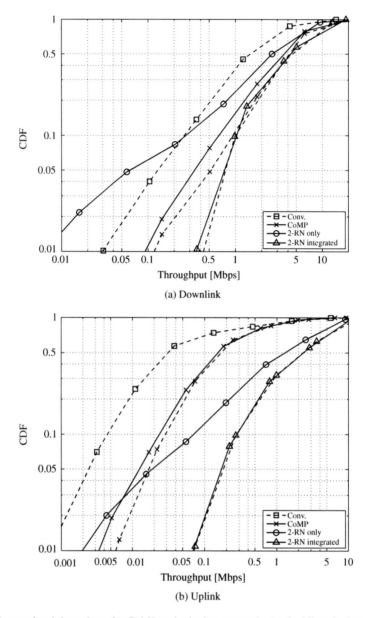

(a) Downlink

(b) Uplink

Figure 7.10 System-level throughput for CoMP and relaying protocols. Dashed lines indicate unlimited inter-site backhaul and solid lines indicate only intrasite cooperation

site surrounded by two tiers of interferers which totals 54 interfering BSs. All RNs are modeled as half-duplex in-band RNs. All further parameters and a simulator description are given in Rost (2009) and Rost et al. (in press). The previous assumptions differ slightly from the IMT-Advanced as well as 3GPP assumptions where an intersite distance of 500 m and a fixed UE density of 10 UEs per cell is assumed. However, this worst case assumptions illustrates the ability of relaying to significantly improve performance even in cases of low-density deployment and high user density.

Figure 7.10 shows the UL and DL performance for the previously introduced approaches. Consider at first the DL performance shown in Figure 7.10(a). It can be deduced that the integrated approach and CoMP cases (denoted respectively in the figure caption by "2-RN integrated" and "CoMP") outperform the direct transmission and the relay-only cases (denoted respectively by "Conv." and "2-RN only"). The relay-only approach is outperformed due to the significant performance loss of UEs close to the BS, which experience a very high path loss towards the RNs. The integrated approach slightly improves the fairness (higher fifth percentile throughput) but does not significantly improve the average throughput. The slightly improved fairness results from the improved channel conditions at the cell edge and the almost equal average throughput follows from the interference cancellation ability of CoMP. However, compare the performance of CoMP with unlimited intersite cooperation (dashed line) and the performance of CoMP with intrasite cooperation without backhaul between individual sites (solid line). While the integrated approach maintains its performance as most of the interference originates from the same site, the CoMP performance significantly drops due to the worse SINR at intersite cell borders.

Now consider Figure 7.10(b), which shows the uplink performance and further emphasizes the conclusions from the discussed downlink results. By contrast to the downlink performance figures, the performance difference between the integrated approach and CoMP is in the range of a factor of 10. This enormous performance difference is a result of the significantly improved path loss between UEs and radio access points, which did not have the same effect in the DL due to the high transmit power at BSs (46 dBm) compared to the transmit power at RNs (37 dBm) and UE (24 dBm). These results demonstrate that beyond IMT-Advanced networks should rather integrate both CoMP and relaying than choosing only one particular technology.

7.4 In-band RNs versus Femtocells

The previous section discussed in-band RNs, which share the same spectral resources for UE-RN and RN-BS links. By contrast, as shown in Chapter 4, femtocells (Chandrasekhar et al. 2008; Knisely et al. 2009) are usually deployed indoors and exploit existing broadband connections as backhaul links towards the core network. This additional deployment of small indoor cells is a cost-efficient alternative to significantly improve the indoor coverage, data rates, and to allow for traffic offloading. Femtocells may also be deployed outdoors (in which case they are called pico cells) and use either wireless out-of-band backhaul towards the assigned BS or a wired backhaul connection towards the core network. Relay Nodes provide more flexibility than indoor and outdoor femtocells, but they suffer from a performance loss as the same spectral resources are used for BS-RN links and RN-UE links.

Nonetheless, RNs and femtocells face the same challenges, that is, intercell and intracell interference coordination and avoidance. In order to assess the performance loss due to the in-band operation of RNs and the intra/intercell interference, Tables 7.3 and 7.4 provide numerical results for the already introduced wide-area deployment and a typical Manhattan street-grid deployment, where we again apply a worst-case scenario, that is, high shadowing between the deployed outdoor-BSs and indoor-UEs due to the high carrier frequency. Consider at first Table 7.3, which shows the mean and the 5th Throughput (TP) for the same deployment as illustrated in Figure 7.8. We compare a deployment of two additional RNs (inband and half-duplex) per macrocell and a deployment of two additional (outdoor) femtocells per macrocell with an unlimited backhaul towards the core network. The same table shows that (outdoor) femtocells improve the average downlink throughput by about 60% and the uplink performance by about 35% compared to inband RNs. This improvement is caused by the increased number of resources available to RNs and BSs for serving UEs. While in the downlink the worst UE performance (represented by the 5th TP) is improved by

Table 7.3 System level throughput for wide-area scenario comparing CoMP, relaying and femtocell deployments. The stated number of RNs and femtocells is per macrocell

	Scenario / Protocol	Mean TP [Mbps]	5th TP [Mbps]
Downlink	Conventional	2.0	0.1
	CoMP	4.4	0.5
	2 (outdoor) RNs	5.0	0.6
	2 (outdoor) femtocells	8.3	1.0
Uplink	Conventional	0.3	$2.3 \cdot 10^{-3}$
	CoMP	0.5	$1.6 \cdot 10^{-2}$
	2 (outdoor) RNs	3.5	0.2
	2 (outdoor) femtocells	4.8	0.2

about 60%, it remains the same in the uplink as the worst UEs suffer from the same insufficient channel conditions.

Table 7.4 considers a typical Manhattan street-grid using the channel models described in WINNER-II (2006) and UEs placed uniformly indoors *and* outdoors. Furthermore, we consider two setups. The first one places two outdoor RNs/femtos per macrocell at the adjacent crossings. In the second setup, two additional (indoor) RNs/femtos are placed in the two adjacent buildings. The resulting setups are shown in Figure 7.11. We can see in Table 7.4 that femtocells with a backhaul connection do not significantly improve the performance for the setup of two additional outdoor access points. However, in the case of additional indoor femtocells the performance is improved by a factor of about four in the case of DL and a factor of about two in the case of UL. Furthermore, only a setup with indoor RNs or femtocells is able to provide high data rates due to the very high path loss from outdoor RNs or femtocells towards indoor UEs. These results are supported by recent field trial results (Fettweis et al. 2010) for relay-based communication in a LTE environment. Femtocells provide significant performance benefits, particularly for indoor UEs, but they also require a broadband backhaul connection, which makes the deployment less flexible. Hence, while femtocells appear to be the right choice for indoor UEs, RNs seem to be the better option for outdoor deployments due to their flexibility and cost-efficient deployment.

Table 7.4 System level throughput for Manhattan-area scenario comparing CoMP, relaying and femtocell deployments. The stated number of RNs and femtocells is per macrocell

	Scenario / Protocol	Mean TP [Mbps]	5th TP [Mbps]
Downlink	Conventional	1.2	$1.1 \cdot 10^{-4}$
	CoMP	1.8	$2.5 \cdot 10^{-4}$
	2 (outdoor) RNs	1.5	$1.4 \cdot 10^{-3}$
	2 (outdoor) femtocells	1.8	$1.8 \cdot 10^{-3}$
	2 (outdoor) + 2 (indoor) RNs	3.8	0.2
	2 (outdoor) + 2 (indoor) femtocells	14.2	0.6
Uplink	Conventional	0.3	$2.3 \cdot 10^{-3}$
	CoMP	0.3	$4.6 \cdot 10^{-5}$
	2 (outdoor) RNs	0.7	$5.8 \cdot 10^{-5}$
	2 (outdoor) femtocells	0.7	$6.9 \cdot 10^{-5}$
	2 (outdoor) + 2 (indoor) RNs	5.6	0.1
	2 (outdoor) + 2 (indoor) femtocells	13.4	0.6

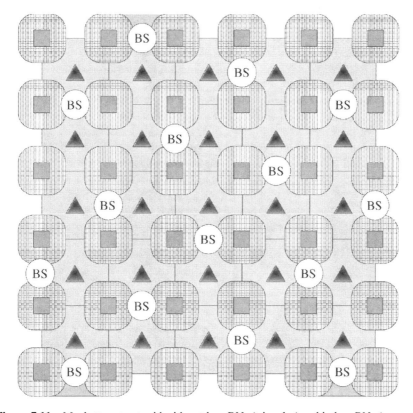

Figure 7.11 Manhattan street grid with outdoor RNs (triangles) and indoor RNs (squares)

7.5 Cooperative Relaying for Beyond IMT-Advanced

One way to introduce diversity in the received signal is to utilize multiple antennas at the transmitter and possibly also at the receiver. An alternative approach is to introduce macro-diversity utilizing cooperative relaying.

> A cooperative relaying system is a relaying system where the information sent to an intended destination is conveyed through various routes and combined at the destination. Each route can consist of one or more hops utilizing the RNs. In addition, the destination may receive the direct signal from the source.

The introduction of cooperative communication systems will increase macro- and/or multipath diversity gains. Such systems offer the possibility of reducing the effective path loss between communicating RN entities, which may benefit the end UE. Cooperative relaying systems are typically limited to only two (or a few) hops. In the literature, several names are in use, such as cooperative diversity (Laneman 2002), cooperative coding (Stefanov and Erkip 2004), and virtual antenna arrays (Döhler et al. 2002).

Cooperative communication can be divided into several categories: cooperative diversity, distributed space-time coding (Laneman and Wornell 2002) and cooperative (channel) coding.

There has been a plethora of cooperative diversity, and distributed space-time coding schemes (Anghel and Kaveh 2006; Ben Slimane and Osseiran 2006; Larsson 2003; Osseiran et al.

2007; Scaglione and Hong 2003; Wei et al. 2004). Just to cite a few: Alamouti diversity based cooperative relaying (Anghel and Kaveh 2006), coherent combining based relaying (Larsson 2003), Relay selection diversity (RSD), and Relay Cyclic Delay Diversity (RCDD).

Another way to achieve cooperative diversity is to use cooperative channel coding. Cooperative coding designates those schemes where the channel coding operations are distributed within the cooperative nodes instead of being conducted at a single node. The idea of extending convolutional codes, turbo coding and Low-Density Parity-Check (LDPC) to a cooperative manner has been proposed since the early 2000s. For instance, the distributed Rate Compatible Convolutional Codes (RCPC) similar to incremental redundancy Automatic Repeat-reQuest (ARQ) was proposed in 2002 by (Hunter and Nosratinia 2002, 2003). Moreover, distributed turbo coding technique was suggested by (Zhao and Valenti 2003) and distributed Space-Time Trellis Codes (STTC) were analyzed in (Rost and Fettweis 2007). Finally distributed LDPC appeared in (Chakrabarti et al. 2007; Razaghi and Yu 2007). In the following we will describe briefly the most known cooperative relaying schemes. It should be noted that cooperative communications in the context of Network Coding (NC) is treated in Chapter 8.

Relay Selection Diversity (RSD)

Relay Selection Diversity is similar to antenna selection diversity. In RSD the BS will select one RN out of a set of RNs belonging to the BS. The BS transmits in the first transmission phase. In the second transmission phase, the selected RN will forward the information from the BS to the UE. Contrary to classical antenna selection where the selected antenna for transmission is based on short-term statistics (e.g. fast fading), the selection criterion for RSD may be done on a slow basis and may consist of the distance and shadow fading gain.

Relay Coherent Combining

Relay Coherent Combining (RCC) was first proposed in Larsson (2003). It involves multiplying the transmitted signal at each RN by a phase that compensates the one introduced by the channel. In fact the effective channel at the UE will be a constructive summation of all the RN signals transmitting to the desired UE. However, this method requires very detailed phase-CSI at the RN.

Distributed Space Time Block Coding

Distributed Space Time Block Code (STBC) is another way to get the spatial diversity by reusing in a distributed manner the STBC initially designed for MIMO schemes. In the first step the BS transmits the information symbols $s_1, \ldots, s_N$ to the RNs. In the second step each RN encodes the symbols linearly into new symbols corresponding to one line of a STBC matrix. Figure 7.12 illustrates this process in the case of two RNs where u_i and v_i $(i = 1, \ldots, K)$ are linear combinations of symbols $s_j(j = 1, \ldots, N)$. When multi-hop RNs are used, the synchronization between RNs can be lost. Note that the STBC properties can be destroyed due to imperfect synchronization between the transmitting antennas. Some recent works propose the use of new distributed STBC that are delay tolerant (Damen and Hammons 2007; Nahas et al. 2010).

A particular case of these distributed STBC is the Relay Alamouti code. Relay Alamouti diversity consists of two distributed RNs that are used to mimic conventional Space Time Transmit Diversity (STTD). For instance, the BS will transmit an even number of OFDM symbols in the first transmission phase then the RN will retransmit these symbols during the second transmission phase. For simplicity let us assume that the duration of the transmission phases is two OFDM symbols,

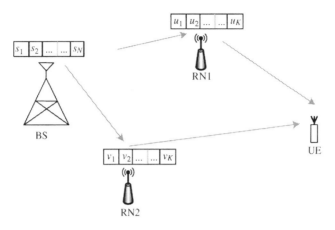

Figure 7.12   Distributed space time block coding with two RNs

for example, s_0 and s_1. For the second transmission phase (of equal duration to the first transmission phase), two RNs attached to the BS are selected. Each of the two RNs is equipped with a single antenna. The two antennas of the RNs will act jointly as in the case of two transmitting antennas for a conventional STTD. The only difference is that each of the antennas is attached to a different antenna system instead of being controlled by the same radio unit. After demodulating the two received symbols the RNs will act jointly as a STTD encoder. While the first RN will resend the same symbols that is, $[u_0, u_1] = [s_0, s_1]$, the second RN will swap the order of the received symbols in addition to conjugating them as is done at the second antenna of a conventional STTD encoder, that is, $[v_0, v_1] = [-s_1^*, s_0^*]$.

Distributed Relay Cyclic Delay Diversity

Distributed Cyclic Delay Diversity (CDD) is also known as RCDD (Osseiran et al. 2007) as it mimics the conventional CDD. The idea is to transmit the same signals from a set of distributed RNs each applying a different time delay (i.e. cyclic shift) and thereby causing multipath fading even when the actual channel is flat. Besides the improved link gain usually obtained from a cooperative relaying system, RCDD introduces frequency selectivity and macro-diversity in OFDM based systems.

Cooperative LDPC coding

Cooperative or distributed LDPC is commonly implemented in a two-hop DF relaying scenario. The encoding is executed in two steps. At first the information bits are encoded at the source (by a generator matrix). Then the RN decodes the received signal. Afterward it computes a new packet of the parity bits (based on the parity-check matrix of the information bit at the source). The new packet is forwarded to the destination. An illustration of a distributed LDPC is shown in Figure 7.13. There has been a large amount of work on the application of LDPC codes to full duplex and half duplex RN (Chakrabarti et al. 2007; Ezri and Gastpar 2006; Hu and Duman 2006; Razaghi and Yu 2007; Wu et al. 2010). In Chakrabarti et al. (2007) and Razaghi and Yu (2007) code designs for irregular coding are proposed. Moreover Ezri and Gastpar (2006) consider the use of independent source and RN codebooks. In Hu and Duman (2006) LDPC codes are applied to the half-duplex RN channel and a random puncturing scheme for LDPC codes was proposed.

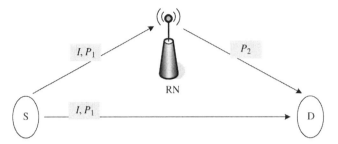

Figure 7.13   Distributed LDPC coding in a 2-hop scenario. S is the source, R the RN and D the destination. *I* is the information matrix. P_1 and P_2 are the parity matrices from the source and RN, respectively

In WINNER+ (2010) a new code design for distributed LDPC coding was devised. In fact, most of the distributed LDPC coding schemes proposed in the literature are based either on serial or parallel code concatenation. From the code design point of view, the serial or parallel concatenation of LDPC codes has intrinsic limitations, mainly because parity-check matrices used for decoding at the RN and the destination are included one in the other, resulting in inappropriate matrix topologies (density on nonzero entries, column and row weight distributions, cycles, etc.). The proposed code design in WINNER+ (2010) aims to create incremental redundancy for LDPC codes, while avoiding both serial and parallel concatenation. It is based on a Split-and-Extend (SpE) approach, and allows the construction of codes with enhanced correction capacity and low integration cost. After decoding the received signal, the RN computes extra parity bits by splitting parity checks corresponding to rows of the parity-check matrix. Then the RN transmits these new parity bits towards the destination. The whole process aims to create a new matrix, whose rows correspond to parity checks involving both old and new parity bits. This new matrix can be used at the destination to jointly decode the received signals from both the source and relay.

It was shown in WINNER+ (2010) that the distributed LDPC based on the SpE approach yields up to 4 dB improvement in terms of SNR compared to repetition coding schemes (i.e. the case where the RN forwards simply the information bits to the destination). The detailed results as well the simulation assumption can be found in WINNER+ (2010, App. C).

7.6   Relaying for beyond IMT-Advanced

Relaying is still an active and promising research area. A few of the promising ideas for future directions are summarized below. Among those ideas is the daunting task of multihop relaying (i.e. more than two hops), which has already been considered in an IEEE standard and is expected to be standardized stin the 3GPP in near future. Moreover the use of relaying in a mobile framework such as high-speed trains is economically very attractive.

7.6.1   Multihop RNs

Hitherto we focused on two-hop RNs. In some scenarios the coverage extension provided by a single RN does not suffice but additional RNs are required. There exist two solutions to control multihop RNs: the centralized and the distributed case. The centralized case implies that the BS manages all resource allocations for all involved RNs while in the distributed case RNs independently assign their resources. One problem in multihop relaying is the difficulty of synchronization due to the unavailability of wired backhaul by construction.

7.6.2 Mobile Relay

Mobile Relay is a generic term indicating that the RN is moving. These RNs can be mounted on top of a car, a bus or a train, where a large number of UEs are located and where a wireless high-data rate network access is required. In this case, the backhaul link must cope with extreme conditions such as frequent handovers and therefore the predictability of the trajectory is a good means to reduce overhead, facilitate handovers and improve the backhaul link.

Mobile Relaying represents an interesting area where real-time positioning information will greatly improve the communication link. On the other hand, these RNs introduce new problems for the owners of cars. On top of a public transportation vehicle one can easily imagine who will own, maintain and operate the RN. However, this is less evident if the RNs are targeted for private vehicles, which gives rise to difficulties similar to those problems experienced with femtocell, that is, such as the question which set of UEs is eligible to access the mobile RN. As long as the vehicle is large enough, it is likely that the transmit power of these RNs will be sufficiently high to support a large number of UEs. On the other hand, mobile RNs complicate the cell planning and long-term interference mitigation in networks.

7.6.3 Network Coding

The RN deployment in the network implies that several points share the same information. Moreover, these points are able to support a high processing complexity. In order to increase further the network gains offered by cooperative relaying schemes, network coding can be used on top of cooperative communication. Hence the naming of cooperative networks. These networks will be presented in the following chapter.

References

3GPP 2010 Further Advancements for E-UTRA Physical Layer Aspects. Technical Report 36.814 v9.0.0, 3rd Generation Partnership Project (3GPP), http://www.3gpp.org/ftp/Specs/2010-03/Rel-9/.

Anghel P and Kaveh M 2006 On the performance of distributed space-time coding systems with one and two non-regenerative relays. *IEEE Trans. Wireless Commun.* **5**(2), 682–692.

Ben Slimane S and Osseiran A 2006 Relay communication and delay diversity for future communication systems. *Proc. VTC 2006 Fall - IEEE 64th Vehicular Technology Conf.*, pp. 1–5.

Chakrabarti A, Baynast AD, Sabharwal A and Aazhang B 2007 Low density parity check codes for the relay channel. *IEEE J. Selected Areas in Commun.* **25**(2), 280–291.

Chandrasekhar, V. Andrews J and Gatherer A 2008 Femtocell networks: a survey. *IEEE Commun. Mag.* **46**(9), 59 –67.

Chen D and Laneman J 2006 Modulation and Demodulation for Cooperative Diversity in Wireless Systems. *IEEE Trans. Wireless Commun.* **5**(7), 1785–1794.

Costa M 1983 Writing on dirty paper. *IEEE Trans. Inform. Theory* **IT-29**(3), 439–441.

Cover T and Gamal AE 1979 Capacity theorems for the relay channel. *IEEE Trans. Inform. Theory* **25**(5), 572–584.

Damen M and Hammons A 2007 Delay-tolerant distributed tast codes for cooperative diversity. *IEEE Trans. Inform. Theory* **53**, 3755–3773.

Dohler M, Lefranc E and Aghvami H 2002 Virtual antenna arrays for future wireless mobile communication systems *ICT 2002*, Beijing, China.

Döttling M, Mohr W and Osseiran A 2009 *Radio Technologies and Concepts for IMT-Advanced*. John Wiley & Sons, Ltd, Chichester.

Etkin R, Tse D and Wang H 2008 Gaussian interference channel capacity to within one bit. *IEEE Trans. Inform. Theory* **54**(12), 5534–5562.

Ezri J and Gastpar M 2006 On the performance of independently designed LDPC codes for the relay channel, *2006 IEEE International Symposium on Information Theory*, pp. 977–981.

Fettweis G, Holfeld J, Kotzsch V, Marsch P, Ohlmer E, Rong Z and Rost P 2010 Field trial results for LTE-Advanced concepts. *Proc. ICASSP 2010 IEEE Int. Conf. Acoust. Speech and Signal Processing*, Dallas, TX.

Figuier L 1868 *Les Merveilles de la science ou description populaire des inventions modernes* vol. 2. Jouvet Furne, Paris.

Hammerström I, Kuhn M and Wittneben A 2004 Cooperative diversity by relay phase rotations in block fading environments, *Proc. SPAWC 2004 - Sig. Proc. Advances in Wireless Commun.*, pp. 293–297.

Han T and Kobayashi K 1981 A new achievable rate region for the interference channel. *IEEE Trans. Inform. Theory* **IT-27**(1), 49–60.

Hu J and Duman T 2006 Low density parity check codes over half-duplex relay channels, *2006 IEEE International Symposium on Information Theory* pp. 972–976.

Hunter T and Nosratinia A 2002 Cooperation diversity through coding, *2002 International Symposium on Information Theory*, p. 220.

Hunter T and Nosratinia A 2003 Performance analysis of coded cooperation diversity, *Proc. ICC 2003 - IEEE Int. Conf. Commun.* vol. 4, pp. 2688–2692.

IEEE 2007 Draft Standard for Local and Metropolitan Area Networks-Part 16: Air Interface for Fixed and Mobile Broadband Wireless Access Systems Multi-hop Relay Specification. Technical Specification P802.16j/D1, Task Group IEEE802.16j.

IEEE 2009 IEEE Standard for Local and metropolitan area networks Part 16: Air Interface for Broadband Wireless Access Systems Amendment 1: Multiple Relay Specification. Standard 802.16j-2009, Task Group IEEE802.16j, http://standards.ieee.org/getieee802/download/802.16j-2009.pdf.

IEEE 2010 IEEE Draft Amendment Standard for Local and Metropolitan Area Networks - Part 16: Air Interface for Broadband Wireless Access Systems – Advanced Air Interface. Standard P802.16m/D8.

Joham M, Utschik W and Nossek J 2005 Linear transmit processing in MIMO communications systems. *IEEE Trans. Signal Processing* (8), 2700–2712.

Knisely D, Yoshizawa T and Favichia F 2009 Standardization of femtocells in 3 GGP. *IEEE Commun. Mag. IEEE* **47**(9), 68–75.

Kramer G, Gastpar M and Gupta P 2005 Cooperative strategies and capacity theorems for relay networks. *IEEE Trans. Inform. Theory* **51**(9), 3037–3063.

Laneman JN 2002 Cooperative Diversity in Wireless Networks: Algorithms and Architectures PhD thesis Massachusetts Institute of Technology Cambridge, MA.

Laneman J, Tse D and Wornell G 2004 Cooperative diversity in wireless networks: Efficient protocols and outage behavior. *IEEE Trans. Inform. Theory* **50**(12), 3062–3080.

Laneman J and Wornell G 2002 Distributed space-time coded protocols for exploiting cooperative diversity in wireless networks, vol. 1, pp. 77–81.

Larsson P 2003 Large-scale cooperative relay network with optimal coherent combining under aggregate relay power constraints *Proc.. Future Telecommunications Conference*, pp. 166–170, Beijing, China.

Marsch P and Fettweis G 2007 A framework for optimizing the uplink performance of distributed antenna systems under a constrained backhaul. *Proc. ICC 2007 - IEEE Int. Conf. Commun.* pp. 975–979, Glasgow, Scotland.

Nahas M, Saadani A and Rekaya G 2010 New delay-tolerant code for distributed antenna *IEEE Personal, Indoor and Mobile Radio Conferencee*, Istanbul, Turkey.

Osseiran A, Logothetis A, Ben Slimane S and Larsson P 2007 Relay cyclic delay diversity: Modeling and system performance. *IEEE International Conference on Signal Processing and Communication (ICSPC07)*, Dubai, UAE.

Pabst R, Walke B, Schultz D, Herhold P, Yanikomeroglu H, Mukherjee S, Viswanathan H, Lott M, Zirwas W, Dohler M, Aghvami H, Falconer D and Fettweis G 2004 Relay-Based Deployment Concepts for Wireless and Mobile Broadband Radio. *Communications Magazine, IEEE* **42**(9), 80–89.

Razaghi P and Yu W 2007 Bilayer low-density parity-check codes for decode-and-forward in relay channels. *Information Theory, IEEE Transactions on* **53**(10), 3723–3739.

Rost P 2009 Opportunities, Benefits, and Constraints of Relaying in Mobile Communication Systems PhD thesis Technische Universitt Dresden Dresden, Germany.

Rost P and Fettweis G 2007 Space-time trellis coding exploiting superimposed transmissions in half-duplex relay networks. *VTC 2007 Fall - IEEE 66th Vehicular Technology Conf.*, Baltimore, USA.

Rost P and Fettweis G. In press. Protocols and performance limits for half-duplex relay networks submitted to IEEE Transactions on Communications.

Rost P, Fettweis G and Laneman J 2009 Opportunities, constraints, and benefits of relaying in the presence of interference, *Proc. ICC 2009 - IEEE Int. Conf. Commun.*, Dresden, Germany.

Rost P, Fettweis G and Laneman J. In press. Energy and cost efficient mobile communication using multi-cell MIMO and relaying submitted to IEEE Transactions on Wireless Communications.

Scaglione A and Hong Y 2003 Opportunistic large arrays: cooperative transmission in wireless multihop ad hoc networks to reach far distances. *IEEE Trans. SP '03* **51**(8), 2082–2092.

Somekh O, Simeone O, Poor V and Shamai S 2010 Cellular systems with non-regenerative relaying and cooperative base stations. *IEEE Trans. Wireless Commun.* **9**(8), 2654-2663.

Stefanov A and E Erkip 2004 Cooperative Coding for Wireless Networks. *IEEE Trans. Commun.* **52**(9), 1470–1476.

Wei S, Goeckel D and Valenti M 2004 Asynchronous cooperative diversity *Conference on Information Sciences and Systems*.

WINNER+ 2010 Celtic Project CP5-026 *Final Innovation Report* (ed. Svensson T. and Zinovief E.). Public Deliverable D1.9, Wireless World Initiative New Radio – WINNER+, April 2010, http://projects.celtic-initiative.org/winner+/index.html (accessed June 2011).

WINNER-II 2006 IST-4-027756 *Test Scenarios and Calibration Issue 2* (ed. Döttling M.). Public Deliverable D6.13.7, Wireless World Initiative New Radio – WINNER, 2006, http://projects.celtic-initiative.org/winner+/index.html (accessed June 2011).

Wu M, Weitkemper P, Wubben D and Kammeyer KD 2010 Comparison of distributed LDPC coding schemes for decode-and-forward relay channels. *International ITG Workshop on Smart Antennas (WSA)*, Bremen, Germany, pp. 127–134.

Xie LL and Kumar P 2004 A network information theory for wireless communication: Scaling laws and optimal operation. *IEEE Trans. Inform. Theory* **50**(5), 748–767.

Xie LL and Kumar P 2005 An achievable rate for the multiple-level relay channel. *IEEE Trans. Inform. Theory* **51**(4), 1348–1358.

Zhao B and Valenti M 2003 Distributed turbo coded diversity for relay channel. *Electronics Letters* **39**(10), 786–787.

8

Network Coding in Wireless Communications

Afif Osseiran, Ming Xiao and Slimane Ben Slimane

This chapter deals with Network Coding (NC) in general and with its application to wireless communications in particular. A short overview of Network Coding in wired and wireless communications is given in section 8.1. Section 8.2, NC describes methods in uplink wireless communications. In section 8.3 the advantage of nonbinary network codes over binary network codes is shown in a multiuser cooperative communication and multirelay scenarios. The application of Network Coding (NC) to broadcast scenarios is then described. Finally, NC trends in wireless communications are given in section 8.5.

8.1 An Overview of Network Coding

In a classical network, data streams originating from a source and intended to a desired destination are routed through intermediate nodes before reaching their final destination. Those intermediate nodes, also called routers, simply route and/or replicate each data stream within the network.

By contract, NC manipulates those data streams at an intermediate node by combining the data from the streams before forwarding it to the destination. NC should not be confused with any kind of source or channel coding in a network executed over a single data stream at an intermediate node.

In order to illustrate the idea of NC, the butterfly example is shown in Figure 8.1 where a communication network is represented by a directed graph. In Figure 8.1, S is a source node transmitting the data streams a and b, both destinated to D1 and D2. The data streams can be routed through R1, R2 and R3 in order to reach the desired destination. In a classical network (shown in Figure 8.1(a)) the node R3 is the bottleneck since by receiving simultaneously a and b, it can only transmit one stream at a time. On the other hand, as shown in Figure 8.1(b), NC will alleviate such a drawback by transmitting $a + b$, the combined (i.e. added) data streams. Consequently, both destinations get a and b at the same time after proper decoding operations. It should be noted that the addition operation, is usually conducted in the finite Galois Field GF(p), where p is a positive integer. In particular, the binary

Mobile and Wireless Communications for IMT-Advanced and Beyond, First Edition.
Edited by Afif Osseiran, Jose F. Monserrat and Werner Mohr.
© 2011 Afif Osseiran, Jose F. Monserrat and Werner Mohr. Published 2011 by John Wiley & Sons, Ltd.

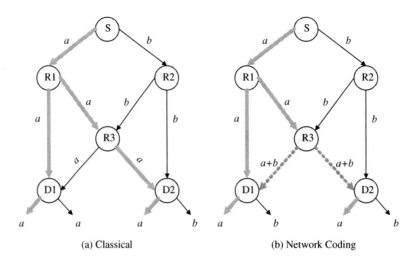

(a) Classical   (b) Network Coding

Figure 8.1   Butterfly network

field[1], GF(2) is the most commonly used field, where the addition operation, XOR, is an exclusive OR operation. Hence, the term XOR is widely used to express the binary addition operation.

8.1.1   Historical Background

The concept of Network Coding was first introduced for satellite communication networks in Yeung and Zhang (1999) and then fully developed in Ahlswede et al. (2000) where initially the term NC was used. Thereafter, the electrical and computer engineering communities took huge interest in it. In particular a plethora of work appeared on the data communication and information theory aspects of NC. For instance, several tutorials on the theory of NC are worth mentioning: Fragouli and Soljanin (2007b), Ho and Lun (2008), Yeung (2008) and Yeung et al. (2005). The application of NC in the engineering field is treated in Fragouli and Soljanin (2007a).

   A beginner trying to acquire basic knowledge in the field of NC will instantly face a large number of widely used terms and keywords within the NC field. Hence, a basic definition and explanation of the most relevant keywords are given in the following:

Multicast means to transmit information from a source node to a specific set of destination/receiver nodes.
Unicast is the special case of multicast with one destination node.
Capacity[a] is the maximum rate at which the source can multicast to all destinations.
Min-cut bound, also called the max-flow min-cut bound, is the upper bound which gives the limit on the amount of rates that can be multicast in the network.

---

[1] A finite field is an algebraic structure, composed of a set of finite elements. Further, the four arithmetic operations, while satisfying a well-defined axiom, can be performed on the elements of the set. Those operations are addition, subtraction, multiplication and division.

[a] The term capacity in the NC literature is mostly used in the sense of the capacity of a graph as is the case in graph theory.

In the literature NC is mainly treated from a graph theory perspective. One of the areas that has attracted a lot of attention is the design of codes that achieve the min-cut bound. Moreover, it was shown that linear NC achieves the min-cut bound capacity (Yeung 2008).

8.1.2   Types of Network Coding

Network Coding can be divided into several categories depending on the network topology, characteristics and the whereabouts (i.e. communication layer) of the NC operation. In particular, NC can be divided into the following families:

Single- or multiple source: indicating if there is a single or multiple source at the origin.
Cyclic or Acyclic: referring to whether there is a directed cycle or not, between the source and the destination.
Analog or Digital: indicating whether the NC operation is conducted on the signal or finite field at the NC node.

From the above classification, it can be seen that each family is composed of two exclusive elements (i.e. a NC can be either cyclic or acyclic). Hence a NC can be identified by three attributes where each attribute corresponds to one element of a family. For example, coded bidirectional relaying (Larsson et al. 2006) with two steps falls under NC having multisource, acylic and digital attributes.

It should be noted that analog NC is also called Physical Network Coding or superposition coding. The digital NC can be executed at any layer (Link, Application etc.).

In the acyclic case all the nodes in the network simultaneously receive all their inputs and produce their outputs (implying a memoryless channel). In the cyclic case there is delay by construction and has to be considered, hence it can be modeled as convolutional codes (Yeung 2008).

8.1.3   Applications of Network Coding

The notion of coding at the packet level, today known as Network Coding, has attracted a lot of attention after the work of Ahlswede et al. (2000). At first NC appeared useful for multicast in wireline communication but soon its utility increased. In fact, because of the broadcast nature of wireless channels, NC turned out to be a very promising scheme for wireless communication applications. Network Coding has the potential to improve energy efficiency, enhance link performance, and improve system throughput in various wireless networks. This is very pleasing as it is clear to network researchers, engineers, and businesses that wireless in its various forms will be the dominant medium of communication in the future.

The most relevant domains in which the NC idea is applied are sketched in Figure 8.2. This section mainly summarizes the applications of Network Coding and its potential in both wired and wireless networks.

Security for Network Coding

A big concern about the use of NC is protection against malicious User Equipments (UEs). Unlike regular transmission, where each block can be digitally signed, in NC each intermediate node produces new network coded blocks. This makes the system vulnerable and it can be subject to a pollution attack. For instance, a malicious node may inject a junk packet into the network. If this junk

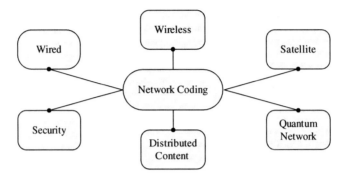

Figure 8.2   Applications of network coding

packet is mixed into the buffer of a node, the buffer will be polluted and the output of the node will become junk. This pollution may easily propagate to the entire network through coding.

In order to solve this pollution problem several counter-measures have been proposed in the literature. This includes homomorphic hash functions (Gkantsidis and Rodriguez 2006), a homomorphic signature scheme (Charles et al. 2006), and a secure random checksums scheme (Gkantsidis et al. 2006).

Application of Network Coding in the Internet

The primary example of the application of NC is in the Internet, both at the network layer (routers in ISPs) and at the application layer (dedicated infrastructure such as content distribution networks and ad hoc networks such as peer-to-peer networks). For NC to be practical, random coding could be used to allow the encoding, among the nodes, to proceed in a distributed manner. Packets need to be tagged to allow the decoding to proceed in a distributed manner (i.e. the nodes decode independently). Buffering is also needed to allow for asynchronous packet arrivals and departures with arbitrary varying rates, delay and loss.

Avalanche is a widely known application that uses NC (Yeung 2007). In a peer-to-peer content distribution network, a server usually splits a large file into a number of blocks. Peer nodes try to retrieve the original file by downloading blocks from the server but also by distributing downloaded blocks among them. To this end, peers maintain connections to a random number of neighboring peers, with which they exchange blocks. In Avalanche, the blocks sent out by the server are random linear combinations of all original blocks. Similarly, peers send out random linear combinations of all the blocks available to them. A node can either determine how many innovative blocks it can transmit to a neighbor by comparing its own and the neighbors matrix of decoding coefficients, or it can simply transmit coded blocks until the neighbor receives the first non-innovative block. The node then stops transmitting to this neighbor until it receives further innovative blocks from other nodes. Coding coefficients are transmitted together with the blocks with little overhead as long as the packets size is large enough.

The application of NC can reduce the file download time because a coded block uploaded by a peer contains information about every block possessed by that peer. A Network Coding-based solution is much more robust in case the server leaves early, as it is more likely that the active peers have all the information necessary for recovering the whole file.

Finally, practical NC appears to be very suitable for video multicast in lossy packet networks (Ramasubramonian and Woods 2009). Compared to traditional routing methods, NC provides significant improvement in the video quality at the receivers. It increases the effective bandwidth seen by the

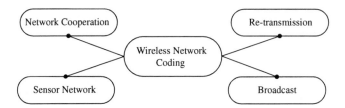

Figure 8.3   Applications of wireless network coding

receivers and reduces the effective loss rate seen by the receivers. NC has shown good performance in other applications such as live media broadcasting, gaming, instant messaging, and distributed storage.

Applications of Network Coding in Wireless Networks

NC has gained a lot of interest within the wireless communications research community (Katti et al. 2006; Li et al. 2003; Popovski and Yomo 2006a,b; Wu et al. 2004). The most relevant areas within wireless networks where NC is applied are shown in Figure 8.3. Moreover, in the following an overview of those promising areas is given.

Unicast Communication: Let us consider the case when a Base Station (BS) wants to transmit a file to a UE within the cell. Let us assume that the BS has N packets to deliver to the UE. With an Automatic Repeat-reQuest (ARQ) protocol of today's wireless networks, each packet needs to be transmitted until an acknowledgement is successfully received at the BS. With an average of d transmissions per packet, $d \times N$ transmissions will be needed to transfer the whole file, which, depending on the reliability of the radio channel, can be quite large. Network Coding, however, provides the ideal solution in this case, with very low complexity. For instance, the BS can transmit random linear combinations of the packets instead of individual packets. In particular, the BS transmits packets of the form $p'_i = \sum_{j=1}^{N} c_{ij} p_j$, where p_j is the jth packet in the file, and c_{ij}s are random coefficients that the BS picks. The UE just waits to correctly receive N such coded packets in order to recover the original packets in the file, by just solving a set of linear equations:

$$\begin{pmatrix} p_1 \\ p_2 \\ \vdots \\ p_N \end{pmatrix} = \begin{pmatrix} c_{11} & c_{12} & \cdots & c_{1N} \\ c_{21} & c_{22} & \cdots & c_{2N} \\ \vdots & \vdots & \ddots & \vdots \\ c_{N1} & c_{N2} & \cdots & c_{NN} \end{pmatrix}^{-1} \begin{pmatrix} p'_1 \\ p'_2 \\ \vdots \\ p'_N \end{pmatrix} \qquad (8.1)$$

Once the UE has decoded the packets, it immediately sends the BS one acknowledgement for the whole file. It is clear that this process will transfer the file much faster than the conventional case and reduces the transfer delay considerably (Fragouli et al. 2007). In fact, the above problem can be considered as a coupon-collector problem. Hence, a gain in the order of $\log_e (N)$ can be achieved (Fragouli and Soljanin 2007a, Chapter 2).

Multicast Wireless Communication: Let us consider the case when a BS wants to multicast a certain file to a group of two UEs within the cell. Say the BS broadcasts the packets p_1, p_2 to the UEs. Since the behavior of the radio channel is different for the two UEs, U_1 and U_2, U_1 may receive only p_1 and U_2 receives only p_2. In this case, the BS has to retransmit both packets to allow both UEs to recover their missing packets. But, instead the BS can broadcast the XOR-ed version on the wireless channel. This single transmission allows both destinations to recover their corresponding

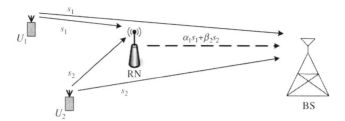

Figure 8.4 Two-user one-relay network

losses, providing efficient reliability. This scheme increases the system throughput with a throughput gain that increases with the number of destinations (UEs) (Chachulski et al. 2007). An opportunistic protocol that exploits this intraflow NC has been developed in Chachulski et al. (2007).

Network Coded Cooperation in Wireless Networks: The form of NC is well suited for cooperation in wireless networks. When such a union (between NC and cooperative networks) occurs, those wireless networks are called *Diversity Network Codes (DNC)*. The cooperation can be obtained via a fixed Relay Node (RN) within the cell, where two or more UEs can cooperate when communicating with the BS. There can also be cooperation between UEs, where each UE forwards the information of the other UE by applying NC. There have been a few pioneer work in that topic, such as Han et al. (2009) and Xiao et al. (2007) consider a scheme for two-user cooperation, in which each UE transmits the binary sum of its own source message and partner messages, resulting in spectrally efficient transmission.

Consider a simple network model where there are two UEs transmitting on the uplink to the BS (see Figure 8.4). Here, the cooperative transmission progresses in two phases. In the first phase, each UE transmits its own data on orthogonal channels while the RN receives and decodes the data of both UEs. In the second phase the RN assists both UEs by Network Coding, by transmitting the XOR-ed version of information from both UEs in the second phase. If any two of the three transmissions succeed, the BS can still recover information data from both. This network coded cooperation scheme provides a diversity order of two for each UE (Chen et al. 2006). If a time division access scheme is used to orthogonalize channels, a total of three time slots are needed for two UEs–see Figure 8.5(b). As can be deduced from Figure 8.5(a), a gain of one time slot is obtained compared to

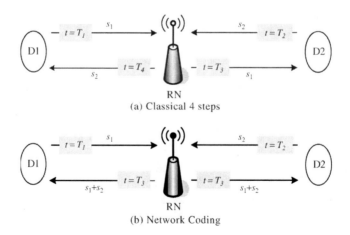

(a) Classical 4 steps

(b) Network Coding

Figure 8.5 Bidirectional relaying

a conventional relay-based cooperation. This scheme can also be generalized to the case of a cellular network with N UEs and M RNs communicating with the BS. In this scheme, each UE still transmits its own data in the first phase on orthogonal channels. In the second phase, a single RN is selected from the M candidates that maximizes the worst instantaneous channel conditions of links from UEs to the RN and from the RN to the BS. The selected RN broadcasts the XOR-ed version of data received from each UE to the BS (Peng et al. 2008). With time-division, a total of $N + 1$ time slots are needed with N time slots in the first phase and one time slot in the second phase to broadcast the XOR-ed message for all the N UEs. Intuitively, the multiplexing gain of the system can be improved by increasing N. However, the diversity is limited to 2. This limitation in diversity gain is due to the fact that, although the coded message can be potentially be helpful for any UE, it can only at most help a UE provided that all the other UEs' data are decoded correctly, no matter what the number of RNs is. Some of these methods that apply NC to cooperative wireless networks will be presented in sections 8.2 and 8.3.

Network-coded cooperation can also occur between UEs and can provide the same diversity-multiplexing tradeoffs. In that case, it takes $2N$ time slots to complete one round of cooperation for N UEs. In the second phase, each UE, acting as a RN now, takes turns to select (randomly) a subset of messages it correctly overheard, and transmits the binary combinations of them. From a coding perspective, the network coded cooperation can be viewed as Low-Density Parity-Check (LDPC) codes applied on the fly. However, this scheme requires a more complex decoding structure and introduces extra overhead to let the BS know how the combinations are formed. In general, NC provides flexibility to cellular networks and can adapt to the lossy nature of wireless channels. The above scenarios are some examples on how to combine NC and cooperative relaying and illustrate some of the benefits of NC in wireless networks. However, more research activities are still needed to take advantage of the full potential of NC and to make it more practical with acceptable complexity.

Physical Network Coding is a smart physical layer technique that transforms the superposition of the electromagnetic (EM) waves as an equivalent NC operation that mixes the radio signals in the air. A form of NC is then created at the physical layer, and works on the EM waves in the air rather than on the digital bits of data packets or channel codes (Zhang et al. 2006). For instance, for the two-way relay channel where two UEs exchange information via one RN, only two time slots are needed for both UE to exchange information (see Figure 8.6), as opposed to three using digital NC (see Figure 8.5(b)) and four using direct transmission (see Figure 8.5(a)). In the first time slot each UE concurrently transmits to the RN. Assuming symbol-level, carrier phase synchronization and channel pre-equalization, the RN does not decode the symbols of the individual UEs. Instead the RN tries to decode and transmit the superposition of the symbols. Physical Network Coding (PNC) has the potential to improve the system throughput performance greatly but this comes at the expense of a loss in diversity gain and strict synchronization rules. In order to relax the synchronization requirements, an Amplify-and-Forward (AF) scheme, in which the RN directly amplifies and forwards the interfered signal to both UEs, was proposed and studied in several works (Ding et al. 2009; Hao et al. 2007; Katti et al. 2007; Popovski and Yomo 2007). Amplify-and-Forward PNC was found to be

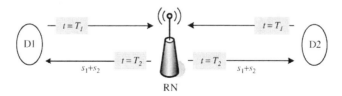

Figure 8.6   Physical network coding

more robust and offers good performance when synchronization is absent. However it suffers from noise amplification even with perfect synchronization.

8.2   Uplink Network Coding

Multiple Access Relay Channel (MARC) (Kramer and van Wijngaarden 2000) refers to a specific topology where a common RN assists several transmitters in relaying their data to a common destination.

In the last few years, implementations of the Multiple Access Relay Channel (MARC) based on NC, also known as cooperative NC, were investigated in the literature. For such a scenario, which is shown in Figure 8.4, and as mentioned before, a total of three transmissions instead of four are needed. In fact four transmissions are needed in a classical Multiple Access Relay Channel (MARC) in order to relay the information of two UEs to the destination. Among other gains, such as the reduction in the transmit power at the RN, this decrease in the number of transmissions can lead to an asymptotic capacity gain of $4/3$ (Hausl et al. 2005; Xiao and Aulin 2009; Yang and Koetter 2007).

8.2.1   Detection Strategies

The detection of the different signals in the NC-based MARC is performed at the destination. This did not receive much attention in the literature. A common (implicit) assumption is that Joint Detection (JD) of the three signals (i.e. the two signals of the UEs and the network coded signal transmitted by the RN) is used at the destination (Yang and Koetter 2007). Joint detection can be implemented using, for example maximum-likelihood decoding (Xiao and Aulin 2009) or iterative decoding (Hausl et al. 2005) among the codewords of different channels. The major drawback of such methods is the cost in terms of high complexity when higher order modulations are used. To circumvent such high complexity, two low-complexity detection schemes, Selection and Soft Combining (SSC) and Majority Vote Detection (MVD), were proposed in Manssour et al. (in press). In the following, JD along with SSC and MVD are described.

Joint Detection

Joint Detection is the optimal and straightforward approach for signal detection in the NC-based MARC. The destination has estimates of three signals (r_1, r_2 and r_x) with different information content and the detector will simply try all possible output alternatives for the different signals before choosing the alternative with the smallest error metric. In JD, the receiver uses jointly the three received samples (r_1, r_2 and r_x) in order to decide on the transmitted symbols (s_1 and s_2).

Selection and Soft Combining

Selection and Soft Combining for NC-based MARC consists of detecting one of the UEs based on its direct transmission, while the other UE is detected based on a combination of its direct transmission with the network-decoded RN transmission.

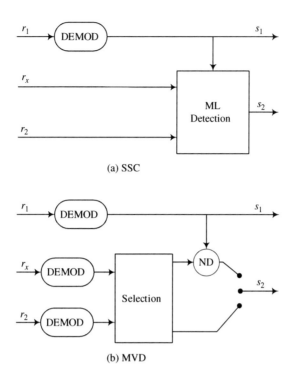

(a) SSC

(b) MVD

Figure 8.7 Selection and soft combining, and majority vote detection strategies for uplink MARC

SSC can be divided into two steps. The first step involves determining the strong UE. This strong UE is then detected solely based on its direct transmission. In the second step the receiver exploits the knowledge of the strong symbol $\hat{s}_1$ by performing a low-complexity JD of the signal transmitted of U_2. An illustration of SSC is shown in Figure 8.7(a), where it is assumed without loss of generality that U_1 is the strong UE.

Unlike the case of JD, the complexity does not increase exponentially as the different symbols are detected separately: the strong UE inherently benefits from selection diversity by being the strong UE, whereas the weak UE benefits from transmit diversity by combining its direct transmission with the network-decoded RN transmission.

Majority Vote Detection

Majority Vote Detection for NC-based MARC consists of using the two strongest links (having the highest Signal-to-Noise Ratios (SNRs)) at the receiver out of the received three, hence the nomenclature Majority Vote Detection.

The detection in this scheme works as follows, the first stage (similar to SSC scheme) in MVD is to determine the strong link. The strong link is detected solely based on its received sample. At the second stage, the receiver uses the second strong link (called the middle link) to obtain an estimate of the second symbol. An illustration of MVD is shown in Figure 8.7(b).

8.2.2   User Grouping

A common assumption in most NC works is that there are two UEs in the system, on which NC is applied. In reality, in a wireless network system there is a set of active UEs in a cell that can be conveniently paired together. Hence, the first obstacle to be faced at a network level is which set of UEs should be selected and grouped to perform the NC operation on. Obviously, a random selection will not yield the optimal system capacity. In fact if UEs are randomly paired then we could end up pairing UEs with *noncomplementary* channel conditions to the RN and BS and consequently losing the advantage provided by NC. Therefore, if both of the paired UEs have a bad channel towards the BS, one of them will be decoded with a low SNR. Similarly, if both UEs have a good channel towards the BS, the capacity would decrease as compared to a direct transmission due to the time division among the UEs and the RN.

> Consequently, grouping UEs with complementary characteristics in a multicell, multiuser scenario is essential in order to ensure a good performance of the NC scheme.

To illustrate how different sets of UEs can possibly be formed, consider the case where a total of four UEs, $U_1, \ldots, U_4$, have data to transmit to the BS and can cooperate via one RN. For simplicity, let us assume that only two UEs are allowed to cooperate at a time, then three sets can be formed

$$S_1 = \{G_1; G_2\} = \{(U_1, U_2); (U_3, U_4)\}$$
$$S_2 = \{G_3; G_4\} = \{(U_1, U_3); (U_2, U_4)\}$$
$$S_3 = \{G_5; G_6\} = \{(U_1, U_4); (U_2, U_3)\}$$

Each set contains two groups of two UEs each; for instance the group G_1 contains the pair U_1 and U_2. As an illustration, the sets S_1 and S_2 are shown in Figure 8.8(a) and 8.8(b), respectively.

In the following, the User Grouping (UG) method proposed in Manssour et al. (2008) is described. The UG algorithm determines which group of pairs should be network coded together so that the designed cost function of the considered set of UEs is optimized. Assume M active UEs in a cell. Let S_l be the set $l = 1, \ldots, L$ where L is the total number of sets. The sets size is by design equal to the number of active UEs that is, $|S_l| = M \ \forall l$. Let $S = \bigcup_{l=1}^{L}(S_l)$ be the union set containing all possible groups of pairs of UEs. For simplicity, assume M is an even number. Hence, it can be easily shown that $|S|$, the number of possible group of pairs, is given by:

$$|S| = (M - 1) \times (M - 3) \times \ldots \times 3 \times 1 \tag{8.2}$$

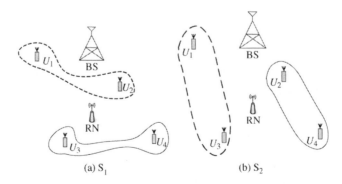

(a) S_1 (b) S_2

Figure 8.8   An example of user grouping for two sets S_1 and S_2, assuming 4 active UE

Let $J(S_l)$ denote the cost function to be optimized. Then the optimal UG should satisfy:

$$S^* = \arg\max_{S_l \in S} J(S_l) \qquad (8.3)$$

where S^* is the UG satisfying the optimization problem. The cost function can be, for instance, based on maximizing the sum-capacity C of a set of active UEs at a certain time. Assuming all transmitting nodes have equal access to the channel and a SSC receiver, the resulting sum capacity for a scheduled pair (U_i, U_j), in unit [bps/Hz], is given by:

$$C = \frac{1}{3}[\log_2(1 + \min(\Gamma_i, \Gamma_{r,i})) + \log_2(1 + \min(\Gamma_j + \Gamma_x, \Gamma_{r,j}))] \qquad (8.4)$$

where Γ_i (resp. Γ_j) is the received Signal to Interference plus Noise Ratio (SINR) for U_i (resp. U_j) at the BS after T_1 (resp. T_2), $\Gamma_{r,i}$ (resp. $\Gamma_{r,j}$) is the SINR for U_i (resp. U_j) at the RN after T_1 (resp. T_2), and Γ_x is the signal received at the BS after T_3 where the RN transmits the network-coded signal. It should be noted that the min operation reflects successful network encoding at the RN. Finally note that the UG method can be easily generalized for groups of UEs greater than two (Osseiran et al. 2011).

8.2.3   Relay Selection

One of the most critical issues in relaying systems is where to deploy the RNs (Timus 2009) and, moreover, which of these RNs should be selected for transmission. In fact, the use of RN can adversely affect the performance as opposed to not using any RNs in the case where the RN selection is done randomly or inappropriately. When applying RN selection in conjunction with NC, the problem becomes even more challenging. This is because the data of both sources to be encoded together should be available at the NC node.

Generic Relay Selection

In the literature, the RN selection algorithms can be generically divided into two categories: opportunistic RN selection and multiple RN selection.

In *opportunistic relay selection*, only one RN of a set of RNs will be selected to assist the transmitting node and different criteria are used to solve the selection problem. The RNs are useful even when they do not actively transmit, as long as they adhere to the opportunistic cooperative behavior and give priority to the best available RN (Bletsas et al. 2006).

In *multiple relay selection* schemes, multiple RNs will be chosen to assist the transmitting node. Different schemes exist for selecting these multiple RNs (Lo et al. 2008). For example, Laneman and Wornell (2003) considers the case where all RNs that can decode the source's transmission, participate in the communication using the distributed space-time codes. Other algorithms promote the use of a threshold to decide on the number of selected RNs (Ban et al. 2007).

Relay Selection for Network Coding

When NC is introduced, the Relay Selection (ReS) scheme should be aware of this operation and select the best RN accordingly. Otherwise, if the ReS choice is simply based on an individual UE without any consideration of the other UE in the pair, this other UE could be detrimentally affected. In fact, the ReS algorithm should be aware of the NC operation and take into account both sources to be network coded when selecting the RN. In Manssour et al. (2009a) an opportunistic ReS for a

NC-based MARC scenario was proposed. Specifically, the proposed Relay Selection considers the distinct cases where NC is performed on the RN that maximizes:

- the capacity of the weak source;
- the capacity of the strong source;
- the sum-capacity of the network coded pair.

This algorithm consists of a simple exhaustive search to find the RN n out of a total of N_R active RNs such that the achievable capacity is maximized. Mathematically, the ReS problem can be stated as:

$$n^* = \arg\max_{n \in \{1,...,N_R\}} C(RN_n) \qquad (8.5)$$

8.2.4 Performance

In Figure 8.9, the performance of the three **detection strategies** is compared by plotting the Bit Error Rate (BER) as a function of SNR assuming that both UEs have equal average SNR (i.e. $\Gamma_1 = \Gamma_2$). The average RN SNR Γ_x is 20 dB . The trend of the BER curves is similar for all detection schemes, with JD slightly outperforming the other two schemes (i.e. SSC and MVD). Similar performance results have been observed while using different values of Γ_x and for higher order modulations (e.g. 8 Phase-Shift Keying (PSK)), confirming that SSC and MVD have a very similar performance to JD (Manssour et al. in press).

Concerning **User Grouping**, in Figure 8.10, the CDF of the capacity of random NC is compared with NC with UG with different set size (sets of four and six, respectively). Simulation results evidence that the bigger the set size the better the performance is achieved by UG. For instance, the normalized mean capacity increases from 1.27 bps/Hz for random Network Coding, to 1.52 bps/Hz for UG of a set of four and to 1.70 bps/Hz for a set of six. Consequently, mean capacity gains of 34% and 16% can be achieved by the application of UG on a search window of six and four UEs

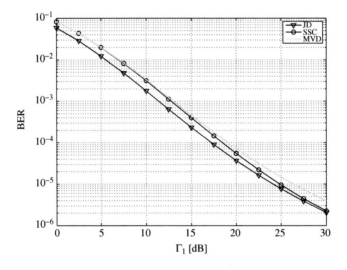

Figure 8.9 Bit Error Rate versus the user SNRs for a RN of $\Gamma_x = 20$ dB

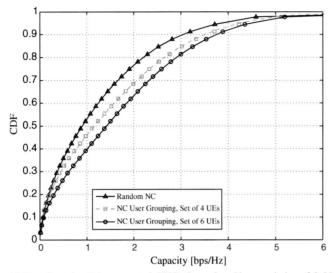

(a) User Grouping (Manssour et al. 2008). Reproduced by permission of © 2008 IEEE

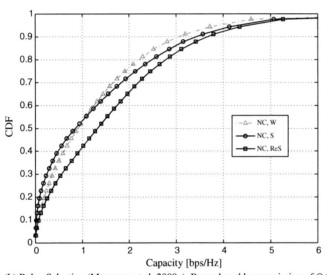

(b) Relay Selection (Manssour et al. 2009a). Reproduced by permission of © 2009 IEEE

Figure 8.10   CDF of the cell capacity for UG and RN selection methods in uplink MARC

respectively, as compared to random NC. Increasing the search window size (i.e. larger sets of UEs) further increases the capacity gains at the expense of higher complexity as the number of possible pairings increases.

Concerning **ReS**, in Figure 8.10(b), the performance of NC in the presence of opportunistic ReS is evaluated. Simulation results shows that substantial capacity gains can be obtained due to NC awareness (denoted NC, ReS in the legend of Figure 8.10(b)). For the simulated scenario, up to 48% gain in the median capacity was achieved as opposed to algorithms not aware of the NC operation. Note that the cases denoted (NC, W) and (NC, S) in the legend designate the capacity when the ReS is based on the weak and strong UE, respectively.

Furthermore, this awareness leads to an increased fairness among UEs. In particular, NC awareness achieves max-min fairness for the majority of the percentiles. It should be noted that applying a joint UG and ReS algorithm will results in up to 75% capacity gain as observed in Manssour et al. (2009a). Moreover, further capacity improvement can be achieved by taking into account an appropriate time division scheme (Manssour et al. 2009b).

8.2.5 *Integration in IMT-Advanced and Beyond*

The integration of MARC topology in Long Term Evolution (LTE) Release 10 (Rel-10) and beyond requires simply the RNs and donor BS to be aware of the NC operation. As a consequence, the proposed detection strategies can be implemented in LTE without any modification. The impact of UG and RN selection for uplink RN NC on signaling depends on the assumed cost function to select the UEs and RNs. At best, the Medium Access Control (MAC) layer (scheduler at the BS) shall be moderately modified since the BS needs to instruct the RN on how to apply the NC operation. Additional signaling (from the RNs to the BSs) will be required in case the quality of the link between of the RNs and the UEs is needed at the BSs.

8.3 Nonbinary Network Coding

This section discusses the design of NC for cooperative and relaying networks with the objective to increase the wireless diversity. In general, binary NC can not exploit networks containing inherently a diversity order higher than 2. In addition the XOR operation may not be optimal, in the sense of asymptotic performance for certain network settings. By contrast, nonbinary codes, exploit such networks and further improve their performance. Mathematically speaking, in nonbinary NC scenarios, the combining operation of the data streams at the NC nodes is based on the finite field GF(n), where n is an integer greater than 2.

Cooperative communication (Hunter et al. 2006; Sendonaris et al. 2003) is an efficient method to combat wireless fading. In cooperative communications, two or more UEs send messages to a common BS in order to help each other convey the desired information to the destination, as shown in Figure 8.11 (assuming $\alpha_1 = \beta_2 = 0$ and $\alpha_2 = \beta_1 = 1$), in a classic two-user cooperative scenario, where the UE, U_1, communicates with the BS. The partner UE, U_2, also receives the message because of the broadcasting property of the wireless medium. Then, U_2 can try to decode it, and, if the decoding is successful, forwards the message of U_1 to the BS (using Decode-and-Forward (DF)). A similar scheme works when U_2 transmits and U_1 relays. Information messages are therefore transmitted to the BS through two independent fading paths: one direct path and one through a RN.

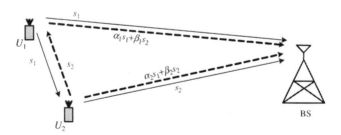

Figure 8.11 Two-user cooperative networks with or without network codes over finite fields

8.3.1   Nonbinary NC based on UE Cooperation

For wireless cooperative networks, carefully designed network codes over finite fields can offer solid performance improvement over previous schemes based on binary codes or without NC. The schematic of nonbinary NC for two-user cooperative communications is illustrated in Figure 8.11, in which network codes are on top of channel coding, to encode relayed and local messages. The NC scheme is time-invariant at each RN (i.e. deterministic codes). In the first time slot, the two source nodes (U_1 and U_2) use proper channel coding to transmit their own messages s_1 and s_2 (systematic blocks) respectively in different orthogonal channels. In the second time slot, if both RNs successfully decode the channel codes, the transmitted messages for U_1 and U_2 are encoded using NC as $\alpha_2 s_1 + \beta_2 s_2$ and $\alpha_1 s_1 + \beta_1 s_2$, respectively. Here, it is assumed that the NC combining operation, is done in the finite field GF(4), constructed based on a minimal polynomial[2]. Further, block-fading channels are assumed, that is, the channels stay constant over the transmission of a codeword at the physical layer but vary independently between successively transmitted blocks[3]. Hence, the variation is independent and identically distributed (i.i.d.). Clearly, in total, the BS receives codewords carrying four different messages: s_1, s_2, $\alpha_1 s_1 + \beta_1 s_2$ and $\alpha_2 s_1 + \beta_2 s_2$, where each message has experienced independent fading. Any two of these four blocks can rebuild the two source blocks s_1 and s_2. If a RN cannot decode correctly, it repeats its own message using the same channel code. Then the BS performs Maximum Ratio Combining (MRC) of these codewords and decodes. Here, perfect error detection is used, by, for example, Cyclic Redundancy Check (CRC). Thus, for all channels, the outputs from channel decoders are either dropped or can be considered error free. Now, the outage probability of the above cooperative system, is shown as a form of a theorem accompanied with the proof in Appendix D.

> **Theorem 8.1.** *Assuming any link (between a transmitter and receiver) experiences Rayleigh fading with an outage probability of P_e, then the outage probability for two cooperating UEs and a BS using nonbinary NC (as shown in Figure 8.11) is given by $3.5 P_e^3$. Hence, the diversity order d is equal to 3.*

The proof of Theorem 8.1 is given in Appendix D.1.

An important observation is that *it is necessary that the UE messages can be reconstructed from any two out of four network codewords to achieve diversity order* 3. Note that successive codewords of the same channels have independent fading. The above nonbinary network codes, are designed to exploit the diversity of independent block fading and cooperation among UEs, that is, time and space diversity are exploited. This cannot be fully accomplished by schemes based on binary NC (i.e. $(\alpha_1, \beta_1) = (\alpha_2, \beta_2) = (1, 1)$). For instance, if the second block of each UE is $s_1 \oplus s_2$ over GF(2), then the received blocks at the BS are s_1, s_2, $s_1 \oplus s_2$ and $s_1 \oplus s_2$. For U_1, outage occurs when two blocks cannot be decoded. These two blocks can be s_1 and s_2 (two $s_1 \oplus s_2$ cannot rebuild the source messages), resulting in a diversity order of 2. Thus, the approach based on non-binary network codes can achieve a higher diversity order, and a performance gain for medium-to-high SNRs.

Although the above example network only considers two-user cooperative networks. A similar principle can also be extended to multiuser networks (Xiao and Skoglund 2010).

---

[2] For example for the case $(\alpha_1, \beta_1) = (1, 2)$ and $(\alpha_2, \beta_2) = (1, 1)$, the minimal polynomial is given by $p(X) = X^2 + X + 1$. Hence, the four elements are the polynomials 0, 1, X and $X + 1$. Consequently, integer notation can be used to denote the field elements, that is, 0, 1, 2 and 3, respectively.

[3] In case the channel variation is slow. Then, a frequency hopping scheme can be used for channel switching among UEs to achieve block fading.

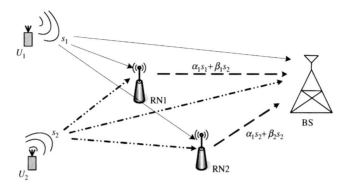

Figure 8.12   A two-user two-relay wireless network

8.3.2   Nonbinary NC for Multiuser and Multirelay

In cooperative networks (Laneman et al. 2004; Tse and Viswanath 2005), the UEs are assumed to cooperate with each other. The assumption relies heavily on the UEs' capabilities. An alternative way would be to use a fixed RN to perform the NC operation in a multi-user multi-relay scenario. To illustrate the idea, a schematic of a two-user, two-relay network is shown in Figure 8.12. In this network the channels among UEs and RNs are orthogonal. In Figure 8.12, if the UEs communicate with the BS, both RNs will receive the corresponding codewords from the UEs. Then, the RNs try to decode. In case the decoding is successful, the RNs forward the information to the BS by applying Network Coding, assuming that the fading coefficients are i.i.d. random variables for different channels but constant for all symbols of one or more codewords of the same channel. Consequently, the corresponding codewords for each UE are transmitted to the BS through three independent fading paths: one direct path and two paths via the two RNs.

Let Global Encoding Kernel (GEK) denote the linear relation between the ingoing and outgoing codewords, that is, $\underline{C} = \underline{G}\,\underline{S}$, where $\underline{C}$ denotes the outgoing codeword at the RN, $\underline{G}$ is the GEK, and $\underline{S} = [s_1, s_2]$ represent the codewords originating at the source. For example, in Figure 8.12, RN1 has the GEK $\underline{G}_2 = [\alpha_2, \beta_2]$, and RN2 the GEK $\underline{G}_1 = [\alpha_1, \beta_1]$, respectively. The vectors $\underline{G}_1$ and $\underline{G}_2$ in the non-binary NC scenario are chosen to be linearly independent in GF(4). If one Source Relay (SR) channel is in outage, the corresponding coding coefficients are equal to zero. Hence, in the case of an outage event on the SR channels, the GEKs are different for the transmitted codewords in the RN.

In case of perfect SR channels, the BS receives four different codewords (i.e. $s_1, s_2, \alpha_1 s_1 + \beta_1 s_2$, and $\alpha_2 s_1 + \beta_2 s_2$). If the RN can only decode one UE's codeword, due to an outage event in one of the SR channels, it then reencodes the information by using the same codeword used at that UE. At the BS, the two codewords of this message are combined via MRC and decoded (i.e. GEKs are $[1, 0]$ or $[0, 1]$). Now, the outage probability for the two-user two-relay scenario, assuming binary NC, is shown as a form of a theorem accompanied with the proof in Appendix D.

Theorem 8.2. *Assuming any link (between a transmitter and receiver) experiences Rayleigh fading with an outage probability of P_e, then the outage probability of any of the UEs (U_1 or U_2), for the two-user two-relay scenario, assuming binary NC (i.e. $(\alpha_1, \beta_1) = (\alpha_2, \beta_2) = (1, 1)$, as shown in Figure 8.12) is given by P_e^2. Hence the diversity order d is equal to 2.*

The proof of Theorem 8.2 is given in Appendix D.2. A similar analysis reveals the outage probability for any of UEs in Figure 8.4 (with only one RN) is $3P_e^2$. Consequently going from one to two RNs (as illustrated in Figures 8.4 and 8.12) improves the UE outage probability from $3P_e^2$ to P_e^2 while providing identical diversity order $d = 2$.

Now, let us derive the outage probability for the two-user two-relay scenario, assuming nonbinary NC.

Theorem 8.3. *The outage probability of any of the UEs (U_1 or U_2) for the two-user, two-relay scenario, assuming nonbinary NC (i.e. $(\alpha_1, \beta_1) \neq (\alpha_2, \beta_2)$, as shown in Figure 8.12) is given by $6P_e^3$. Hence the diversity order d is equal to 3.*

The proof of Theorem 8.3 is given in Appendix D.2. The above scheme can be generalized for an arbitrary number of M UEs and N RNs. The generalized scheme (Xiao et al. in press), is called Maximum Diversity Network Codes (MDNC), and achieves a (lower bound) diversity order of $N + 1$ for all UEs.

8.3.3 Performance

In Figure 8.13, simulation results for cooperative communication without Network Coding, and cooperation based on binary and nonbinary NC (called also DNC) are given. Here regular LDPC codes are used as physical layer channel codes. In Figure 8.13, reciprocal interuser channels are considered. The codes have 200 input bits and 400 output coded bits. Each column of the parity check matrix has three 1s and the other elements are 0s. Binary Phase Shift Keying (BPSK) signals are used for transmission. In the same figure, the outage probabilities are also shown. From the figure, the improvements by using nonbinary NC in cooperative communications are clear. The improvement is pronounced in the medium-to-high SNRs.

In Figure 8.14 simulation results of the two-user two-relay scenario are presented. Further, we compare binary NC and nonbinary NC, in terms of outage probability or FER versus SNR. Channel

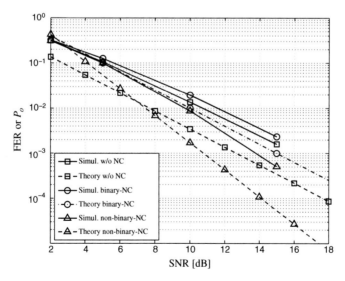

Figure 8.13 FER (i.e. simulation) or outage probabilities (i.e. theory) versus SNR for UE cooperative networks

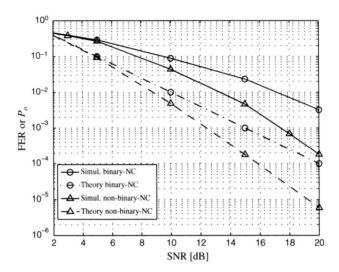

Figure 8.14 FER (i.e. simulation) or outage probabilities (i.e. theory) versus SNR for two-user two-relay networks with network coding

codes, regular $(3, 6)$ LDPC codes of block length of 400 coded bits, are used. Figure 8.14 also shows exact outage probabilities, where there are clear gaps between theoretical outage probabilities and simulation results. The main reason is that outage probabilities were calculated with the assumption of unconstrained Gaussian channel inputs, while in the simulations BPSK signals are assumed. However, as can be seen, the diversity orders are predicted correctly by the analytical results. From the figure, one can also see that analytically obtained outage probabilities and simulations closely match in the high SNR regime. As shown in Figures 8.14 and 8.13, the nonbinary NC yields a gain of 4.2 dB and 2.6 dB, respectively at a FER of 10^{-3} compared with binary NC.

8.3.4   Integration in IMT-Advanced and Beyond

The implementation of NC for UE cooperation, and multiuser multirelay cooperation will impact the signaling and architecture of LTE. In particular:

- the frame structure should be modified as two uplink phases are required;
- additional signaling is needed as the cooperating UEs must listen to each other;
- the MAC layer should be modified to take into account the user cooperation;
- at the RNs, coding in a higher finite field needs extra computational costs;
- at the BS, more complex decoding algorithms is needed.

For the NC based on the UE cooperation method, it is assumed that the UEs are transmitting and receiving at the same time. In a Time Division Duplex (TDD) scenario it will be challenging to implement this due to the dynamic range problem in the transceiver and an extra time slot is needed. In a Frequency Division Duplex (FDD) scenario, it implies stringent and fast frequency swapping (frequency hopping) between the transmitter and receive carrier, and an extra frequency band is needed to relay the information of the partner. For the multiuser multirelay the BS should select the UEs and the RNs conducting the NC operation. Hence, the MAC layer should be modified. Finally this scheme incurs higher costs in terms of the network topology because it assumes two RNs.

8.4   Network Coding for Broadcast and Multicast

Wireless Digital Broadcasting (DB) applications such as multimedia real-time broadcasting are becoming increasingly popular because the digital format allows for quality improvements as compared to the traditional analogue broadcast. Although these applications are currently mainly used through digital TV, more and more applications are available in wireless cellular systems (e.g. LTE Multimedia Broadcast Multicast Service (MBMS)). In a typical DB scenario, a BS broadcasts common information to a set of UEs through wireless channels. In DB systems, error control strategies should be introduced to improve the reliability and QoS, such as delay.

The broadcast is based on packet-level transmission, where packets are subject to channel noise, fading and interference at physical layer. Channel error correction may not be perfect, especially when delay or processing complexity is limited. However, assuming perfect error detection at higher layers, a received packet at a UE is either error-free or discarded as erroneous. Consequently, the higher layer broadcast transmission from the BS to the set of UEs can be modeled as a broadcast packet-erasure (or block-erasure) channel as shown in Figure 8.15, where the i-th UE is designated by U_i.

Block-erasure channel coding has been investigated in Guillén i Fàbregas (2006) and Lapidoth (1994) where Maximum-Distance Separable (MDS) codes are shown to be optimal in terms of error probability. Therefore coding schemes applied across a sequence of broadcast information packets are typically considered for error control. For instance, ARQ (Djandji 1994) is a widely used error control for packet-level transmission. In an ARQ scenario, a retransmission is initiated by any UE with an erased packet. When the number of UEs increases in a broadcast scenario, ARQ becomes increasingly inefficient, both in terms of feedback and retransmissions.

To improve the system efficiency, the use of NC (Ho et al. 2006; Koetter and Médard 2003; Li et al. 2003) during the retransmission phase has been proposed for wireless broadcasting (Larsson et al. 2010; Nguyen et al. 2009; Sundararajan 2009; Xiao et al. 2008). In most of these schemes, packets lost by different UEs are jointly encoded with a suitable network code, leading to a reduction in the total number of transmitted blocks required for retransmission. In (Xiao et al. 2008), an analytical approach is proposed to determine the improvements in efficiency obtained by Network Coding, while a lower bound on the transmission overhead is developed in (Nguyen et al. 2009). A specific NC scheme for two UEs is further proposed in (Nguyen et al. 2009). Furthermore, in (Larsson and Johansson 2006) and (Larsson et al. 2010), NC is combined with ARQ and hybrid ARQ schemes for

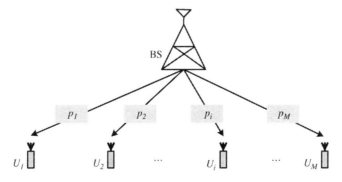

Figure 8.15   A wireless broadcast system with M UEs, and block-erasure channels with p_i

unicast transmission, respectively. The analysis of (Larsson et al. 2010) is mainly from information theory perspective. In addition, the assumed codes are not optimal. Finally, in (Sundararajan 2009), feedback is considered for NC from queuing theory perspective (i.e. traffic engineering). However, the propagation channel impairments were not taken into account.

In this section, a brief introduction of a new NC method for wireless broadcasting is given. In general, the application of NC leads to higher efficiency than traditional ARQ schemes. The benefits of using NC for broadcasting can be seen from the following simple example. In a broadcast session a set of N information blocks s_i, $i = 1, 2, ..., N$ has to be broadcast from a BS to a set of $M \geq 2$ UEs. Since they are at different locations, the M BS-to-UE block-erasure channels are assumed to be independent with block-erasure probabilities p_i, $i = 1, 2, ..., M$, respectively. To facilitate the analysis, the transmission process is divided into two phases: the information transmission phase and the retransmission phase. In the information transmission phase, the BS broadcasts N information blocks, and during the transmission, some blocks are lost over the respective BS-to-UE block-erasure channels. Each UE feeds back a packet with indices of the erased blocks. Feedback is considered as instantaneous and error free. A corresponding *error matrix* **E** is evaluated at the BS to indicate the block-erasure status of the UEs. The dimension of **E** is $M \times N$, where $e_{i,j} = 1$ if the j-th block of U_i is erased; otherwise, $e_{i,j} = 0$.

Assuming $M = 2$ and $N = 6$, one snapshot of **E** can be, for instance:

$$\mathbf{E} = \begin{pmatrix} 1 & 0 & 0 & 1 & 0 & 1 \\ 0 & 1 & 1 & 0 & 0 & 1 \end{pmatrix} \tag{8.6}$$

Clearly, when traditional ARQ is applied, each individual erased block for any UE will be retransmitted separately. Thus, blocks s_1, s_2, s_3, s_4, and s_6 are retransmitted separately, making a total of five retransmitted blocks.

However, if NC is used, only three (encoded) blocks, $s_1 \oplus s_2$, $s_3 \oplus s_4$ and s_6 are needed, assuming that in the information broadcast period, U_1 and U_2 have correctly received s_2, s_3, s_5 and s_1, s_4, s_5, respectively. Then, after correctly receiving the retransmissions packets, both UEs can retrieve the respective erased blocks through simple modulo-2 addition. Clearly, transmission efficiency is improved by using even a simple NC scheme.

8.4.1 Efficient Broadcast Network Coding Scheme

Now we discuss a more efficient coding scheme (Lu et al. 2010) for a more general network. In order to measure the system efficiency, a normalized overhead η can be defined as

$$\eta \triangleq \frac{X}{N} \tag{8.7}$$

Here X denotes the number of blocks sent from the BS until termination of the broadcast session.

Consider n_i blocks are erased on the BS-to-U_i link, $i = 1, 2, ..., M$, during the information transmission phase. Let $\hat{n} = \max_i n_i$ and $\hat{\imath} = \arg\max_i n_i$. It follows that if $\hat{n} > 0$ then a new round of retransmissions is required. Since the UE, $U_{\hat{\imath}}$ must receive at least $\hat{n}$ blocks in order to recover all N information blocks, $\hat{n}$ is a lower bound on the total number of required retransmissions before the termination condition is satisfied. The following NC scheme is partly motivated by this observation.

As part of the coding scheme, the concept of *column groupings* of the error matrix is used. In fact, it is a coding constraint aiming to reduce complexity and delay. The columns of the error matrix are grouped into a minimum number, $k \geq \hat{n} > 0$, of submatrices such that each row in a column grouping has at most one 1. Since each column in the error matrix corresponds to an information block, the column groupings correspond to *encoding sets*, C_ℓ, $\ell = 1, 2, ..., k$, containing the indices of the respective information blocks in a grouping. The information blocks whose indices are in the same encoding set, will be jointly XOR-ed into one single block. The single block will used for retransmission. For example, if $C_1 = \{1, 2\}$, then the encoded block is $s_1 \oplus s_2$. Since each UE has at most one erased information block within an encoding set, the erased blocks can be retrieved easily.

The proposed NC retransmission scheme can be summarized in the following three steps.

1. **Initialization**: Determine $\hat{n}$ and $\hat{\imath}$, and denote the corresponding erased blocks as $\mathbf{S}_{m_1}$, $\mathbf{S}_{m_2}, ..., \mathbf{S}_{m_{\hat{n}}}$. Initialize $k = \hat{n}$ and k coding sets as $C_\ell = \{m_\ell\}$, $\ell = 1, 2, ..., k$.
2. **Index Allocation**: In this step, the erased blocks of the remaining UEs are allocated to the encoding sets C_ℓ, $\ell = 1, 2, ..., k$ if possible. Obviously, if all erased blocks can be allocated to the $\hat{n}$ encoding sets (and encoded into $\hat{n}$ codewords), the lower bound is achieved when the first round of retransmissions is successful for all UEs. The remaining erased blocks are sorted in descending order, according to the number of encoding sets. In the first round, these erased blocks are allocated subject to the column grouping constraints. The remaining erased blocks are allocated to an eligible encoding set. If there are no eligible encoding sets available for a particular block, a new encoding set is generated.
3. **Retransmission**: All blocks assigned to a particular encoding set are jointly network encoded through modulo-2 addition and transmitted.

The encoding constraint enforced by the column groupings rule simplifies the decoding process and minimizes delay with no loss of throughput performance. A UE retrieves an erased block for each received retransmitted block through a simple modulo-2 addition. The operation of the NC scheme is illustrated with the following example. The error matrix is assumed to be

$$
\mathbf{E} = \begin{pmatrix}
1 & 1 & 0 & 0 & 0 & 1 & 0 & 1 & 0 & 0 \\
1 & 1 & 0 & 0 & 1 & 0 & 0 & 1 & 1 & 0 \\
1 & 0 & 1 & 0 & 1 & 0 & 0 & 0 & 1 & 1 \\
1 & 0 & 0 & 1 & 0 & 1 & 0 & 0 & 1 & 1 \\
1 & 0 & 0 & 1 & 0 & 1 & 0 & 0 & 0 & 1
\end{pmatrix}
\tag{8.8}
$$

Using the proposed encoding approach, the encoding sets are $C_1 = \{1\}$, $C_2 = \{2, 3, 4\}$, $C_3 = \{5, 6\}$, $C_4 = \{8, 10\}$ and $C_5 = \{9\}$, leading to five encoded blocks, s_1, $s_2 \oplus s_3 \oplus s_4$, $s_5 \oplus s_6$, $s_8 \oplus s_{10}$ and s_9 for retransmission. Under the assumption that the retransmissions are received correctly at all UEs, the number of retransmissions required is five, whereas six retransmissions are required for the scheme in Xiao et al. (2008), and nine retransmissions are required for traditional ARQ.

The above retransmission method is quite powerful. In a fact, for $M = 2$, the scheme is optimal. For instance, if U_1 loses n_1 packets, and U_2 loses n_2 packets with $n_1 \geq n_2$. One can show that n_1 retransmission packets are sufficient. Further, when $M > 2$, numerical results show Lu et al. (2010) that the number of retransmission packets is quite close to that of the minimum requirement (i.e. $\hat{n}$, the maximum number of lost packets among all UEs).

8.4.2   Performance

The performance of different NC schemes is compared with the performance of traditional ARQ. Specifically, the impact of M on the normalized overhead η is considered. Assume that the links have unequal erasure probabilities. Let p_1, the erasure rate on link 1, be given by $p_1 > p$; and let all other links have identical erasure rate for example, $p_i = p$ for $i = 2, 3, ..., M$.

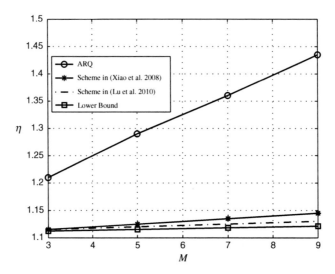

Figure 8.16 The impact of M on the normalized overhead η for divers methods. $N = 100$, $p_1 = 0.1$ and $p_i = 0.05$ for $i \neq 1$ (Lu et al. 2010). Reproduced by permission of © 2010 IEEE

In Figure 8.16, the normalized overhead (given by Equation 8.7) is shown as a function of M. It can be observed that the proposed scheme in Lu et al. (2010) enjoys substantially better performance compared to ARQ. Also, the performance of the proposed scheme in Lu et al. (2010) outperforms that of Xiao et al. (2008). For instance, for $M = 5$ (resp. $M = 9$) the overhead in Lu et al. (2010) is improved by 13% (resp. 22%). Further, the results of Lu et al. (2010) are close to the lower bound performance designated by $\hat{n}$, namely, the maximum number of packet erasures among all UEs.

8.5 Conclusions and Future Directions

The relative performance gains of the most promising Network Coding schemes in Wireless Networks that were covered in this chapter are summarized in Table 8.1. These gains are mostly relative to a conventional two-hop relaying scenario. Where this is not the case, it is mentioned under the comment field of Table 8.1.

Although the publications on NC have been extensive during the last five years, NC has still to reach wireless standards such as LTE-Advanced (LTE-A) and 802.16m. Nonetheless timid proposals have been initiated (Alcatel-Lucent 2009). Moreover, the idea of NC has been applied to areas not treated within this chapter such as Multiple-Input Multiple-Output (MIMO) (Manssour et al. 2010), the coordinated multipoint scenario (Du et al. in press) and more recently to retransmissions.

Table 8.1 Performance gains of Network Coding Methods relative a conventional relaying system in Wireless Networks

NC Method	Gain	Metric	Comments
User Grouping and Relay Selection	70%	Throughput	Relative to Random NC
Nonbinary codes	up to 4.2 dB	SNR	at 10^{-3} FER
Efficient Broadcast NC	22%	Overhead	Relative to ARQ
Coded Bidirectional	100%	Throughput	see (Tarokh 2009, Ch. 11)

The research community continues to have a high interest in the application of NC to wireless communications. The plethora of publications on that topics has not solved all the related issues. Hence there are many opportunities which are still unexplored. More specifically, some areas with a high potential for future research are:

- jointly optimizing source coding, channel coding and NC;
- optimal use of NC for retransmission under a unicast scenario;
- application of NC to a multicell context, for instance in a Coordinated Multipoint transmission or reception (CoMP) scenario;
- investigation of whether the NC operation should be applied at the signal or bit or symbol level;
- designing resource allocation in NC-based wireless networks.

Finally NC can be applied to emerging areas such as cognitive radio and power/energy efficiency.

References

Ahlswede R, Cai N, Li SY and Yeung R 2000 Network information flow. *IEEE Trans. Information Theory* **IT-46**(4), 1204–1216.

Alcatel-Lucent 2009 Applications of network coding in LTE-A. Technical Report R1-090774, 3GPP Third Generation Partnership Project, Working Group RAN1, meeting 56, Athens, Greece, http://ftp.3gpp.org/.

Ban TW, Jung BC, Sung DK and Choi W 2007 Performance analysis of two relay selection schemes for cooperative diversity *Personal, Indoor and Mobile Radio Communications, 2007. PIMRC 2007. IEEE 18th International Symposium on*, pp. 1–5.

Bletsas A, Shin H, Win M and Lippman A 2006 Cooperative diversity with opportunistic relaying *Wireless Communications and Networking Conference, 2006. WCNC 2006. IEEE*, vol. 2, pp. 1034–1039.

Chachulski S, Jennings M, Katti S and Katabi D 2007 Trading structure for randomness in wireless opportunistic routing *In Proc. of ACM SIGCOMM 2007*.

Charles D, Jain K and Lauter K 2006 Signatures for network coding *Information Sciences and Systems, 2006 40th Annual Conference on*, pp. 857–863.

Chen Y, Kishore S and Li JT 2006 Wireless diversity through network coding. *IEEE WCNC Proceedings*.

Ding Z, Leung K, Goeckel D and Towsley D 2009 On the study of network coding with diversity. *IEEE Trans. Wireless Commun.* **8**(3), 1247–1259.

Djandji H 1994 An efficient hybrid ARQ protocol for point-to-multipoint communication and its throughput performance *IEEE Trans. Infor. Theory* **40**, 1459–1473.

Du J, Xiao M and Skoglund M. In press. Cooperative Network Coding Strategies for Wireless Relay Networks with Backhaul. *IEEE Trans. Commun.*

Fragouli C and Soljanin E 2007a Network coding applications. *Foundations and Trends in Networking* **2**(2), 135–269.

Fragouli C and Soljanin E 2007b Network coding fundamentals. *Foundations and Trends in Networking* **2**(1), 1–133.

Fragouli C, Katabi D, Markopoulou A, Medard M and Rahul H 2007 Wireless network coding: opportunities and challenges *Military Communications Conference, 2007. MILCOM 2007. IEEE*, pp. 1–8.

Gkantsidis C and Rodriguez P 2006 Cooperative security for network coding file distribution *INFOCOM 2006. 25th IEEE International Conference on Computer Communications. Proceedings*, pp. 1–13.

Gkantsidis C, Miller J and Rodriguez P 2006 Anatomy of a p2p content distribution system with network coding *IPTPS 06*, pp. 27–28.

Guillén i Fàbregas A 2006 Coding in the block-erasure channel. *IEEE Trans. Inform. Theor.* **52**, 5116–5121.

Han Z, Zhang X and Poor VH 2009 High performance cooperative transmission protocols based on multiuser detection and network coding. *IEEE Trans. Wireless Commun.* **8**(5), 2352–2361.

Hao Y, Goeckel D, Ding Z, Towsley D and Leung KK 2007 Achievable rates for network coding on the exchange channel *Military Communications Conference, 2007. MILCOM 2007. IEEE*, pp. 1–7.

Hausl C, Schreckenbach F, Oikonomidis I and Bauch G 2005 Iterative network and channel decoding on a tanner graph *Proc. IEEE 43rd Allerton*.

Ho T and Lun D 2008 *Network Coding: An Introduction*. Cambridge University Press, New York.

Ho T, Médard M, Koetter R, Karger D, Effros M, Shi J and Leong B 2006 A random linear network coding approach to multicast. *IEEE Trans. Inform. Theor.* **52**(10), 4413–4430.

Hunter TE, Sanayei S and Nosratinia A 2006 Outage analysis of coded cooperation. *IEEE Trans. Inform. Theor.* **52**(2), 375–391.

Katti S, Gollakota S and Katabi D 2007 Embracing wireless interference: analog network coding *SIGCOMM '07: Proceedings of the 2007 Conference on Applications, Technologies, Architectures, and Protocols for Computer Communications*, pp. 397–408. ACM, New York, NY, USA.

Katti S, Kattabi D, Hu W, Rahul HS and Médard M 2006 The importance of being opportunistic: Practical network coding for wireless environments. *SIGCOMM'06*.

Koetter R and Médard M 2003 An algebraic approach to network coding. *IEEE/ACM Trans. Networking* pp. 782–795.

Kramer G and van Wijngaarden AJ 2000 On the white gaussian multiple-access relay channel *Proc. 2000 IEEE Int. Symp. Inform. Theory*, p. 40, Sorrento, Italy.

Laneman J and Wornell G 2003 Distributed space-time-coded protocols for exploiting cooperative diversity in wireless networks. *Information Theory, IEEE Transactions on* 49(10), 2415–2425.

Laneman JN, Tse D and Wornell GW 2004 Cooperative diversity in wireless networks: Efficient protocols and outage behavior. *IEEE Trans. Inform. Theor.* 50, 3062–3080.

Lapidoth A 1994 The performance of convolutional codes on the block erasure channel using various finite interleaving techniques. *IEEE Trans. Inform. Theor.* 48, 1688–1698.

Larsson P and Johansson N 2006 Multi-User ARQ *Vehicular Technology Conference, 2006. VTC 2006-Spring. IEEE 63rd*, vol. 4, pp. 2052–2057.

Larsson P, Johansson N and Sunell KE 2006 Coded bi-directional relaying *Vehicular Technology Conference, 2006. VTC 2006-Spring. IEEE 63rd*, vol. 2, pp. 851–855.

Larsson P, Smida B, Koike-Akino T and Tarokh V 2010 Analysis of network coded HARQ for multiple Unicast Flows *Communications (ICC), 2010 IEEE International Conference on*, pp. 1–6.

Li SY, Yeung RW and Cai N 2003 Linear network coding. *IEEE Trans. Information Theory* **IT-49**(2), 371–381.

Lo C, Vishwanath S and Heath R 2008 Relay subset selection in wireless networks using partial decode-and-forward transmission *Vehicular Technology Conference, 2008. VTC Spring 2008. IEEE*, pp. 2395–2399.

Lu L, Xiao M, Rasmussen L, Skoglund M, Wu G and Li S 2010 Efficient network coding for wireless broadcasting *Proc. IEEE WCNC*.

Manssour J, Ahsin T, Osseiran A and Ben Slimane S. In press. Detection strategies for cooperative network coding: Analysis and performance. *Physical Communication, Elsevier*.

Manssour J, Osseiran A and Ben Slimane S 2008 Wireless network coding in multi-cell networks: Analysis and performance *Signal Processing and Communication Systems, 2008. ICSPCS 2008. 2nd International Conference on*, pp. 1–6.

Manssour J, Osseiran A and Ben Slimane S 2009a Opportunistic relay selection for wireless network coding *Communications (MICC), 2009 IEEE 9th Malaysia International Conference on*, pp. 102–106.

Manssour J, Osseiran A and Ben Slimane S 2009b Time allocation in wireless network coding *12th Symposium on Wireless Personal Multimedia Communications (WPMC 09)*, Sendai, Japan.

Manssour J, Osseiran A and Ben Slimane S 2010 High-rate redundant space-time coding. *Journal of Electrical and Computer Engineering* **2010**(324138), 1247–1259.

Nguyen D, Tran T, Nguyen T and Bose B 2009 Wireless broadcasting using network coding. *IEEE Trans. Veh. Technol.* **58**, 782–786.

Osseiran A, Xiao M, Ben Slimane S, Skoglund M and Manssour J 2011 Advances in Wireless Network Coding for IMT-Advanced & Beyond *2nd International Conference on Wireless Communications, Vehicular Technology, Information Theory and Aerospace and Electronic Systems(WIRELESS VITAE 2011)*, Chennai, India.

Peng C, Zhang Q, Zhao M, Yao Y and Jia W 2008 On the performance analysis of network-coded cooperation in wireless networks. *IEEE Trans. Wireless Commun.* **7**(8), 3090–3097.

Popovski P and Yomo H 2006a The anti-packets can increase the achievable throughput of a wireless multi-hop network *Proc. IEEE International Conference on Communication (ICC2006)*, Istanbul, Turkey.

Popovski P and Yomo H 2006b Bi-directional amplification of throughput in a wireless multi-hop network *IEEE 63rd Vehicular Technology Conference (VTC)*, Melbourne, Australia.

Popovski P and Yomo H 2007 Physical network coding in two-way wireless relay channels. *IEEE International Conference on Communications (ICC2007)* Glasgow, Scotland, pp. 707–712.

Ramasubramonian A and Woods J 2009 Video multicast using network coding *SPIE VCIP*, pp. 1–8.

Sendonaris A, Erkip E and Aazhang B 2003 User cooperation diversity. part i and part ii. *IEEE Trans. Commun.* **51**(11), 1927–1948.

Sundararajan JK 2009 *On the role of feedback in network coding* PhD thesis Massachusetts Institute of Technology, http://dspace.mit.edu/handle/1721.1/54230.

Tarokh V 2009 *New Directions in Wireless Communications Research*. Springer, Heidelberg.

Timus B 2009 Studies on the Viability of Cellular Multihop Networks with Fixed Relays PhD thesis Royal Institute of Technology (KTH) Stockholm, Sweden.

Tse D and Viswanath P 2005 Fundamentals of Wireless Communication. Cambridge University Press.

Wu Y, Chou PA and Kung SY 2004 Information exchange in wireless networks with network coding and physical-layer broadcast *39th Annual Conference on Information Sciences and Systems (CISS)*, Baltimore, MD.

Xiao L, Fuja T, Kliewer J and Costello D 2007 A network coding approach to cooperative diversity. *IEEE Trans. Inform. Theor.* **53**(10), 3714–3722.

Xiao M and Aulin T 2009 Optimal decoding and performance analysis of a noisy channel network with network coding. *IEEE Trans. Commun.* **57**, 1402–1412.

Xiao M and Skoglund M 2010 Multiple-user cooperative communications based on linear network coding. *IEEE Trans. Commun.* **58**(12), 3345–3351.

Xiao M, Kliewer J and Skoglund M. In press. Design of network codes for multiple-user multiple-relay wireless networks. *IEEE Trans. on Wireless Commun.*

Xiao X, Lu-Ming Y, Wei-Ping W and Shuai Z 2008 A wireless broadcasting retransmission approach based on network coding *Circuits and Systems for Communications, 2008. ICCSC 2008. 4th IEEE International Conference on*, pp. 782–786.

Yang S and Koetter R 2007 Network coding over a noisy relay: a belief propagation approach. *IEEE ISIT* **Nice, France**, pp. 801–804.

Yeung R and Zhang Z 1999 Distributed source coding for satellite communications. *IEE Trans. Inform. Theor.* **45**(4), 1111–1120.

Yeung RW 2007 Avalanche: A network coding analysis. *Commun. in Inform. Systems J.* **7**(4), 353–358.

Yeung RW 2008 *Information Theory and Network Coding*. Springer Publishing Company, Heidelberg.

Yeung RW, Li SYR, Cai N and Zhang Z 2005 Network coding theory. *Commun. Inf. Theory* **2**(4), 241–381.

Zhang S, Liew S and Lam P 2006 Hot topic: physical-layer network coding *MobiCom '06: Proceedings of the 12th Annual international Conference on Mobile Computing and Networking*, pp. 358–365.

9

Device-to-Device Communication

Klaus Doppler, Cássio B. Ribeiro and Pekka Jänis

9.1 Introduction

One aspect of the design of IMT-Advanced systems that has not received sufficient attention so far is the emergence of high data-rate local services. Such local services can provide the high data rates needed to consume rich multimedia services through mobile computers such as tablets, laptops, netbooks and smart phones.

As an example, one may consider the case where a media server is put up at a music concert from which visitors can download promotional material using a Device-to-Device (D2D) connection. Currently, only Wireless Local Area Network (WLAN) or Bluetooth operating on a license-exempt band could be used to setup a direct connection to the media server. On the other hand, only a licensed band can guarantee a controlled interference environment, which increases the reliability of the data transfers. Hence local service providers might prefer to pay a small amount of money to gain access to licensed spectrum when the license-exempt bands become crowded. Cellular operators may offer such cheap access to spectrum with controlled interference enabled by D2D communication as underlay to the cellular network. The licensed spectrum may be used as the only resource for communication or it may be complemented by license-exempt spectrum.

Figure 9.1 illustrates a D2D network underlaying a cellular network. The communication is facilitated by a cellular Base Station (BS), which allows the media server and the UE to communicate directly while keeping some control over the link to limit the interference to the cellular receiver. The D2D operation itself can be transparent to the user who simply enters a Uniform Resource Locator (URL), the network would detect traffic to the media server and hand it over to a D2D connection. The user does not have to configure a WLAN Access Point (AP) or to perform Bluetooth pairing, which can be tedious, especially if a secure connection is required.

Compared to other local connectivity solutions based on, for example, Bluetooth or WLAN, the D2D communication supported by a cellular network offers additional compelling advantages. First the network can advertise local services available within the current cell. Hence, for automated service discovery, the UEs do not have to constantly scan for available WLAN APs or Bluetooth devices. This is especially advantageous when considering that the constant scanning of Bluetooth devices or WLAN APs is often switched off by users to reduce the power consumption. Secondly, the cellular

Mobile and Wireless Communications for IMT-Advanced and Beyond, First Edition.
Edited by Afif Osseiran, Jose F. Monserrat and Werner Mohr.
© 2011 Afif Osseiran, Jose F. Monserrat and Werner Mohr. Published 2011 by John Wiley & Sons, Ltd.

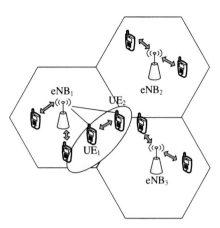

Figure 9.1 Device-to-Device (D2D) communication underlaying the cellular network with eNB maintaining some control over the UE in D2D communication. The shaded area marks potential interference to the cellular communication

network can distribute encryption keys to both D2D UEs so that a secure connection can be established without manual pairing of UEs or entering encryption keys.

Typically, users want to share a file with another user. Instead of selecting a nondescriptive Bluetooth device name or a WLAN Service Set Identifier the cellular network can link a device identifier to the user identity. Instead of a device identifier the user simply selects the other user to send a file, which greatly improves the user experience of local sharing.

Section 9.2 presents the state of the art on D2D communication. In section 9.3 we introduce the technology enablers for D2D communication as an underlay network to the cellular network. We introduce interference management mechanism for underlay D2D communication that allows for secondary usage of the spectrum with limited disturbance to the cellular network. Further, we introduce the core network functionalities for transparent D2D setup and a mode-selection algorithm to ensure reliable D2D communication with high data rates. Finally, we give future research directions in section 9.4.

D2D communication is also referred to as Machine-to-Machine (M2M) communication, direct communication, Peer-to-Peer communication or ad hoc communication.

9.2   State of the Art

9.2.1   In Standards

Several wireless standards have addressed the need for D2D operation in the same band as the BS, AP or central controller. Examples of such standards are WLAN networks based on IEEE 802.11 standards, Terrestrial Trunked Radio (TETRA) and HIPERLAN. In all these examples D2D communication is assumed to occur on dedicated resources, which limits the spectral efficiency of the combined cellular and D2D communication. Using the simple scenario illustrated in Figure 9.1 we explain how these standards implement the D2D communication.

IEEE 802.11

The ad-hoc/Independent Basic Service Set (IBSS) mode of WLANad-hoc communication can be used for setting up direct communication between UEs. It is based on relatively old standards

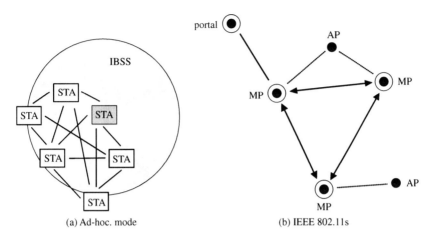

Figure 9.2 Direct communication modes in WLAN. STA is UE in WLAN terminology. Mesh points (MP) can be both in a mesh network and associated to an AP

(1999/2003). Hence, it does not support enhancements such as Enhanced Distributed Channel Access (EDCA) but it works for D2D communication with a limited number of UE and coverage area. The coverage area is defined by the UE that transmits the beacon, which can be located at the edge of the ad hoc network (see Figure 9.2(a)).

Carrier Sense Multiple Access With Collision Avoidance (CSMA/CA) is used to coordinate the different transmissions. A UE senses the medium and if it is free, it transmits. In this operation mode the AP (eNB in WLAN terminology) does not have a direct possibility to control the D2D links. On the other hand, in infrastructure mode the AP is involved in setting up the D2D connections and it can limit the number of active D2D connections. IEEE 802.11z defines a Tunneled Direct Link Setup (TDLS) through the AP. The AP facilitates both the link setup and the key handshake to setup a secure connection. By not forwarding the direct link setup messages the AP can perform an "admission control" for D2D links. The basic Direct Link Setup (DLS) operation is illustrated as follows:

- A station sends a <DLS request> to the AP.
- The AP forwards the <DLS request> to the targeted station, if the stream is allowed according to policies and the targeted station is associated to the AP.
- The targeted station accepts the connection and send a <DLS response>.
- The AP forward the <DLS response> after which the direct link becomes active.

Recently, IEEE 802.11s has defined mesh points that can communicate directly to other mesh points. At the same time they can be associated with an AP as illustrated in Figure 9.2(b). This can be seen also as co-existence of D2D communication and AP-based communication. In contrast to the ad hoc mode, IEEE 802.11s supports advanced power save and Quality of Service (QoS) mechanisms. Furthermore, all nodes in the mesh network transmit a beacon, and the coverage area of the mesh network can be much larger than in the ad hoc mode.

Hiperlan 2

In Hiperlan 2 (ETSI 2002) UE1 would send a resource request over a slot, for example several Orthogonal Frequency Division Multiplexing (OFDM) symbols for direct communication with UE2 to the central controller (eNB in HIPERLAN terminology). After receiving a resource grant, UE1 transmits to UE2 in the granted slots within the direct link phase in the Medium Access Control (MAC) frame. If UE2 wants to transmit to UE1 it has to reserve slots as well, except in acknowledged mode, where the central controller reserves also slots for the acknowledgements of the other UE. Further, it is possible for an UE to request a fixed slot allocation, that is, selected slots are allocated to a UE for multiple frames. The main drawbacks are that the allocation is always for a single UE and the central controller as well as other UE in direct mode cannot transmit at the same time. This does not make efficient use of the available radio resources. Each UE has to reserve slots for each and every transmission, which results in high signaling load. Using fixed slot allocation sets a hard limit to the number of direct links in the subnet. Furthermore, only full OFDM symbols can be reserved, which can be inefficient in a system with large amount of subcarriers.

TETRA

TETRA is a set of standards (ETSI 2007) developed by the European Telecommunications Standards Institute (ETSI) that describes a mobile radio communication system targeted primarily at the needs of public emergency services (police, fire service, ambulance), government agencies, transport services, military, etc. The first TETRA standard was published in 1995. Contrary to legacy trunking services, TETRA provides a D2D communication mode between UEs being outside the coverage of trunking infrastructure. This scenario is called Direct Mode Operation (DMO) in TETRA standards (ETSI 2006a,b).

There are four basic operational schemes within the TETRA DMO:

- Back-to-back: used for standard UE to UE communication, when UEs are out of coverage of Trunked Mode Operation (TMO) or trunking infrastructure is down, overloaded or inaccessible. It also offers covert operations, which cannot be monitored.
- DMO Repeater: extends the coverage of DMO with special UEs called repeaters and can be used when direct communication between UEs is impossible (due to long distance for instance).
- DMO Gateway: provides interconnection between DMO and TMO, which extends the range of TMO and possibly reduces costs of infrastructure (especially during the early stage of network rollout by operators).
- Dual watch: periodic scanning of the other mode (DMO or TMO), which means that UEs communicating in one mode monitor signalling of the other mode in the background.

Several frequency channels are reserved purely for D2D communication. However, a fixed allocation of channels to D2D communication reduces the resources available for the eNB-UE communication.

9.2.2 In Literature

Underlay communication has been studied extensively in the research field of cognitive radio (Haykin 2005; Mitola and Maguire 1999). Cognitive radio systems aim to utilize the "white spaces" in the spectrum locally where the primary service is not present. The primary service in our case is the cellular communication. In order to protect the primary service, secondary systems without cooperation with the primary system need to operate very conservatively and avoid frequency channels and bands where they detect activity of the primary system. Typically, they would not detect white spaces on a band where a cellular system operates even if the network load is low.

A network architecture supporting UE cooperation and D2D transmission among small clusters of UEs in a cellular network was proposed in Chen and Katz (2009) and Fitzek et al. (2009). Hsieh and Iinatti (2005) discuss the network architecture for D2D augmented content distribution in a cellular network. The benefits from coordinating the use of D2D and cellular links is shown in Hsieh and Sivakumar (2004), where the focus is on mobile Internet usage with D2D relaying and no local traffic.

Handing short-range links from cellular mode to direct mode communication was proposed in Adachi and Nakagawa (1998) in order to reduce the battery consumption of UEs. In Adachi and Nakagawa (2000) the capacity of a hybrid system incorporating spread spectrum modulation and both direct and cellular mode communications was studied.

While the work presented in this chapter was being carried out, other research groups were studying D2D communication as an underlay to the cellular Uplink (UL). In Kaufman and Aazhang (2008) a similar power control mechanism to that presented in section 9.3.2 was proposed. The power control mechanism limits the D2D interference to the cellular communication. In Huang et al. (2009) the reuse of cellular UL resources has been compared to using dedicated resources and the tradeoff between link density and outage probability is analyzed by means of stochastic geometry. Omiyi and Haas (2004) propose an interference coordinating MAC protocol for a system that permits cell-edge ad-hoc Time Division Duplex (TDD) communication underlying the otherwise underutilized Frequency Division Duplex (FDD) UL band. Popova et al. (2007) also propose a D2D underlay for cellular FDD UL for more efficient content distribution. In the latter, the interference to cellular UL is limited by choosing very low transmit power for the D2D transmitters.

One problem in designing a D2D concept is mode selection between cellular mode and D2D mode in either dedicated or shared cellular resources. The work in Jänis et al. (2009c) evaluates the potential capacity gains from enabling D2D communication when the communication mode is optimally selected between available options. The results of Hakola et al. (2010) indicate that a simple mode selection scheme is sufficient if the D2D link distance is small, in the order of one-hundredth of cell radius. Otherwise, interference awareness in mode selection is crucial. A similar study is carried out in Koskela et al. (2010), where D2D communication within small device clusters is considered. The potential gain from interference-aware mode selection has also been pointed out in Yu et al. (2009), where the Signal to Interference plus Noise Ratio (SINR) distributions have been derived for Downlink (DL) and UL resource sharing as a function of the D2D pair location in the cell.

A system where dedicated time slots are reserved for D2D communication is investigated in Bennis et al. (2008). A clustering-based scheduling algorithm is proposed for maximizing the amount of simultaneous D2D links per timeslot and cell. Simulations in a wide area network indicate that significant capacity gains are available by enabling direct D2D communication.

9.3 Device-to-Device Communication as Underlay to Cellular Networks

In this section we introduce a D2D communication mode to LTE-Advanced (LTE-A), and describe the session setup and management. Further, we introduce interference coordination and D2D mode selection as enabler for D2D underlay communication.

The D2D underlay communication shares the resources with the cellular network. In peer-to-peer communication between devices in very close proximity, the operator network need not be involved in the actual data transport except for signaling of the session setup, charging, and policy enforcement. Otherwise, interference coordination is needed to ensure that the interference to the cellular receiver is limited and, together with the mode selection, a reasonable D2D throughput is achieved.

As illustrated in section 9.3.1, the eNB maintains the radio resource control for a UE in D2D communication and assigns resources to the D2D communication. It can assign free (dedicated) cellular UL and/or DL resources to the D2D communication or UL and DL resources that are reserved for cellular use (reuse). The interference coordination adapts to the type of resources allocated.

9.3.1 Session Setup

Long Term Evolution (LTE) systems with the System Architecture Evolution (SAE) (Holma and Toskala 2009) operate fully in the packet-switched domain using Internet protocols. The session setup happens in the user plane using the Session Initiation Protocol (SIP). The SAE architecture includes the Mobile Management Entity (MME) and the Packet Data Network (PDN) gateway, which together take care of the UE context, setting up the SAE bearers, IP tunnels, and IP connectivity between the UE and the serving PDN gateway. SAE provides connectivity to the Internet, where a SIP Application Server (AS) is found by a discovery procedure or operator assignment. The session setup can be initiated by a SIP invite message to the SIP AS. After successful session setup, any two or more devices (UE, or UE and servers) may communicate over the Internet.

In the SAE architecture, the node in the network that is aware of wide-area (global) IP addresses is the gateway (serving PDN gateway). The gateway keeps a routing table that enables IP routing from/to the Internet. The gateway is able to route IP packets to the proper eNB serving the active destination UE. The gateway is able to detect potential D2D traffic since it actually processes the IP headers of the data packets and it knows by which eNB the UE is served. Potential D2D traffic is any flow between UE that are served by the same or neighboring cells. The gateway earmarks packets of a potential D2D traffic flow as depicted in Figure 9.3 (Doppler et al. 2009b).

As a next step, the eNB, serving the flow earmarked as D2D flow, requests the UE to make measurements to check if the D2D devices are in communication range and selects the D2D communication mode (discussed in section 9.3.4). If direct communication is selected, the *eNB sets up a D2D radio bearer directly between the two UEs so that they communicate using D2D communication resources.* In cases where the UEs are in neighboring cells, the eNBs serving the UEs have to coordinate D2D measurements and the D2D bearer setup over the X2 interface (Holma and Toskala 2009).

Even if the D2D connection setup is successful, the eNB still maintains the SAE bearer between the UE and the gateway for cellular communications. Furthermore, the eNB maintains the radio resource control for both cellular and D2D communication and the UE can continue to communicate with the Internet. The D2D connection setup, by detecting IP traffic between nearby devices, is transparent to the user, that is, the user does not have to initiate a D2D session explicitly. The automatic switching between cellular and D2D connection should be reliable and seamless to guarantee user satisfaction.

Alternatively, explicit D2D session setup signaling can be used (Doppler et al. 2009b). A specific address format can separate a D2D SIP session request from a generic SIP session request. The new address format of D2D SIP could simply be username@realm.D2D_keyword, where the former part (username@realm) is a well-known SIP Uniform Resource Indicator (URI), and the latter part (D2D_keyword) is a novel extension to let the SAE handle the local D2D session in a special way. In this approach, the application (or the UE) at the requesting UE needs to decide whether to prefer initiation of a D2D session or a regular session. Figure 9.4 depicts the additional functionalities that are required for D2D communications in the current LTE architecture. *In SAE, the MME negotiates with the Serving/PDN gateways to get IP addresses to the UE. The MME thus acts as a binder between the IP addresses, subscription information and SAE network identification. All this justifies the delivery of a D2D session initiation request (like SIP invite) to the MME.* In this case UE1 calls UE2 by SIP invite message using the .D2D_keyword extension with the URI of UE2. The SIP invite message is

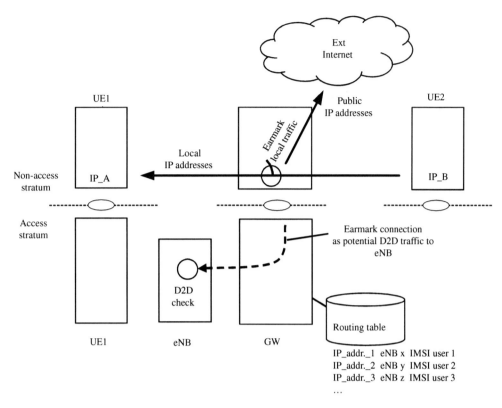

Figure 9.3 The gateway earmarks local traffic to indicate potential D2D traffic to the eNB. The eNB can then check if the devices corresponding to these packets can set up a D2D connection (Doppler et al. 2009b). Reproduced by permission of © 2009 IEEE

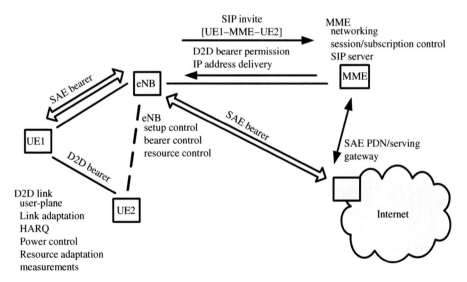

Figure 9.4 D2D communications in the SAE architecture. Added functionality regarding D2D communications is indicated for UE, eNB and MME. SAE bearers are shown for reference (Doppler et al. 2009a). Reproduce by permission of © 2009 IEEE

encapsulated in a Non-Access Stratum (NAS) control-plane message. In the SAE architecture, the MME receives all NAS messages from the served UE and it is responsible for the mobility management, that is, it knows by which eNB a UE is served. The MME can be enhanced by a light SIP handler to keep track of the SIP addresses of the UEs inside the tracking area. This can be done at the initial access, when the UE registers to the network and the MME assigns the Temporary Mobile Subscriber Identity (TMSI) to the UE that is used for all control of the UE in the cellular network.

After handling the SIP message and detecting that the D2D UEs are in the same or neighboring cells, the MME will request a setup of a D2D radio bearer from the serving eNB (or serving eNBs for D2D communication happening in an area of multiple cells). Again the eNB requests measurements from the UEs and will select the D2D mode based on these measurements (discussed in section 9.3.4). If direct communication is selected, the MME could then initiate an IP address delivery for the D2D terminating UEs. IP addresses for D2D communications could be created with a local subnet scope, similar to a local breakout solution. The IP-like connectivity over the D2D link offers seamless operation to the higher layer protocol stack (TCP/IP and UDP/IP) in the UE and it eases the mobility procedures between cellular and D2D networking. The D2D communication can follow the LTE operation principles with some adaptations, for example on the definition of identifiers.

The UE in D2D communication can stay in the connected state and receive data during the D2D transmission. Therefore, *the D2D transmission needs to have mechanisms to coordinate its operation with the cellular communication. The Discontinuous Reception (DRx) mechanism in LTE can be used for that purpose.* During DRx off periods D2D communication will take place. This might require that the eNB align the DRx cycles of the D2D UEs. Secondly, during DRx on periods, a UE involved in a D2D link can monitor the allocations of the other UE and, if there is no allocation, continue with D2D communication. If a UE wants to initiate a cellular communication it can request a transmission gap to announce its absence from the D2D communication for a certain time period.

9.3.2   D2D Transmit Power

The most straightforward way of reducing the D2D interference toward cellular receivers is to limit the D2D transmit power. This is a natural step toward better performance under the assumption of short-distance D2D links in comparison to cellular links. When the D2D link is sharing cellular UL resources the D2D power setting is aided by predicting the resulting interference at the eNB. Such interference-awareness is harder to accomplish on cellular DL resources.

Transmit Power when Sharing Cellular Uplink Resources

During cellular UL transmission, the eNB is the victim receiver of interference from D2D transmitters. Since the UE in D2D connection is still controlled by the serving eNB, it can limit the maximum transmit power of the D2D transmitters. In particular, it can use the cellular power control information for the devices involved in D2D communications.

The transmit power of the D2D transmitter is reduced by a backoff value from the transmit power determined by the cellular power control when the D2D transmitter reuses cellular UL resources. No backoff is required when the D2D transmitter uses dedicated (otherwise free) resources. The eNB can additionally apply power boosting for the UL transmission of a cellular UE to ensure that the UL SINR of the cellular UE meets the target SINR. The boosting is dependent on the backoff value, as described in Jänis et al. (2009c).

Assume that the cellular UL and D2D transmit power are P_1 and P_2, respectively. The UL channel gain is c_1 and the interference channel gain from the D2D transmitter to the eNB is c_2. Thus the UL cellular UE signal to (intra-cell D2D) interference plus noise ratio, SINR_{UL}, is

$$\text{SINR}_{\text{UL}} = \frac{P_1 c_1}{P_2 c_2 + \sigma^2} \qquad (9.1)$$

where σ^2 is the noise level. Assume that power control has a nominal target received power P (and thus target Signal to Noise Ratio (SNR) of P/σ^2). On a UL resource where there is D2D transmission the target power levels of the cellular transmission and D2D interference are modified as follows. The D2D transmission is backed off by a factor β and the UL cellular UE transmission is boosted by a factor α, such that

$$P_2 c_2 = P/\beta$$
$$P_1 c_1 = \alpha P$$

Uplink power boosting may be applied to counter the D2D interference such that the resulting SINR_{UL} equals the power control target $\text{SNR}_{\text{target}} = P/\sigma^2$. This is attained by choosing

$$\alpha = \frac{\text{SNR}_{\text{target}}}{\beta} + 1 \qquad (9.2)$$

Transmit Power when Sharing Cellular Downlink Resources

The actual location of cellular receivers in the DL depends on the short-term scheduling decisions of the eNB. Hence, the victim receiver at a given set of resources can be any of the served UEs. After setting up a D2D connection, the eNB can set the maximum D2D transmit power to a predetermined value. The maximum D2D transmit power will be higher when a D2D connection uses dedicated (free) resources than when it reuses (reserved) cellular resources.

A suitable D2D transmit power limit can be found by long-term observations of the impact of different D2D power levels on the quality of cellular links. In addition, the eNB can observe the link-quality feedback from its served UEs. If it observes a degradation in the UEs link quality, it can reduce the transmit power of D2D transmitters (Axnäs et al. 2007).

9.3.3 Multiantenna Techniques

Downlink Precoding to Reduce Interference

In case of D2D communication reusing the cellular DL the main interference originates from the eNB. A multiantenna eNB may use some of its degrees of freedom to help the D2D communication by suppressing the interference generated towards the D2D receiver (Jänis et al. 2009b).

The remaining degrees of freedom are used to design the precoder such that the signal quality towards the cellular receiver(s) that it transmits to is increased.

Let us assume the system in Figure 9.5, where the eNB has N_1 transmit antennas and the cellular UEs have M_1 receive antennas. The D2D transmitters and receivers have N_2 and M_2 antennas, respectively. The eNB then restricts the transmitted signal to the nullspace of the eNB-to-D2D channel

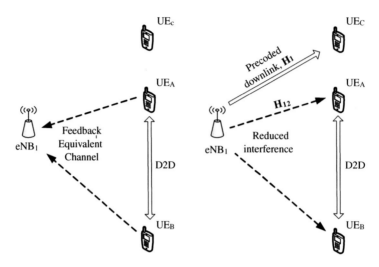

Figure 9.5   Interference suppression towards D2D by DL precoding

as follows. The Singular Value Decomposition (SVD) of the interference channel can be found as $\mathbf{H}_{12} = \mathbf{U}_{12}\Lambda_{12}\mathbf{V}_{12}^H$. Assuming $N_1 > M_2$ the matrix $\mathbf{H}_{12}$ is wide and we partition $\mathbf{V}_{12}$ as

$$\mathbf{V}_{12} = [\mathbf{v}_{12}(1)\dots\mathbf{v}_{12}(M_2)\mathbf{v}_{12}(M_2+1)\dots\mathbf{v}_{12}(N_1)] \qquad (9.3)$$

Now we define $\mathbf{V}_{12}^{\perp} = [\mathbf{v}_{12}(M_2+1)\dots\mathbf{v}_{12}(N_1)]$ as the basis of the interference-free subspace. Note that $\mathbf{V}_{12}^{\perp}{}^H\mathbf{V}_{12}^{\perp} = \mathbf{I}$ and $\mathbf{H}_{12}\mathbf{V}_{12}^{\perp} = \mathbf{0}$. We can now form a projection matrix Π and a projected cellular DL channel $\tilde{\mathbf{H}}_1$ given by:

$$\Pi = \mathbf{V}_{12}^{\perp}\mathbf{V}_{12}^{\perp}{}^H \text{ and } \tilde{\mathbf{H}}_1 = \mathbf{H}_1\Pi \qquad (9.4)$$

The eNB is free to use any Multiple-Input Multiple-Output (MIMO) transmission scheme by designing transmitter weights $\tilde{\mathbf{W}}_1$ for $\tilde{\mathbf{H}}_1$. Finally, the cellular DL precoder is recognized as

$$\mathbf{W}_1 = \Pi\tilde{\mathbf{W}}_1 \qquad (9.5)$$

This precoding approach facilitates many different transmission schemes, as essentially any MIMO transmission scheme can be designed for the DL transmission over channel $\tilde{\mathbf{H}}_1$, including beamforming, spatial multiplexing, and open-loop diversity (Hottinen et al. 2003). Note that there cannot be more than $N_1 \times M_2$ nonzero singular values in the equivalent DL channel and hence the proposed scheme effectively reduces the available degrees of freedom in the DL transmission. The corresponding gain in D2D link quality more than compensates for this.

In case the D2D link utilizes a lower rank transmission than the number of receive antennas at the D2D receiver, it is possible to enhance the DL precoder by reducing the dimensionality of the interference subspace. In this case, the D2D receiver may feedback the equivalent eNB-D2D channel after receiver processing for the D2D link, that is, $\mathbf{Q}_2^H\mathbf{H}_{12}$, where $\mathbf{Q}_2$ is the receiver for the D2D link. This procedure is illustrated in Figure 9.5.

In case there are multiple D2D receivers sharing the DL transmission band, the eNB may avoid generating interference to that set of D2D receivers by stacking the eNB-to-D2D channels into a matrix prior to SVD when finding the interference-free subspace. This can be used, for example, in case the eNB has allocated time-frequency resources for a D2D pair, but does not have information about the timing of D2D upstream and downstream transmissions. The dimensionality of the stacked

D2D channel increases quickly, so this is mostly useful in combination with the closed-loop scheme for lower rank D2D transmissions described above.

Advanced Receivers

Device-to-device receivers equipped with multiple receive antennas can suppress the DL interference further. To be able to suppress the interference from the eNB, for example, a Minimum Mean Square Error (MMSE) receiver could be used by the D2D UEs. This implies that the receiver has to estimate not only the D2D channel but also the channel to the eNB, which may be aided by a suitable reference signal structure and synchronous D2D and cellular communication given the fact that the transmissions occur in the same cell.

Numerical Evaluation of Multiantenna Techniques

We have studied the interference-aware DL precoding mechanism in an interference-limited local area indoor scenario illustrated in Figure 9.6. Nine omnidirectional eNBs serve an area of 100×100m where a set of walls form small rooms, corridor-like longer rooms and a large open area in the center. Similar elements are typically found in shopping malls and office areas. The wireless propagation is modeled according to the Wireless World Initiative New Radio (WINNER) II channel model A1 for office/indoor scenarios (WINNER-II 2008). Cellular UEs are uniformly distributed in the area and the same amount of D2D pairs are generated with the restriction that a D2D link must be within a single room. We consider a synchronized LTE-A cellular network operating on a 100 MHz band using TDD. The band is split into five sub-bands of 20 MHz to keep backwards compatibility with 3GPP LTE. The D2D pairs reuse the cellular DL resources while UL resources have no D2D communication. We investigated the performance of the DL precoding scheme with 4×2 antenna configuration. The D2D transmitter is utilizing single stream transmission, while an eNB utilizes

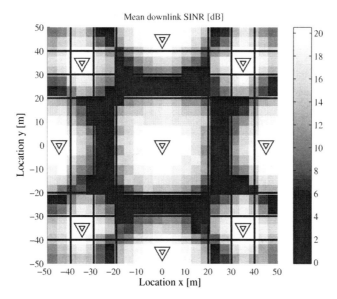

Figure 9.6 Indoor test scenario. The triangles represent the locations of eNBs and the black horizontal and vertical lines represent walls. The color indicates the mean DL SINR without D2D interference as a function of location (Jänis et al. 2009a). Reproduced by permission of © 2009 IEEE

four-antenna transmission. Eigenbeamforming (EBF) with genie-aided rank adaptation on the cellular DL was simulated as a reference, including the case where no D2D links are present. The interference suppression precoder was simulated without equivalent channel knowledge (rank two interference channel) and with equivalent channel knowledge (rank one interference channel). These precoders are called IS and ISCL, respectively. The DL employed MMSE receivers with the assumption of white interference covariance, while the D2D receiver was Maximum Ratio Combining (MRC).

Figure 9.7(a) shows the SINR CDF of the DL and D2D transmissions. As can be seen from the figure, there is a significant improvement in D2D SINR of around 15 dB. As can be expected the D2D SINR is not different between the IS and ISCL cases since both effectively prevent intracell

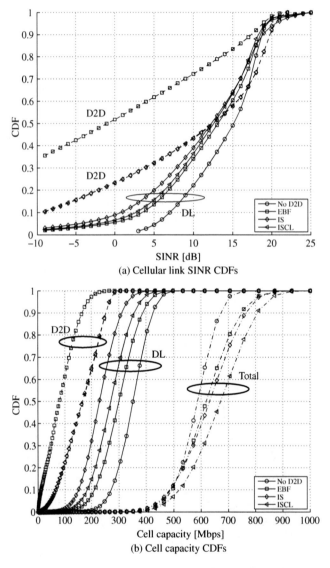

(a) Cellular link SINR CDFs

(b) Cell capacity CDFs

Figure 9.7 Interference-aware DL precoding (Jänis et al. 2009b): (a) Link SINR CDFs for DL (solid lines) and D2D (dashed lines) transmissions and (b) cell capacity CDFs for DL transmissions (solid lines), D2D transmissions (dashed lines), and total cell capacity (DL+UL+D2D, dash-dot lines). Reproduced by permission of © 2009 IEEE

eNB-to-D2D interference. The D2D transmitters induce a reduction in median cellular DL SINR of around 3 dB. Here the D2D transmission power was assumed to be 0 dBm while that of the eNBs is 18 dBm per sub-band.

The D2D, DL, and total cell capacity CDFs are presented in Figure 9.7(b). The D2D capacities are significantly lower than the cellular DL capacity due to the lower transmission power and single-stream transmission without precoding. The median cellular DL capacity reduces by 14% due to D2D transmissions when no interference avoidance precoder is used. In the case of ISCL and IS this reduction increases to 23% and 34%, respectively. The median D2D capacity is, however, increased by more than 200% in the IS and ISCL cases compared to EBF. The net effect is that the overall cell capacity rises by 9% when ISCL is utilized compared to plain EBF. Note that the reference case of total cell capacity without D2D communication is not fair due to the fact that in this case the D2D communication should be relayed through the eNB. That is, the result corresponds to the case where there is no D2D traffic at all. Finally, we present results for D2D UEs equipped with two antennas in comparison to single antenna reception. The difference from the preceding case is that here we assume the intracell interference covariance to be known at the D2D receiver. Figure 9.8 presents the average received SINR of the D2D UEs. The D2D pairs reuse cellular DL resources and use a transmit power of 0 dBm which is 24 dB below the eNB transmit power to keep the SINR degradation below the required 3 dB at the tenth percentile of the difference from SINR CDF. However, with such a low transmit power almost 60% of the single antenna D2D receivers achieve an SINR of less than 0 dB, which would not allow D2D communication with reasonable data rates. Using multiple antennas at the D2D receivers can improve the SINR experienced by D2D pairs. With a MRC receiver, about 10% more D2D pairs can operate at reasonable SINRs. With an MMSE receiver the amount of D2D connections having a SINR below 0 dB can be reduced to below 30%.

However, even with interference-aware DL precoding and advanced multiantenna receivers, not all the D2D connection pairs can achieve a minimum SINR of 0 dB, which would allow direct communication. In particular, D2D pairs close to the eNB experience a low SINR and should get dedicated resources without interference from the eNB or the eNB could act as a relay node. There is therefore a need for interference-aware resource allocation, mode selection, and D2D admission-control schemes to allocate D2D communication in a cellular network.

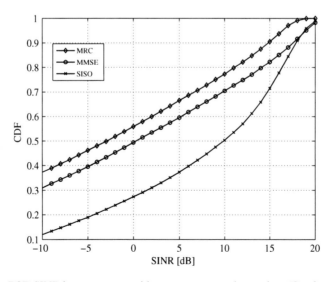

Figure 9.8 D2D SINR improvements with two antennas at the receiver (Osseiran et al. 2009)

9.3.4 Radio Resource Management

Interference-aware Resource Allocation

In addition to limiting the interference to cellular UEs by reducing the transmit power, the eNB can use interference-aware scheduling (Jänis et al. 2009a) to reduce the interference. It schedules cellular UEs that are well isolated in propagation from a D2D connection preferably on the same resources as the D2D connection. For example, the eNB might schedule indoor D2D connections together with outdoor cellular UEs. In this way the interference to D2D receivers that reuse cellular UL resources and the interference generated by D2D transmitters to cellular receivers in DL is reduced.

Interference-aware resource allocation requires that the eNB can acquire local awareness of the channel gains between cellular and D2D UE in its cell. Here we have identified two options: first, the D2D UEs measure the received interference power of cellular UEs UL transmissions. The power measurement can be done from UL pilots if the receiver synchronizes with the transmission, or by simply measuring the total received power during the UL transmission. If the D2D UE is able to decode the UL grants of the cellular UE in the cell, it can use this information to report the interfering UE to the eNB, which requires a signaling of the Cell Radio Network Temporary Identifier (CRNTI) used in the cell to the D2D UE. Second, the cellular UE could be requested to transmit sounding signals and the D2D UEs would then report the interference experienced to the eNB. Both options are backwards compatible and do not require any changes to the cellular UEs.

Mode Selection

Device-to-device communication can be facilitated as direct communication between the UE or relayed by the eNB as in regular cellular communication. In the case of direct communication we distinguish between the usage of dedicated (free) resources or the reuse of resources reserved for cellular communication.

The optimal D2D mode selection strategy does not only depend on the quality of the D2D link and the quality of the link between D2D UE and the eNB. It largely depends on the interference situation, that is, the position of the D2D receiver relative to the cellular eNB when reusing DL resources and to the cellular UE when reusing UL resources. In a multicell scenario the interference from other cells will also affect the decision.

In real networks the choice between direct communication and cellular mode will also depend on the load situation of the cell. The achievable throughput for D2D communication in cellular mode will be lower when the cell is fully loaded and the eNB will allocate fewer resources to D2D communication on dedicated resources.

A mode selection strategy that takes all these aspects into account has been proposed in Doppler et al. (2010). The eNB decides whether the D2D communication receives dedicated resources, reuses the same resources as the cellular communication, or operates in cellular mode. In order to aid the decision the eNB additionally needs information about the expected interference for different D2D communication modes. The SINR for cellular communication can be obtained from procedures presently used in cellular networks, that is, UE channel state information feedback in DL, and from previous UL communication.

We propose the following algorithm to obtain the necessary information at the eNB and to perform the mode selection:

1. D2D UEs send probing signals to each other with power known/set by eNB and estimate the received signal powers.
2. D2D UEs estimate interference plus noise power with/without own eNB signal present in DL.
3. D2D UEs estimate interference plus noise power in UL with/without UEs transmitting in own cell.
4. D2D UEs send the obtained information to the eNB to support mode selection.
5. eNB decides on the amount of resources in dedicated and cellular mode it would allocate to the D2D UEs in UL/DL based on cellular load.
6. eNB decides on the maximum transmit power the D2D UEs can use for different direct modes.
7. eNB estimates the expected SINR for each communication mode.
8. eNB estimates the expected throughput based on SINR and available amount of resources for each communication mode and selects communication mode with the highest throughput.

The algorithm is described for a D2D pair but it is straightforward to extend it to more than two UEs engaged in D2D communication. Note that even though the signaling load of the proposed selection algorithm is significant, we expect it to be feasible, because low mobility and a limited amount of active D2D connections are expected in the local services scenario. For the direct communication modes the eNB can assign dedicated resources or allow the D2D pairs to reuse the same resources as the cellular communication, which results in different interference situations. If it is not known which of the UE is the transmitter and receiver, the worse SINR is used for the mode selection.

In DL, the expected interference can be estimated at the D2D UEs by measuring the signal strength of the serving eNB and interfering eNBs. In UL, the D2D UEs observe the cellular UL communication and estimate the average interference with and without cellular UEs transmitting in their own cell. Information about cellular UEs transmitting in their own cell can be obtained for example from the UL grants broadcasted by the eNB or the eNB can signal unused UL resources to the D2D UEs. However, in order to ensure a quick D2D connection setup, the D2D UEs will not be able to observe the interference on UL resources for a long period. Nearby cellular UEs might be silent during the observation period but activate during the D2D communication causing packet losses for the D2D communication.

The eNB then decides on the transmit power that can be used for the direct communication mode. The maximum transmit power can be higher when the D2D communication receives dedicated resources because it does not interfere with any cellular intracell communication. The maximum tolerable D2D transmit power will also be different depending on whether UL or DL resources are chosen as illustrated in section 9.3.2. Based on the selected D2D transmit power and the reported interference for each mode, the eNB estimates the SINR of each communication mode. Together with the amount of resources that the D2D pair would receive in each communication mode, the eNB can estimate the expected D2D throughput in each of the modes and select the mode with the maximum throughput. In cellular mode, the D2D rate is limited by the link (UL or DL) with lower throughput. The expected throughput can be obtained by using the Shannon formula or by using a lookup table that contains the expected throughput depending on the SINR, the available modulation and coding schemes, the amount of resources (code block length) and the available multiantenna techniques in the system.

Numerical Evaluation of Resource Management Techniques

We studied the proposed interference coordination mechanism in the scenario described in section 9.3.3 and illustrated in Figure 9.6. In these simulations the system utilizes a 100–160 MHz band using TDD. The band is split into $N = 5$ or $N = 8$ sub-bands of 20 MHz. Each cellular UE

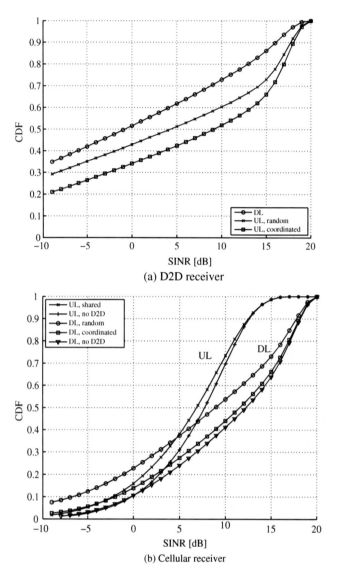

Figure 9.9 SINR improvements with interference-aware resource allocation (Jänis et al. 2009a). The upper figure shows the gain in D2D SINR from scheduling D2D receivers with cellular UL transmitters that are max-imally separated. Similarly, the lower figure shows the gain in cellular DL SINR from scheduling cellular DL receivers with D2D transmitters that are maximally separated. Reproduced by permission of © 2009 IEEE

and D2D pair is assigned either randomly or using interference-aware scheduling to one out of N subbands. In this case the D2D pairs reuse both cellular UL and cellular DL resources.

The results in Figure 9.9 present the empirical CDF of the resulting cellular and D2D SINR, respec-tively, for $N = 8$ subbands. The cellular DL SINR shows the gain from separating D2D transmitter and DL cellular receiver. This gain is higher in the lower percentiles since the low SINR values are mainly induced by a nearby D2D transmitter, which is effectively removed by interference-aware resource allocation. A similar effect can be seen for the D2D SINR sharing cellular UL resources. Figure 9.9 also illustrates that the D2D SINR is higher when reusing UL resources than when reusing

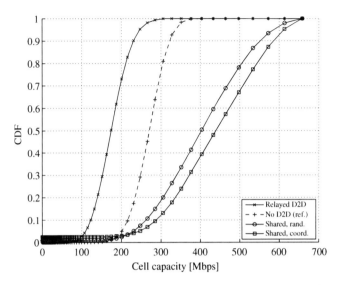

Figure 9.10   Capacity improvement results (Jänis et al. 2009a). Reproduced by permission of © 2009 IEEE

DL resources. In the case of reusing DL resources the D2D transmit power has to be restricted to protect cell-edge UEs of other cells that cannot benefit from interference-aware resource allocation. The cellular UL SINR and the D2D SINR when D2D is in DL phase are not affected by interference-aware resource allocation. Therefore the corresponding curves are left out from the figures.

The aggregate cell capacity in Figure 9.10 reveals that the median cell capacity increases threefold when direct D2D links are enabled versus relaying the D2D traffic via the eNB. This large gain is due to two factors. Firstly, the D2D links are shorter than the UE to eNB links because the D2D pairs are assumed to be in the same room. Secondly, when the traffic is routed through the eNB it needs to be transmitted twice, where the effective link capacity is the minimum of the corresponding UL and DL capacities, which leads to substantial cell capacity reduction. The gain from interference-aware resource allocation over random allocation is an additional 10%. It should be noted, however, that the interference-aware resource allocation effectively reduces the outage of cellular DL UEs (see Figure 9.9).

The results in (Jänis et al. 2009a) indicate that most of the gain is obtained already at $N = 4$ sub-bands. Thus, the intracell D2D-to-DL and UL-to-D2D interference can be mitigated with rather small amount of multiuser diversity. This is a very encouraging result from a system design point of view, because the signaling related to obtaining local awareness grows substantially with the number of UEs.

Next, we evaluate the mode-selection algorithm for D2D communications in the same indoor scenario as in Figure 9.6 with $N = 5$ sub-bands. 180 cellular UEs are randomly located resulting in 20 cellular UEs on average in each cell. Ten D2D pairs have been generated for each cell. Both UEs forming a D2D pair are located in the same room to model local communication. We assume a full buffer traffic model both in cellular UL and DL as well as for D2D communication, resulting in a fully loaded cellular network.

Five D2D pairs are randomly assigned to share UL and the other five to share cellular DL resources and the interference-aware resource allocation of section 9.3.4 is used to allocate a single D2D pair to each sub-band in UL and DL. The allocation maximizes the minimum pathloss between D2D and cellular UEs sharing the same sub-band. The mode-selection algorithm described above determines whether a pair reuses cellular resources, receives dedicated resources or communicates in cellular

Table 9.1 Comparison of mode selection performance with different available modes for D2D communication (Doppler et al. 2010). The "Direct Comm. ratio" is the ratio of D2D pairs communicating directly over the total number of D2D pairs. Reproduced by permission of © 2010 IEEE

Available Modes	Direct comm. ratio [%]	Cell. TP per cell [Mbps]	D2D TP per cell [Mbps]	D2D TP <100 kbps ratio [%]
Cell.	0	174	38	10
Reuse	100	257	92	51
Dedicated	100	262	27	37
Dedicated/Cell.	60	202	35	11
Reuse/Cell.	62	211	101	23
Reuse/Dedicated/Cell.	68	210	107	17

mode. After the decision is done, a D2D pair stays for 2 s in this mode and then a new D2D pair is generated.

The amount of resources are evenly allocated among the UEs. Without a D2D pair in dedicated mode each cellular UE gets $1/n_{cell}$ of the available resources whereas n_{cell} denotes the number of cellular UEs in the cell. If there is a D2D pair in dedicated mode the D2D communication receives $1/(1 + n_{cell})$ of the resources and the remaining resources are shared evenly among the cellular UEs. D2D pairs that reuse the same resources as the cellular communication transmit on the whole sub-band.

The transmit power values have been obtained using the algorithm described in section 9.3.2. The cellular UEs are prioritized over the D2D communication that shares the same band.

> The SINR degradation in UL and DL is required to be less than 3 dB at the fifth percentile of the cellular SINR CDF. In DL this is achieved by setting the D2D transmission power to 0 dBm when reusing the same DL resources within a cell and to 5 dBm when receiving dedicated resources compared to 24 dBm maximum transmit power of the eNB and cellular UE. In UL this is achieved by a backoff value of 5 dB compared with the transmit power determined by the cellular power control. A backoff value of 5 dB ensures that the SINR degradation for 95% of the cellular UE is less than 3 dB.

Table 9.1 compares the average cellular and D2D throughput of the proposed mode selection algorithm, under various cases[1], to other possible strategies. The combined cell throughput (Cell. + D2D throughput (TP)) increases by 50% compared with a network without D2D communication where all local traffic is handled through the cellular network (Cell.). As the eNB does not need to spend resources on D2D communication, the cellular throughput also increases for the proposed mode selection scheme. Hence, it is highly beneficial to introduce D2D communication to offload local traffic from the network. A simple mode selection strategy that forces all D2D UEs to reuse the same resources as the cellular communication (Reuse) achieves a higher cell throughput but 51% of the D2D pairs experience a close to zero throughput. Assigning dedicated resources to all D2D pairs (Dedicated) instead of reusing the same cellular resources in a cell still leaves 37% of the D2D pairs with

---

[1] Three cases for the proposed mode selection algorithm, are investigated: "**Reuse/Dedicated/Cell.**", "**Dedicated/Cell.**" and "**Reuse/Cell.**". In the first case the mode-selection algorithm is fully executed. In the "Dedicated/Cell." case (resp. "Reuse/Cell."), the algorithm is constrained to select either a dedicated (resp. reuse) mode or a cellular mode.

close to zero throughput. *The number of D2D pairs with less than 100 kbps throughput reduces to 17% for the proposed mode selection algorithm with all modes available (i.e. Reuse/Dedicated/Cell).* Please note that we only limited the D2D pairs to be in the same room but we did not limit the distance between these pairs, that is, the D2D distance can be up to 50 m. The investigated scenario is clearly interference limited and with the full buffer assumption even for all D2D pairs in cellular mode (Cell) 10% of the D2D pairs experience very low throughput, which is similar for cellular UEs in DL. The main difficulty in the mode selection originates from the uncertainty when sharing the cellular UL because the location of the cellular interferer can change after the mode selection if a nearby UE is activated which has been idle during the UL SINR estimation period. As a solution, the D2D pair should use a fallback mode to switch to cellular communication when the direct communication fails. The amount of D2D pairs with very low throughput is also effectively reduced by allowing a cellular mode in addition to assigning dedicated resources to the D2D pairs (Dedicated/Cell). However in this case the D2D throughput is only one-third of the proposed full mode selection offering Reuse, Dedicated and Cellular mode. Reusing the spectrum of the cellular UE ensures high throughput for direct communication of D2D pairs with short link distance, Dedicated mode allows direct communication with reduced interference for D2D pairs that are not able to communicate in Reuse mode and the Cellular mode is needed for D2D pairs that cannot communicate directly.

Please note, that a path-loss based mode selection as proposed in Frlan (2000) would have selected a direct communication mode in all cases for such an interference limited local area scenario with D2D pairs located in the same room.

The dependence of the selected mode on D2D UE location is illustrated in Figure 9.11(a) and Figure 9.11(b), when sharing the cellular UL and DL, respectively. For DL, the selection is plotted for the D2D receiver because the interference from the eNB has a large influence on the selection. In UL the selection is plotted for the D2D transmitter because the D2D link quality will largely depend on its transmit power, which is determined by the power control procedure described above. As shown in the figure, in UL the reuse of cellular resources is the dominant mode except close to the eNB where the UE transmit power is restricted. Similarly in DL the cellular resources are not reused close to the eNB but the cellular D2D mode is mostly selected. In Figure 9.12 we evaluate the throughput of the D2D user as a function of the D2D link distance and compare it to a case where all D2D traffic is relayed by the eNB. Especially for short distances of less than 5m the D2D throughput can be more than quadrupled.

In Figure 9.13 we present results for a case where the D2D pair is not generated in the same room. The results show that the D2D throughput reduces drastically in this case. Hence, D2D communication underlaying a cellular network will be limited to local (proximity) traffic (in the same room) and the applications utilizing D2D communication should be designed accordingly. Nevertheless, the sum throughput (cellular plus D2D) may increase largely (up to 65% compared to the sum traffic when all D2D traffic is relayed by the network) when limiting D2D communication to UEs in the same room. On the other hand, the throughput of the cellular UEs decreases significantly (35% compared to a scenario without D2D UEs) if all D2D traffic has to be relayed as the D2D pairs require both UL and DL resources.

9.4 Future Directions

In this chapter we have outlined the integration of D2D communication into the LTE-A architecture. We have presented the main technology enablers and we have presented performance results for underlay D2D communication. Our results clearly show that D2D underlay communication is feasible without excessive interference to the cellular network. Furthermore, direct communication offers up to five times the data rate of cellular communication.

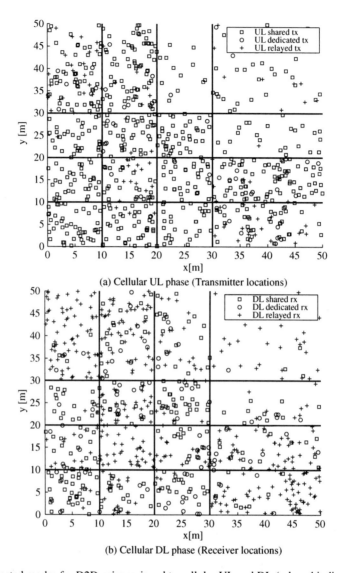

Figure 9.11   Selected modes for D2D pairs assigned to cellular UL and DL (relayed indicates cellular mode)

There are still some research questions that need to be solved to enable D2D underlay communication and eNB-controlled D2D communication in general. The solution presented in this chapter was limited to a single eNB that controls the D2D connections. However, two UEs might be located in different cells and the resource allocation and mode selection needs to be coordinated across the cells. The coordination can use the X2 interface between eNBs and it should be compatible with intercell interference coordination mechanisms that are defined in the Third Generation Partnership Project (3GPP).

The cellular network can facilitate the D2D session setup to be transparent to the user. The approaches to session setup presented in this chapter are limited to UEs served either by the same PDN gateway or the same MME, that is, UEs served by the same operator. This may work well for machine-to-machine type communication, where an operator offers the whole solution and provides the services to all the UEs. However, this is not the case for direct communication between UEs or

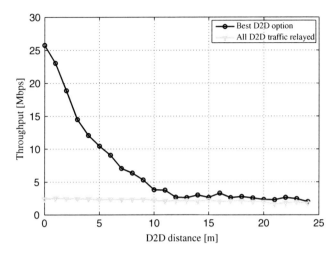

Figure 9.12 D2D throughput as a function of the D2D link distance when choosing the best sharing option and when relaying all D2D traffic through the network (Doppler et al. 2009b). Reproduced by permission of © 2009 IEEE

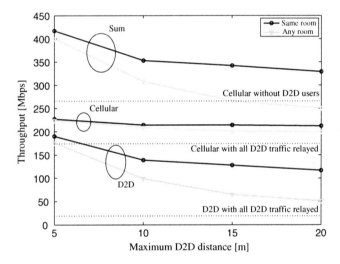

Figure 9.13 Throughput of cellular UEs and D2D UEs in a cell for D2D pairs with and without restriction to be in the same room. The D2D link distance has been limited to maximum D2D distance (Doppler et al. 2009b). Reproduced by permission of © 2009 IEEE

for service providers that want to offer their services to a wide range of customers. It is atypical for the whole circle of friends or customers to be served by the same operator and the same technology. Therefore, a solution that works across different operators would greatly enhance the user acceptance of D2D communication.

Moreover, future research should identify options to further increase the data rates of direct communication. Promising directions could be related to carrier aggregation, which has been standardized for LTE Release 10 (Rel-10) in 3GPP, see Chapter 2 for more details. Further, D2D communication could utilize additional bands that are not in use by cellular systems or license-exempt bands. However, these solutions need to ensure that the UE can still stay connected to the cellular network even though the D2D communication may take place on another band at the same time.

Another trend in D2D communication is the formation of local communities, as for example in the Nokia Instant Community (Nokia Conversations 2010) concept. Nokia Instant Community uses the ad hoc WLAN function of UEs to exchange messages with other UEs in the vicinity. The communities should form automatically without user intervention. On the other hand, the instant community operation should not significantly decrease the operation time of the UE to be accepted by the user. Hence, the duty cycle should be limited. It should be possible to adapt the solution to very sparse environments with few UEs and crowded environments like a stadium where thousands of UEs might share the same space. In a crowded place the discovery of other UEs will be a challenge, as well as keeping the duty cycle low. New approaches to self-organization, synchronization of the UEs and the usage of the OFDMA principles already in the discovery phase may be possible enablers to be studied (Doppler et al. 2011).

Finally, D2D communication as presented in this chapter could be energy efficient. For instance it could provide a way of implementing a light-weight Home eNB. The light-weight Home eNB will be idle when the members of its closed subscriber group are not close by. The wide area eNB will activate both the light weight Femto eNB and a UE in the vicinity and facilitate the D2D connection setup. The interference coordination mechanisms proposed in this chapter will effectively protect the wide area network from harmful interference.

References

Adachi T and Nakagawa M 1998 Battery consumption and handoff examination of a cellular ad-hoc united communication system for operational mobile robots *IEEE International Symposium on Personal, Indoor and Mobile Radio Communications (PIMRC)*, Boston, MA, USA.

Adachi T and Nakagawa M 2000 Capacity analysis for a hybrid indoor mobile communication system using cellular and ad-hoc modes *IEEE 11th International Symposium on Personal, Indoor and Mobile Radio Communications, 2000*.

Axnäs J, Furuskär A and de Bruin P 2007 Apparatus for limiting peer-to-peer communication interference U.S. Patent WO/2007/055623.

Bennis M, Middleton G and Lilleberg J 2008 Efficient resource allocation and paving the way towards highly efficient imt-advanced systems. *Wirel. Pers. Commun.* **45**(4), 465–478.

Chen T and Katz MD 2009 Cooperative architecture for cellular-short-range combined mesh networks *Mobimedia '09: Proceedings of the 5th International ICST Mobile Multimedia Communications Conference*.

Doppler K, Mika P R, Jänis P, Ribeiro C and Hugl K 2009a Device-to-device communications; functional prospects for LTE-Advanced networks *Proc. ICC 2009 - IEEE Int. Conf. Commun.*, Dresden, Germany.

Doppler K, Ribeiro C and Kneckt J 2011 Advances in D2D communications: energy efficient service and device discovery radio *2nd International Conference on Wireless Communications, Vehicular Technology, Information Theory and Aerospace & Electronic Systems(WIRELESS VITAE 2011)*, Chennai, India.

Doppler K, Rinne M, Wijting C, Ribeiro C and Hugl K 2009b Device-to-device communication as an underlay to LTE-Advanced networks. *IEEE Commun. Mag.* **47**(12), 42–49.

Doppler K, Yu C, Ribeiro C and Jänis P 2010 Mode selection for Device-to-device communication underlaying an LTE-Advanced Network *Proc. WCNC 2010 - IEEE Wireless Commun. and Networking Conf.*, Sydney, Australia.

ETSI 2002 BRAN; HIPERLAN2 Type 2; Data Link Control (DLC) Layer; Part 4: Extension for Home Environment. Technical Report TS 101 761-4, v1.3.2, European Telecommunications Standards Institute (ETSI).

ETSI 2006a Terrestrial Trunked Radio (TETRA); Technical requirements for Direct Mode Operation (DMO); Part 2: Radio Aspects. Technical Report ETSI EN 300 396-2 version 1.3.1, European Telecommunications Standards Institute (ETSI).

ETSI 2006b Terrestrial Trunked Radio (TETRA); Technical requirements for Direct Mode Operation (DMO); Part 3: Mobile Station to Mobile Station (MS-MS) Air Interface (AI) protocol. Technical Report ETSI EN 300 396-3 version 1.3.1, European Telecommunications Standards Institute (ETSI).

ETSI 2007 Terrestrial Trunked Radio (TETRA); Voice plus Data (V+D); Part 2: Air Interface (AI). Technical Report ETSI EN 300 392-2 version 3.2.1, European Telecommunications Standards Institute (ETSI).

Fitzek FH, Katz M and Zhang Q 2009 Cellular controlled short-range communication for cooperative p2p networking. *Wirel. Pers. Commun.* **48**(1), 141–155.

Frlan E 2000 Direct communication wireless radio system United States Patent 6047178.

Hakola S, Chen T, Lehtomäki J and Koskela T 2010 Device-to-Device (D2D) communication in cellular network – performance analysis of optimum and practical communication mode selection *Wireless Communications and Networking Conference (WCNC), 2010 IEEE*.

Haykin S 2005 Cognitive radio: brain-empowered wireless communications. *IEEE J. Select. Areas Commun.* **23**(2), 201–220.

Holma H and Toskala A 2009 *LTE for UMTS OFDMA and SC-FDMA Based Radio Access* first edn. John Wiley & Sons, Ltd, Chichester.

Hottinen A, Tirkkonen O, and Wichman R 2003 *Multi-antenna Transceiver Techniques for 3G and Beyond*. John Wiley & Sons, Ltd, Chichester.

Hsieh H and Sivakumar R 2004 On using peer-to-peer communication in cellular wireless data networks. *IEEE Trans. on Mobile Comput.* **3**(1), 57–72.

Hsieh R and Iinatti J 2005 Persistent bidirectional peer traffic in fix-network augmented broadband wireless access *Proc. International Workshop on Wireless Ad-hoc Networks (IWWAN 2005)*.

Huang K, Lau V and Chen Y 2009 Spectrum sharing between cellular and mobile ad hoc networks: transmission-capacity trade-off. *IEEE J. Select. Areas Commun.* **27**(7), 1256–1267.

Jänis P, Koivunen V, Ribeiro CB, Doppler K and Hugl K 2009b Interference-avoiding MIMO schemes for device-to-device radio underlaying cellular networks *IEEE International Symposium on Personal, Indoor and Mobile Radio Communications (PIMRC)*, Tokyo, Japan.

Jänis P, Koivunen V, Ribeiro C, Korhonen J, Doppler K and Hugl K 2009a Interference-aware resource allocation for device-to-device radio underlaying cellular networks *Proc. VTC 2009 Spring – IEEE 69th Vehicular Technology Conf.*, Barcelona, Spain.

Jänis P, Yu C, Doppler K, Ribeiro C, Wijting C, Hugl K, Tirkkonen O and Koivunen V 2009c Device-to-Device communication underlaying cellular communication systems. *Int. J. Commun. Netw. Sys. Sci.* **2**(3), 169–178.

Kaufman B and Aazhang B 2008 Cellular networks with an overlaid device to device network *Proc. IEEE Asilomar Conf. on Signals, Systems and Computers*, Pacific Grove, CA, USA.

Koskela T, Hakola S, Chen T and Lehtomäki J 2010 Clustering concept using device-to-device communication in cellular system *Wireless Communications and Networking Conference (WCNC), 2010 IEEE*.

Mitola J and Maguire G 1999 Cognitive radio: making software radios more personal. *Pers. Commun. IEEE* **6**(4), 13–18.

Nokia Conversations 2010 Nokia Instant Community Gets you Social, - http://conversations.nokia.com/2010/05/25/nokia-instant-community-gets-you-social/ (accessed January 2011).

Omiyi P and Haas H 2004 Maximising spectral efficiency in 3g with hybrid ad-hoc utra tdd/utra fdd cellular mobile communications *Spread Spectrum Techniques and Applications, 2004 IEEE Eighth International Symposium on*.

Osseiran A, Doppler K, Ribeiro C, Xiao M, Skoglund M and Manssour J 2009 Advances in device-to-device communications and network coding for IMT-Advanced *Proc. ICT Mobile and Wireless Communications Summit 2009*, Santander, Spain.

Popova L, Herpel T and Koch W 2007 Enhanced downlink capacity in umts supported by direct mobile-to-mobile data transfer *NETWORKING'07: Proceedings of the 6th International IFIP-TC6 Conference on Ad Hoc and Sensor Networks, Wireless Networks, Next Generation Internet*.

WINNER-II 2008 IST-4-027756 *WINNER II Channel Models*. Public Deliverable D1.1.2, Wireless World Initiative New Radio – WINNER, April 2008, – http://projects.celtic-initiative.org/winner+/index.html (accessed June 2011).

Yu CH, Tirkkonen O, Doppler K and Ribeiro C 2009 On the performance of device-to-device underlay communication with simple power control *Vehicular Technology Conference, 2009. VTC Spring 2009. IEEE 69th*, pp. 1–5.

10

The End-to-end Performance of LTE-Advanced

Marc Werner, Valeria D'Amico, David Martín-Sacristán, Jose F. Monserrat, Per Skillermark, Ahmed Saadani, Krystian Safjan and Hendrik Schöneich

This chapter addresses the end-to-end performance of one particular IMT-Advanced system, namely the Third Generation Partnership Project (3GPP) LTE-Advanced (LTE-A). This technology moves forward from 3GPP Long Term Evolution (LTE) Release 8 (Rel-8), which is currently starting to be deployed in selected countries, including some of the innovations described in the previous chapters. By LTE-A we refer to Release 10 (Rel-10) and beyond of the 3GPP LTE standard. This version of LTE was submitted to the International Telecommunication Union–Radiocommunication Sector (ITU-R) for evaluation and obtained the official International Mobile Telecommunications Advanced (IMT-Advanced) label. The Wireless World Initiative New Radio + (WINNER+) project evaluated LTE-A in its role as official IMT-Advanced Evaluation Group.

The chapter starts with a description, in section 10.1, of the ITU-R process for defining IMT-Advanced radio access technologies. The ITU-R evaluation scenarios and performance requirements are characterized. Then, some key features of LTE Rel-10 are introduced in section 10.2, clarifying which of the technologies described in the previous chapters are included in this system. The performance of LTE-A in International Telecommunication Union (ITU) scenarios, according to the 3GPP self-evaluation as well as extensive simulations by the WINNER+ Evaluation Group, is evaluated and analyzed in section 10.3. The channel model and simulator calibration activities in the WINNER+ Evaluation Group, which led to a consolidated set of evaluation results, are described in sections 10.4 and 10.5. The chapter ends with an outlook on the further IMT-Advanced process.

10.1 IMT-Advanced Evaluation: ITU Process, Scenarios and Requirements

In 2008, the ITU-R Working Party 5D (WP 5D) started its formal process for the identification of "Fourth Generation (4G)" radio access technologies to be qualified as IMT-Advanced systems. The IMT-Advanced label from ITU-R is especially important from a radio regulatory perspective because the operation of a mobile communication system in certain frequency bands might be

Mobile and Wireless Communications for IMT-Advanced and Beyond, First Edition.
Edited by Afif Osseiran, Jose F. Monserrat and Werner Mohr.
© 2011 Afif Osseiran, Jose F. Monserrat and Werner Mohr. Published 2011 by John Wiley & Sons, Ltd.

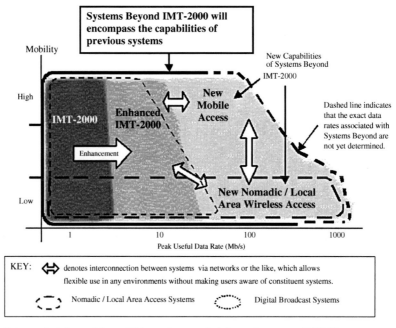

Figure 10.1   Illustration of capabilities of IMT-2000 and IMT-Advanced (ITU-R 2003).
Reproduced by permission of © 2008 ITU

restricted to IMT-Advanced systems in some areas. This was the case, for example, for Universal
Mobile Telecommunication System (UMTS) bands in Europe, which were reserved for International
Mobile Telecommunications 2000 (IMT-2000) systems. ITU-R describes IMT-Advanced systems as
"[...] *mobile systems that include the new capabilities of IMT that go beyond those of IMT-2000.
Such systems provide access to a wide range of telecommunication services including advanced mo-
bile services, supported by mobile and fixed networks, which are increasingly packet-based [...]*"
(ITU-R 2008a). This definition is technology agnostic and service oriented. However, IMT-Advanced
systems are based on new radio interfaces to differentiate themselves from mere IMT-2000 (Third
Generation (3G)) enhancements. The only specific technical performance target for 4G established
from the beginning was a peak data rate of 100 Mbps for high mobility and 1 Gbps for low mobility.
   Report ITU-R M.1645 (ITU-R 2003) describes the trends towards an increased demand for wireless communication (in particular, an increased number of users, higher data rates, and an increased
quality of service) and translates the increased demand into new technical requirements for 4G systems. Figure 10.1 (the so-called ITU van diagram) illustrates the relation between IMT-2000 and
systems beyond IMT-2000 (now called IMT-Advanced).

10.1.1   ITU-R Process for IMT-Advanced

In order to identify IMT-Advanced Radio Interface Technologies (RITs), a stepwise process was
defined by the ITU-R. The process is summarized in the following list and applies to the period from
2008 to 2011:

1. Issuance of a circular letter: invitation for submission of IMT-Advanced proposals.
2. Development of candidate RITs and Set of RITs (SRIT).
3. Submission and reception of the proposals.
4. Evaluation of RITs/SRITs by evaluation groups.
5. Review and coordination of outside evaluation activities.
6. Review to assess compliance with minimum performance requirements.
7. Consideration of evaluation results, consensus building and decision.
8. Development of radio interface recommendation(s).

In accordance with the first step, in March 2008, the ITU invited all its member states and radiocommunication sector members in a circular letter (ITU-R 2008b) for a submission of technology proposals of RITs for IMT-Advanced. The circular letter also referred to supporting documentation in which the process and materials required for submission were explained. All candidate proposals had to be accompanied by a self-evaluation of the RIT according to the IMT-Advanced performance requirements described in Report ITU-R M.2134 (ITU-R 2008c), also known as "IMT.TECH" among the evaluation groups. These performance requirements are reproduced in section 10.1.3. Report ITU-R M.2135 (ITU-R 2009c), known as "IMT.EVAL", specified further guidelines for evaluation of RITs for IMT-Advanced, such as evaluation methods, simulation procedures, test environments and deployment scenarios, antenna characteristics and, in particular, a dedicated spatial channel model (on the basis of the WINNER-II models (Döttling et al. 2009; WINNER-II 2008) and 3GPP Spatial Channel Model (SCM) (3GPP 2009c)) to be used in all evaluations. Details about the IMT.EVAL evaluation scenarios are described in section 10.1.2 and the channel model implementation and calibration are described in section 10.4.

By October 2009, a number of IMT-Advanced RIT proposals were submitted to ITU-R in accordance with the circular letter. All submissions were based on either the 3GPP system LTE-A, or on IEEE 802.16m, that is, Worldwide Interoperability for Microwave Access (WiMAX). The evaluation groups therefore had to assess only two different system proposals (both of which, however, came in different duplexing configurations, that is, Frequency Division Duplex (FDD) and Time Division Duplex (TDD) variants). The submissions were accompanied by extensive self-evaluations as required by the ITU-R process.

Some evaluation groups were registered in ITU-R during the development phase of candidate proposals. They were asked to assess any combination of performance requirements for any of the system proposals. The WINNER+ evaluation group set its focus on the LTE-A proposal while also reviewing the WiMAX proposal, and keeping track of its evaluation. In the following, a list of all registered evaluation groups is given:

- ARIB Evaluation Group
- ATIS WTSC
- Canadian Evaluation Group (CEG)
- Chinese Evaluation Group (ChEG)
- ETSI
- Israeli Evaluation Group (IEG)
- Russian Evaluation Group (REG)
- TCOE India
- TR-45
- TTA PG707
- UADE, Instituto de Tecnología (Argentina)
- WiMAX Forum Evaluation Group (WFEG)

- Wireless Communications Association International (WCAI)
- WINNER+

During the evaluation period, several workshops were held by both 3GPP and Institute of Electrical and Electronics Engineers (IEEE) for the external evaluation groups in order to discuss the evaluation status and provide answers to arising questions. Moreover, the evaluation groups cooperated in the implementation of IMT.EVAL recommendations. For instance a calibration of channel model implementations was achieved between the WINNER+ and Chinese evaluation groups.

The evaluation of the LTE-A system proposal according to steps 4-6 of the IMT-Advanced process mentioned above is described in the subsequent sections of this chapter.

10.1.2   Evaluation Scenarios

According to the evaluation rules of ITU-R (ITU-R 2009c), IMT-Advanced candidate proposals need to fulfill a set of 13 minimum performance requirements in four specific test environments that reflect future use cases of IMT-Advanced systems. Each environment is associated with a deployment scenario that specifies the simulation setup, for example intersite distance, carrier frequency, maximum transmit powers and channel model. The following test environments are defined:

- Indoor: indoor environment targeting isolated cells at offices and/or in hotspots based on stationary and pedestrian users.
- Microcellular: urban microcellular environment with higher user density focusing on pedestrian and slow vehicular users.
- Base coverage urban: urban macrocellular environment targeting continuous coverage for pedestrian up to fast vehicular users.
- High speed: macrocellular environment with high-speed vehicles and trains.

Each test environment focuses on a specific application for the candidate RIT/SRITs and is accompanied by specific values of the performance criteria to be met by the RIT/SRITs. For each of these test environments, at least one deployment scenario was defined to be used for the performance evaluation of candidate RIT/SRIT (see Table 10.1 and Figure 10.2).

The **deployment scenarios** given in (ITU-R 2009c) are:

- **Indoor Hotspot (InH)**: small isolated cells at offices or hotspot areas. Targets high user throughput and high user density. All users are pedestrians. Two base stations operating at 3.4 GHz with an omnidirectional antenna setup are mounted on the ceiling of a long hall with adjacent offices.

Table 10.1   Test environments and deployment scenarios

Test environment	Deployment scenario
Indoor	Indoor Hotspot (InH)
Microcellular	Urban Microcell (UMi)
Base Coverage Urban	Urban Macrocell (UMa)
	Suburban Macrocell (SMa) – optional
High Speed	Rural Macrocell (RMa)

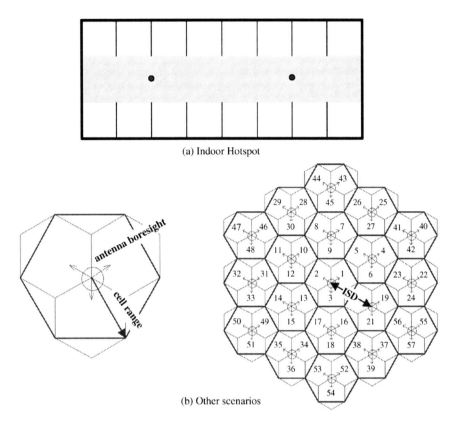

Figure 10.2   Deployment scenarios (ITU-R 2009c). Reproduced by permission of © 2008 ITU

- **Urban Microcell (UMi)**: high traffic and user density for city centers and dense urban areas. Outdoor and outdoor to indoor propagation characteristics for pedestrian users are assumed. Continuous hexagonal deployment is used with three sectors per cell and below rooftop antenna mounting. Base stations operate at 2.5 GHz and have an intersite distance of 200 m.
- **Urban Macrocell (UMa)**: targets ubiquitous coverage for urban areas. A similar hexagonal deployment is used with larger intersite distance of 500 m and antennas mounted clearly above rooftop. Non-line-of-sight or obstructed propagation conditions are common for this scenario. Only vehicular users at moderate speed are assumed, suffering from an additional outdoor to in-car penetration loss. Base stations operate at 2 GHz.
- **Rural Macrocell (RMa)**: similar to UMa, but targets larger cells with support for high-speed vehicular users. Base stations have an intersite distance of 1732 m and operate at 800 MHz, which is more suitable for large cells.
- **Suburban Macrocell (SMa)**: this is an optional scenario for the same test environment as of the UMa scenario. The key difference is an increased intersite-distance of 1299 m and a mix of indoor and high-speed vehicular users.

10.1.3   Performance Requirements

In Report M.2134 (ITU-R 2008c), ITU-R specifies minimum performance requirements that candidate radio technologies must fulfill. The evaluation criteria are grouped by their evaluation method, which is either inspection, analytical or simulation.

Evaluation by inspection only requires evaluation groups to check if the candidate proposal addresses and meets the requirement. Inspection requirements are:

- Bandwidth: the candidate systems must support scalable bandwidth allocations up to and including 40 MHz. Furthermore, proponents are encouraged to support higher bandwidths of up to 100 MHz. It must be possible to demonstrate that the systems support at least three different bandwidths allocation including the minimum and maximum value for the candidate system.
- Intersystem handover: the candidate systems must support intersystem handover between the candidate IMT-Advanced system and at least one IMT-2000 system.
- Deployment possible in at least one of the identified International Mobile Telecommunications (IMT) bands.
- Channel bandwidth scalability.
- Support for a wide range of services: candidate systems must be able to support multiple service classes such as background, streaming, interactive and conversation.

Analytical evaluation involves some calculations to determine whether the candidate meets the minimum requirement. The analytical requirements are:

- Peak spectral efficiency: gross data rate offered by the physical layer of the candidate technology. This criterion allows to estimate the overhead introduced by the physical layer.
- Control plane latency: this allows for estimation of call setup duration. It is measured by the state transition time, for example the time interval between the idle and active mode of a User Equipment (UE).
- User plane latency: a minimum transmission time for IP-Packets through the radio access network is required.
- Intrafrequency and interfrequency handover interruption time: candidates are required to support seamless handovers between cells of the system. Therefore a minimum handover interruption time is required.

Some aspects of IMT-Advanced candidate systems cannot be investigated analytically and need to be addressed by simulation. These requirements are:

- Cell spectral efficiency: IMT-Advanced systems should provide their users with high data rates. The assigned spectrum must be utilized efficiently.
- Cell-edge user spectral efficiency: high data rates must be provided to users, while at all times a minimum data rate should be available to cell edge users. Cell spectral efficiency and cell-edge user spectral efficiency are to be determined in the same simulation runs.
- Mobility: the candidate system should be able to operate at UE speeds of up to 350 km/h. This is evaluated by link-level simulations.
- Voice over IP (VoIP) capacity: IMT-Advanced systems should not only be able to support high data rates but also a large number of users. The VoIP capacity is used to evaluate the maximum load of users – with rather low traffic demands – that can be supported.

Tables 10.2, 10.3, and 10.4 provide the numerical values of inspection, analytical, and simulative performance requirements set by ITU-R for IMT-Advanced candidates.

The subsequent sections provide an overview of the evaluation work that has been carried out in WINNER+ to analyze the LTE-A performance with respect to the three groups of requirements. The detailed assessment results are reported in a WINNER+ deliverable (WINNER+ 2010).

Table 10.2 IMT-Advanced performance requirements to be assessed by inspection

Parameter [unit]	Required value
Bandwidth [MHz]	up to 40
Intersystem handover	supported
Deployment possible in at least one of the identified IMT bands	possible
Support for a wide range of services	All service classes defined in (ITU-R 2009c) in at least one test environment
Channel bandwidth scalability	Support of at least three bandwidth values

Table 10.3 IMT-Advanced performance requirements to be assessed by analytical calculation

Parameter [unit]	DL/UL	Required value
Peak spectral efficiency [bps/Hz/cell]	DL	15
	UL	6.75
Control plane latency [ms]	n.a.	<100
User plane latency [ms]	n.a.	<10
Intrafrequency handover interruption time [ms]	n.a.	27.5
Interfrequency handover interruption time within a spectrum band [ms]	n.a.	40
Interfrequency handover interruption time between spectrum bands [ms]	n.a.	60

Table 10.4 IMT-Advanced performance requirements to be assessed by simulation

Parameter [unit] Simulation method	Test environment	DL/UL	Required value
Cell spectral efficiency [bps/Hz/cell]	Indoor	DL	3
System level		UL	2.25
	Microcellular	DL	2.6
		UL	1.8
	Base coverage urban	DL	2.2
		UL	1.4
	High speed	DL	1.1
		UL	0.7
Cell edge user spectral efficiency	Indoor	DL	0.1
[bps/Hz/cell]		UL	0.07
System level	Microcellular	DL	0.075
		UL	0.05
	Base coverage urban	DL	0.06
		UL	0.03
	High speed	DL	0.04
		UL	0.015
Mobility Traffic channel link data rates	Indoor	UL	1.0
[bps/Hz]	Microcellular	UL	0.75
System and link level	Base coverage urban	UL	0.55
	High speed	UL	0.25
Number of supported VoIP users	Indoor	both	50
[active users/sector/MHz]	Microcellular	both	40
System level	Base coverage urban	both	40

10.2   Short Introduction to LTE-Advanced Features

In 2008, 3GPP held two "3GPP IMT-Advanced Workshops" (Network 2008a,b). The goal of these workshops was to investigate the main changes that could be brought forward to evolve the Evolved Universal Terrestrial Radio Access (E-UTRA) radio interface as well as the Evolved Universal Terrestrial Radio Access Network (E-UTRAN) (3GPP 2010a) in the context of IMT-Advanced. In particular, the LTE-A study item was initiated in order to study the development of LTE based on a new set of requirements. This initiative collected operators' and manufacturers' views in order to develop and test innovative concepts that will satisfy the needs of next-generation communications. The requirements were gathered in 3GPP TR 36.913 (3GPP 2009b). The resulting technical report was published in June 2008 and a liaison was sent to ITU-R covering the work in 3GPP RAN on LTE-A towards IMT-Advanced. Finally, 3GPP contributed to the ITU-R towards IMT-Advanced via its proposal "3GPP LTE Rel-10 & Beyond (LTE-Advanced)" (ITU-R 2009a).

The new technical features of LTE-A are defined in 3GPP TR 36.814 (3GPP 2010b). The main technical features are:

- **Support of wider bandwidth**
  Carrier aggregation, where two or more component carriers are aggregated, is considered for LTE-A. Each carrier is assigned a bandwidth up to 20 MHz, in order to support Downlink (DL) transmission bandwidths larger than 20 MHz, for example, up to 100 MHz. A UE may simultaneously receive or transmit one or multiple component carriers depending on its capabilities.
- **Extended multiantenna configurations**
  Extension of LTE DL spatial multiplexing up to eight layers is considered. For the UL, spatial multiplexing to up to four layers is supported.
- **Coordinated multipoint transmission or reception (CoMP) transmission and reception**
  This feature is considered as a tool to improve the coverage of high data rates, the cell-edge throughput and/or to increase system throughput. It implies dynamic coordination among multiple geographically separated transmission points. However, for LTE-A Rel-10, there will be no new communication link for support of CoMP in the inter-eNodeB (eNB) mode, that is, through the X2 standardized interface between eNBs.
- **Relaying functionality**
  Relaying is supported for LTE-A as a tool to improve, for example, the availability of high data rates, group mobility, temporary network deployment, the cell-edge throughput and/or to provide coverage in new areas.

For more details details about the LTE-A features the readers are referred to section 6.2 (for CoMP), to section 9.3 (for extended multi-antenna configurations), and to section 7.2 (for relaying functionality).

10.2.1   The WINNER+ Evaluation Group Assessment Approach

The WINNER+ project produced consistent research work on optimization of the radio interface concepts for IMT-Advanced systems, based on the heritage of the activities carried out in the former European Union framework program 6 project Wireless World Initiative New Radio (WINNER). In addition to developing enabling technologies for the WINNER+ system concept, in November 2008, WINNER+ registered as an independent evaluation group towards ITU-R in order to drive the IMT-Advanced development process.

The evaluation procedure is designed in such a way that the overall performance of the candidate RITs/SRITs may be fairly and equally assessed on a technical basis. Based on the available expertise in IMT-Advanced radio technology concepts and in link- and system-level simulation tools, the WINNER+ Evaluation Group targeted the 3GPP LTE Rel-10 & Beyond (LTE-Advanced) proposal (ITU-R 2010b). The WINNER+ group evaluated all 13 minimum requirements for IMT-Advanced systems according to (ITU-R 2008c) (see section 10.1) by means of analytical, inspection and simulation activities in order to perform a full evaluation of the LTE-A candidate technology.

For simulation purposes, in order to guarantee the reliability of the results, evaluated characteristics were assessed by a plurality of partners. During the course of the work, emphasis was placed on reflecting realistic behavior of the system under consideration by modeling nonideal aspects including, for example, effects of channel estimation errors, channel quality indicator (CQI) measurement errors and feedback delay as well as a correct modeling of the overhead in the system. Simulators of different partner organizations were calibrated in order to provide consistent results.

10.3 Performance of LTE-Advanced

10.3.1 3GPP Self-evaluation

Together with the submission of LTE Rel-10 as an IMT-Advanced technology proposal, 3GPP also performed and submitted an evaluation of the LTE Rel-10 technology. Such an evaluation, in which the technology proponent evaluates its own radio interface proposal to verify that it meets the requirements, is referred to in ITU-R as a self-evaluation. The 3GPP self-evaluation covers the FDD RIT and the TDD RIT and is part of the submissions of LTE Rel-10 to ITU-R. It is also available in the 3GPP technical report (3GPP 2009a, Section 16).

In short, the 3GPP self-evaluation report covers all the technical, spectrum and service requirements. The overall conclusion is that both the FDD RIT and the TDD RIT fulfill all the requirements in the four test environments and, hence, LTE Rel-10 meets all the IMT-Advanced performance requirements.

Technical requirements are evaluated by means of inspection, analytical evaluation, or simulations, whereas the spectrum and service requirements are evaluated by means of inspection only. As for inspection and analytical evaluations, 3GPP summarized the necessary data and provided explanatory information regarding how LTE Rel-10 performs in the respective areas. As for the evaluations that are performed by means of simulations, several organizations contributed to this work and the 3GPP self-evaluation report provides the average values of all samples together with the number of samples.

Table 10.5 summarizes the 3GPP self-evaluation results for the different technical requirements. In many of the evaluations, such as, for example, cell spectral efficiency and VoIP capacity, 3GPP reported results for different antenna configurations and transmission schemes. In addition, for the DL cell spectral efficiency and cell-edge user spectral efficiency evaluations 3GPP performed evaluations for different Physical Downlink Control CHannel (PDCCH) overhead assumptions ($L = 1, 2$, or 3 Orthogonal Frequency Division Multiplexing (OFDM) symbols per subframe). For the cases where different options were considered, Table 10.5 provides a range of figures.

All the details of the 3GPP self-evaluation can be found in (3GPP 2099a). However, some further information about the results of the 3GPP cell spectral efficiency and cell-edge spectral efficiency evaluations is given here. In the DL, 3GPP performed evaluations for different PDCCH overhead assumptions and for a few antenna setups and transmission schemes, for example, Single-User (SU)-Multiple-Input Multiple-Output (MIMO), Multi-User (MU)-MIMO, CS/CB-CoMP, and JP-CoMP. In most cases, four base station transmit antennas were utilized but a few evaluations were also performed assuming eight transmit antennas. FDD and TDD perform relatively similar in most cases, except for JP-CoMP in the base coverage urban scenario where the TDD option seems to take advantage of the available channel reciprocity. Even with the higher PDCCH overhead of $L = 3$ OFDM

Table 10.5 Outcome of the 3GPP self-evaluation of technical requirements (ITU-R 2009a). © 2009 European Telecommunications Standards Institute

Requirement [unit]	Test environment	DL/UL	FDD	TDD
Cell spectral efficiency [bps/Hz/Cell]	Indoor	DL	4.1–6.6	4.1–6.7
		UL	3.3–5.8	3.1–5.5
	Microcellular	DL	2.8–4.5	2.7–4.6
		UL	1.9–2.5	1.9–3.0
	Base cov. urban	DL	2.4–3.8	2.4–3.7
		UL	1.5–2.1	1.5–2.7
	High speed	DL	1.8–4.1	1.6–4.0
		UL	1.8–2.3	1.8–2.6
Cell-edge user spectral efficiency [bps/Hz/Cell]	Indoor	DL	0.19–0.26	0.19–0.24
		UL	0.23–0.42	0.22–0.39
	Microcellular	DL	0.087–0.15	0.085–0.10
		UL	0.073–0.086	0.068–0.079
	Base cov. urban	DL	0.066–0.10	0.067–0.10
		UL	0.062–0.099	0.062–0.097
	High speed	DL	0.057–0.13	0.049–0.12
		UL	0.082–0.13	0.080–0.15
Peak spectral efficiency [bps/Hz/Cell]	n.a.	DL	16.3	16.0
		UL	8.4	8.1
Control plane latency [ms]	n.a.	n.a.		50
User plane latency [ms]	n.a.	n.a.		9.5
Mobility traffic channel link data rate LOS/NLOS [bps/Hz]	Indoor	UL	3.15 / 2.56	3.11 / 2.63
	Microcellular	UL	1.42 / 1.21	1.48 / 1.14
	Base cov. urban	UL	1.36 / 1.08	1.36 / 0.95
	High speed	UL	1.45 / 1.22	1.38 / 1.03
Intrafrequency handover interruption time [ms]	n.a.	n.a.	10.5	12.5
Interfrequency handover interruption time [ms]	n.a.	n.a.	10.5	12.5
Interfrequency handover interruption time between spectrum bands [ms]	n.a.	n.a.	10.5	12.5
VoIP capacity [users/MHz/cell]	Indoor	n.a.	131–140	130–137
	Microcellular	n.a.	75–80	74–74
	Base cov. urban	n.a.	68–69	65–67
	High speed	n.a.	91–94	86–92
Bandwidth [MHz]	n.a.	n.a.		100
Channel bandwidth scalability	n.a.	n.a.	One component carrier supports 1.4, 3, 5, 10, 15, and 20 MHz. By aggregating multiple component carriers, transmission bandwidth up to 100 MHz are supported.	

symbols both the FDD and the TDD RITs fulfill the DL requirements, which are given in Table 10.4. The most challenging test environment seems to be the base coverage urban scenario.

In the Uplink (UL), most of the spectral efficiency and cell-edge spectral efficiency evaluations were performed using Rel-8 Single Input Multiple Output (SIMO) transmission, Rel-8 MU-MIMO or SU-MIMO (2x4). As in the DL, the evaluations show that LTE Rel-8 (and consequently Rel-10) fulfills all the requirements in all test environments and the performance of the FDD RIT and the TDD RIT is very similar. An observation is that, in the UL, the 3GPP self-evaluation indicates that all requirements are also fulfilled with LTE Rel-8 technology.

10.3.2   Simulative Performance Assessment by WINNER+

This section presents the LTE-A assessment made by the WINNER+ group. First the cell and cell-edge spectral efficiencies are addressed for the different test environments for UL, DL, FDD RIT and TDD RIT. Next, the performance evaluation results related to mobility are shown. The traffic channel link data rates and mobility classes are considered. Finally, the VoIP capacity results are presented. The detailed results as well the simulation assumptions are described in (WINNER+ 2010). Furthermore the analytical and inspection performance assessment by WINNER+ can be found in Appendices E.1.2 and E.1.1, respectively.

Cell and Cell Edge Spectral Efficiency

The requirements and simulation results for the cell spectral efficiency are summarized for the FDD and TDD RIT in Table 10.6. Cell-edge spectral efficiency results are summarized in Table 10.7. These results are averaged using different simulators and the same antenna configurations. The simulations were conducted using the common set of parameters summarized in Table 10.8. Detailed simulation results and assumption are described in (WINNER+ 2010). Simulation results show that the requirements (see Table10.4) are achieved for all environments by using different MIMO features of LTE-A.

Mobility

The channel link data rates for FDD and TDD RIT for different environments are shown and compared to the ITU-R Requirements in Table 10.9. The results for FDD are the average of results using

Table 10.6   Cell spectral efficiency (CSE) results for FDD and TDD RIT

Test environments	Indoor	Microcellular	Base coverage urban	High speed
Requirement DL [bps/Hz/Cell]	3	2.6	2.2	1.1
CSE for FDD	4.10	2.88	2.38	3.15
	4×2 SU-MIMO	4×2 MU-MIMO	4×2 MU-MIMO	4×2 MU-MIMO
CSE for TDD	4.92	2.75	2.31	2.67
	4×2 MU-MIMO	4×2 MU-MIMO	4×2 MU-MIMO	4×2 MU-MIMO
Requirement UL [bps/Hz/Cell]	2.25	1.80	1.4	0.7
CSE for FDD	6.06	2.59	2.94	2.38
	1×4 MU-MIMO	2×4 SU-MIMO	2×4 BF	2×4 SU-MIMO
CSE for TDD	5.67	2.35	1.81	2.19
	1×4 MU-MIMO	2×4 SU-MIMO	2×4 SU-MIMO	2×4 SU-MIMO

Table 10.7 Cell edge spectral efficiency (CESE) results for FDD and TDD RIT

Test environments	Indoor	Microcellular	Base coverage urban	High speed
Requirement DL [bps/Hz/Cell]	0.1	0.075	0.06	0.04
CESE for FDD	0.173	0.089	0.067	0.091
	4 × 2 SU-MIMO	4 × 2 MU-MIMO	4 × 2 MU-MIMO	4 × 2 MU-MIMO
CESE for TDD	0.149	0.085	0.067	0.083
	4 × 2 MU-MIMO	4 × 2 MU-MIMO	4 × 2 MU-MIMO	4 × 2 MU-MIMO
Requirement UL [bps/Hz/Cell]	0.07	0.05	0.03	0.015
CESE for FDD	0.442	0.127	0.092	0.117
	1 × 4 MU-MIMO	2 × 4 SU-MIMO	2 × 4 BF	2 × 4 SU-MIMO
CESE for TDD	0.413	0.114	0.084	0.106
	1 × 4 MU-MIMO	2 × 4 SU-MIMO	2 × 4 SU-MIMO	2 × 4 SU-MIMO

Table 10.8 Models and assumptions

Parameter	Value
Bandwidth	10 MHz DL + 10 MHz UL for FDD, 10 MHz for TDD, double bandwidth for InH.
Scheduler	DL: Proportional Fair in Time and Frequency.
Receiver type	MMSE with intercell interference suppression capabilities in DL and UL.
Network synchronization	Synchronized.
Antenna configuration at base station	(a) Uncorrelated co-polarized (used for InH DL/UL and UMi DL baseline): co-polarized antennas separated four wavelengths. (b) Correlated co-polarized (used otherwise) 0.5 wavelengths between antennas.
Antenna configuration at UE	Vertically polarized antennas 0.5 wavelengths separation at UE.
Channel estimation	Nonideal channel estimation.

different assumptions, so that the mean value does not represent the performance of one particular system setup. The obtained results show that the requirements are fulfilled. The support of mobility classes for the TDD and FDD RITs are summarized in Table 10.10 and the requirements are also achieved for all the environments.

VoIP Capacity

The VoIP capacity simulation results for FDD RIT and TDD RIT are described in Table 10.11. Different assumptions were made in the underlying simulations, so that the mean values do not represent

Table 10.9 Traffic channel link data rates for FDD and TDD RIT

Test environments	Indoor	Microcellular	Base coverage urban	High speed
Requirement UL [bps/Hz]	1	0.75	0.55	0.25
Data rates for FDD	> 2	1.25	1.35	1.47
Data rates for TDD	> 2	1.2	1.3	1.4

Table 10.10 Mobility class support for FDD and TDD RIT

Test environments	Indoor	Microcellular	Base coverage urban	High speed
Required mobility classes	Stationary, pedestrian	Stationary, pedestrian, Vehicular (up to 30 km/h)	Stationary, pedestrian, vehicular	High speed vehicular, vehicular
Required mobility classes supported in FDD RIT	Yes	Yes	Yes	Yes
Required mobility classes supported in TDD RIT	Yes	Yes	Yes	Yes

Table 10.11 VoIP capacity for FDD and TDD RIT

Test environments	Indoor	Microcellular	Base coverage urban	High speed
Requirements [active users/sector/MHz]	50	40	40	30
Capacity for FDD	148	83	66	94
Capacity for TDD	139	70	65	78

the performance of one particular system setup. It is shown that the requirements are achieved for all environments.

10.3.3 LTE-Advanced Performance in the Rural Indian Open Area Scenario

The ITU-R defined evaluation criteria but was open for adding new evaluation environments and this opportunity was exploited in ITU-R (2010a), which proposed an additional deployment scenario that is referred to as the so-called Indian Rural environment. The "Rural Indian Open Area" deployment scenario assumes a large-cell coverage. The parameters of the scenario like carrier frequency, UE antennas height, intersite distance, etc., may take several different values, for example, the intersite-distance is 30 km to 50 km which corresponds to typical distances between villages in India. In this scenario UEs are in fixed positions with rooftop directional antennas.

The "Rural Indian Open Area" scenario is interference limited, so there is very limited sensitivity to frequency bands, antenna heights, and intersite distance. In Table 10.12 system performance in this scenario, as simulated by WINNER+, is summarized and compared with other scenarios. The results obtained by TCOE India and WINNER+ evaluation groups confirm that the requirements for this test scenario are met.

10.4 Channel Model Implementation and Calibration

10.4.1 IMT-Advanced Channel Model

The IMT-Advanced channel model defined by the ITU-R is a stochastic and geometric model that allows creating a bidirectional radio channel ITU-R (2009c). Although it is a geometric-based model that knows the exact location of transmitting and receiving elements, it does not specify the position of the scatterers. Only ray directions are known.

Table 10.12 LTE-A performance in rural Indian open area compared with performance in other test scenarios

Test scenario	MIMO scheme	Spectral efficiency [bps/Hz/cell]	
		Cell	*Cell edge*
Rural Indian Open Area	DL 4 × 2 SU-MIMO	2.41	0.071
(ISD = 40 km, f = 2300 MHz)	UL 4 × 2 SU-MIMO	2.64	0.124
Base Coverage Urban	DL 4 × 2 MU MIMO	2.38	0.067
(ISD = 500 m, f = 2 GHz)	UL 2 × 4 BF	2.94	0.092
High Speed	DL 4 × 2 MU-MIMO	3.15	0.091
(ISD = 1732m, f = 800 MHz)	UL 2 × 4 SU-MIMO	2.38	0.117

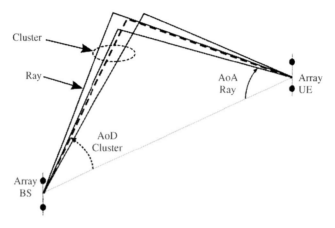

Figure 10.3   IMT-Advanced MIMO channel

Figure 10.3 shows a transmitter and receiver and all existing rays between them. Moreover, this figure represents the concept of cluster, or propagation path – in space, time and angle – that consists of a set of rays affected by nearby scatterers. The figure also includes the concept of Angle of Departure (AoD) and Angle of Arrival (AoA), both at cluster and ray level.

Being a stochastic model, channel parameters are derived from a set of statistical distributions obtained from channel measurements. These measurements were made in extensive trials covering different propagation scenarios in Line of Sight (LoS) and Non Line of Sight (NLoS) conditions. Statistical distributions are defined, for example, in terms of delay, Delay Spread (DS), Shadow Fading (SF) and crosspolarization ratio.

In the IMT-Advanced channel model there are two different sets of channel parameters. The first one is related to the large-scale parameters, such as SF and path loss. The second one concerns small-scale parameters, including angle of arrival and departure of the rays.

In order to generate channel samples between one transmitter and one receiver, mobility and exact location of both ends must be known. Based on this, all large-scale parameters are generated, followed by the small scale parameters. All these parameters are kept constant during the whole simulation. Channel samples are obtained by adding the contributions of all involved rays.

As explained in section 10.1.2, there are five different scenarios and thus channel models: indoor hotspot, urban microcell, urban macrocell, suburban macrocell and rural macrocell. Different models are characterized by different parameters of the statistical distributions used to generate the channel

Table 10.13 Parameters of the IMT-Advanced channel model

Scenario	DS $log10(s)$		ASD $log10(deg)$		ASA $log10(deg)$		SF [dB]	K Factor [dB]	
	μ	σ	μ	σ	μ	σ	σ	μ	σ
InH LoS	−7.70	0.18	1.60	0.18	1.62	0.22	3	7	4
InH NLoS	−7.41	0.14	1.62	0.25	1.77	0.16	4	−	−
UMi LoS	−7.19	0.4	1.20	0.43	1.75	0.19	3	9	5
UMi NLoS	−6.89	0.54	1.41	0.17	1.84	0.15	4	−	−
UMi OtoI	−6.62	0.32	1.25	0.42	1.76	0.16	7	−	−
SMa LoS	−7.23	0.38	0.78	0.12	1.48	0.20	4	9	7
SMa NLoS	−7.12	0.33	0.90	0.36	1.65	0.25	8	−	−
UMa LoS	−7.03	0.66	1.15	0.28	1.81	0.20	4	9	3.5
UMa NLoS	−6.44	0.39	1.41	0.28	1.87	0.11	6	−	−
RMa LoS	−7.49	0.55	0.90	0.38	1.52	0.24	4	7	4
RMa NLoS	−7.43	0.48	0.95	0.45	1.52	0.13	8	−	−

samples. Table 10.13 summarizes the main values of these distributions for each model, distinguishing among the following sight conditions: LoS, NLoS and Outdoor to Indoor (OtoI).

The first parameter, DS, is the delay spread . For each link, a normal random number must be generated with mean μ and standard deviation σ. The resulting number is the delay spread in logarithmic units (dB). Note that DS is, in linear scale, in the order of 10^{-7}s, which is well aligned with the classical tapped delay line models. The lower values of DS are for the Urban Microcell (UMi) model, whereas highest ones correspond with Rural Macrocell (RMa).

Angle Spread Departure (ASD) and Angle Spread Arrival (ASA) parameters measure the ray dispersion between the output of the transmitter and the input of the receiver, respectively. Concerning SF values, these are similar for all scenarios and higher in case of NLoS. Finally, the K factor applies only for LoS as it represents the power ratio among line-of-sight ray and the others.

Coordination between evaluation groups was strongly recommended by ITU-R to facilitate comparison and consistency of results and to ease the understanding of differences in evaluation results achieved by the independent evaluation groups. Indeed, the divergence in the results obtained in the evaluation of the same system is a common problem encountered in all fora where researchers coming from different bodies try to provide their contributions to the progress of science and technology. A possibility for overcoming this situation is the comparison of different approaches using the same calibration process and benchmark data. In this framework, and in order to simplify the IMT-Advanced assessment, ITU-R approved a number of documents describing the evaluation process, requirements and evaluation criteria. In particular, ITU-R (2009c) contains the detailed simulation assumptions and the evaluation methodologies of IMT-Advanced. The M.2135 document represents a significant calibration effort that intends to ensure proper harmonization of the tools used by the external evaluation groups for performance evaluation of the IMT-Advanced technologies.

ITU-R (2009c) is mainly focused on the definition of the reference scenarios for system-level simulations including large- and small-scale parameters of the channel model. The new stochastic geometric model proposed by the ITU-R is far from being simple to implement. Several implementations were freely offered, but the main problem is that these implementations are not coherent and do not provide the same output statistics. Without a proper calibration of the channel model implementation it would have been not feasible to build up a consistent evaluation of candidates. This is why channel calibration has been so important from the beginning.

Table 10.14 Large-scale assumptions

Cell selection	1 dB Handover margin				
Feeder loss	2 dB				
BS antenna tilt	InH	UMi	UMa	RMa	SMa
	N.A	$12°$	$12°$	$6°$	$6°$

The following subsections intend to describe the channel model calibration effort made in WINNER+ (2010). Moreover, this action aimed to share the experience, information and benchmark data with other evaluation groups in order to foster the required coordination and unification of results.

The calibration data presented in this section focuses on one scenario, the Urban Macrocell (UMa), serving as an example of the general procedure. More information and results can be found at the WINNER+ IMT-Advanced evaluation web page.[1] Calibration is split in two main parts: large-scale and small-scale parameters.

10.4.2   Calibration of Large-Scale Parameters

For large-scale calibration, multicell system level simulations are required. Given that this kind of simulator is also to be used for evaluating the IMT-Advanced requirements of cell spectral efficiency, cell-edge user throughput, VoIP capacity and mobility, large-scale parameter calibration is also useful in the sense of detecting potential simulator incoherences among contributors. Some important properties of the system simulations are determined by the environment description in ITU-R (2009c), including the propagation and channel models. The metrics used in this calibration are the path gain and the wideband Signal to Noise plus Interference Ratio (SNIR), which are essentially technology independent, and hence calibration of these metrics can be performed using just a few additional assumptions compared to what is given in (ITU-R 2009c). The path gain is defined as the average signal attenuation between a UE and its serving base-station cell. The measure includes distance attenuation, shadowing and antenna gains (both at the base station and at the UE) while the effects from fast fading are excluded. The path gain may hence be defined as the difference between the (average) received power and the (average) transmitted power.

$$\text{Pathgain} = P_{rx} - P_{tx}[dB] \tag{10.1}$$

The DL wideband SNIR is the (average) power received from the serving cell in relation to the (average) received power from all other cells plus noise. For a UE connected to the base station cell i the geometry (G) is defined as:

$$G = \frac{P_{rx,i}}{\sum_{\forall j \neq i} P_{rx,j} + N_0} \tag{10.2}$$

where $P_{rx,j}$ is the received power from the base station cell j and N_0 is the noise power.

In addition to the evaluation principles and assumption in ITU-R (2009c) and the channel model clarifications that followed, Table 10.14 summarizes other assumptions that have been used to derive the path gain and wideband SNIR distributions.

---

[1] http://projects.celtic-initiative.org/winner+/

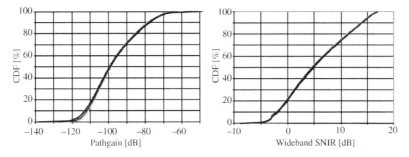

Figure 10.4 Path gain and wideband SNIR distributions in the UMa scenario

Figure 10.4 presents calibration results from up to seven WINNER+ partners. Some very small differences appear in terms of path gain, but these differences decrease when considering the wideband SNIR distribution.

10.4.3 Calibration of Small-Scale Parameters

In the small-scale fading characteristics the delay spread and the angular spread at the base station and at the UE are included. For simplicity, the small-scale fading calibration assumes omnidirectional antennas at both the base station and the UE. If other antenna patterns are assumed, for example a directional antenna pattern at the base station, the results will be different. Moreover, the calibrations are performed separately for LoS, NLoS and OtoI propagation conditions. OtoI propagation is relevant only in the UMi scenario. For calibration of the angular spread for LoS propagation channels it is important to account for the correction under ITU-R (2009b) Section 3. Now assume that each propagation channel comprises N clusters and that each cluster comprises M rays. Assume further that the delay of ray m in cluster n is denoted $\tau_{n,m}$ and that the associated power is denoted $p_{n,m}$. In case of LoS propagation the LoS ray is here included as a separate cluster for which, according to (ITU-R 2009c), only the first ray in the cluster has a non-zero power.

To calculate the delay spread, the average delay is first calculated according to:

$$\bar{\tau} = \frac{\sum_{n=1}^{N} \sum_{m=1}^{M} \tau_{n,m} \cdot p_{n,m}}{\sum_{n=1}^{N} \sum_{m=1}^{M} p_{n,m}} \tag{10.3}$$

Then, the Root-Mean-Square (RMS) delay spread σ_τ is calculated as follows:

$$\sigma_\tau = \sqrt{\frac{\sum_{n=1}^{N} \sum_{m=1}^{M} (\tau_{n,m} - \bar{\tau})^2 \cdot p_{n,m}}{\sum_{n=1}^{N} \sum_{m=1}^{M} p_{n,m}}} \tag{10.4}$$

For the angular spread we use the circular angular spread σ_{AS} as defined in (3GPP 2009c, Annex A), where the angular spread is the minimum spread over different linear shifts Δ. One small addition is used here, however. Before calculating $\theta_{n,m,\mu}(\Delta)$ we wrap the quantity $\mu_\theta(\Delta)$ into the interval $[-\pi, \pi]$ according to Equation (10.5). This step is not explicitly stated in (3GPP 2009c).

$$\mu_\theta(\Delta) = \begin{cases} 2\pi + \mu_\theta(\Delta) & \text{if } \mu_\theta(\Delta) < -\pi \\ \mu_\theta(\Delta) & \text{if } |\mu_\theta(\Delta)| \leq \pi \\ \mu_\theta(\Delta) - 2\pi & \text{if } \mu_\theta(\Delta) > \pi \end{cases} \tag{10.5}$$

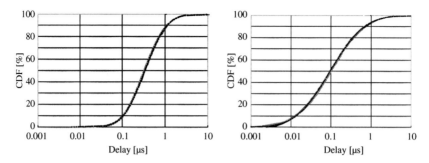

Figure 10.5   RMS delay spread for UMa NLoS (left plot) and LoS (right plot)

The RMS delay spread (σ_τ) and the circular angular spread (σ_{AS}) at the base station and at the UE are calculated for a large number of radio links and in the calibrations depicted in Figure 10.5 the corresponding distributions are compared.

10.5   Simulator Calibration

System-level simulator calibration is the process that aims the alignment of the simulation tools. The alignment can be achieved by changes to the simulator driven by the feedback acquired from the comparison with results from other simulator tools or literature results. The purpose of the calibration is to be sure that the different simulation tools produce comparable outputs.

A review of system-level simulator building blocks shows that calibration is not a trivial task due to the system complexity and the high number of degrees of freedom in modeling and implementation. In general, system-level simulations are focused on modeling the operation of network components based on the performance of multiple links between network nodes and UEs. Resource management is to be modeled as well as effects of radio resources limitations in terms of capacity and coverage. Typical system-level simulators (Döttling et al. 2009, Chapter 12) comprise the following main building blocks:

- user and network deployment modules;
- radio channel model;
- link-to-system interface;
- traffic generator;
- scheduler.

In practice, there is always room for differentiation of any given module operation modeling. Deviations can originate from partial standardization that requires taking additional, sometimes implicit, assumptions or can originate from bug detection in implementation. In order to minimize differences in simulators, the use of particular system configurations with stable, well defined features is preferred. During the IMT-Advanced evaluation process, for example, the LTE Rel-8 basic configuration (3GPP 2010b) served as such a reference. It is important to understand that the priority for the calibration step is not to achieve the best performance figures but to obtain results aligned among calibration partners.

System-level calibration is reasonable if there is confidence that the employed channel model is already calibrated. On top of this confidence there is an opportunity to perform tests that involve as many simulator building blocks as possible to achieve high coverage of the calibration range.

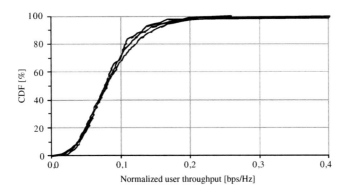

Figure 10.6 CDF of normalized user throughput in UL UMi deployment scenario. Results originated from three organizations

When planning the calibration there is also a need to decide on indicators that should be compared during the process. It is reasonable to choose performance indicators that are most important from user experience point of view. The normalized user throughput is an example of metric to compare.

The claim that results come from a calibrated simulator is true when this simulator was calibrated in a particular deployment scenario. Claim of calibration in one deployment scenario does not automatically mean calibration in any other deployment scenario.

The WINNER+ Evaluation group performed a calibration of the available simulation tools following the aforementioned approach. The simulators were calibrated in UL and DL. In Figure 10.6 an example of a calibration result for UL Urban Micro deployment scenario ITU-R (2009c) is shown.

Numerical characterization of calibration accuracy might help to track progress towards achieving simulator alignment and also allows indicating good and weak calibrated scenarios. Quantitative characterization of calibration accuracy is one of the factors that can support a decision to recognize simulators as calibrated. However, it is difficult to provide a rationale for a classifier based on a partcular numerical limit, which will indicate the state of calibration or noncalibration. The decision on an exact numerical value that will classify simulators as calibrated or not would be arbitrary. That is why a quantitative characterization of calibration accuracy is only an element of support for the final decision which is done by the researcher taking multiple indicators into account.

10.6 Conclusion and Outlook on the IMT-Advanced Process

As a result of the external IMT-Advanced candidate evaluations, there were 13 independent evaluation reports apart from the WINNER+ report (ITU-R 2010b) submitted to ITU-R. The WINNER+ simulation results provided in this chapter have confirmed that the 3GPP LTE Rel-10 proposal satisfies all the IMT-Advanced requirements. In fact, despite differences in implementation details, all reports confirmed the proponents' claims that both LTE-A and WirelessMAN-Advanced (the WiMAX-based proposal, part of the IEEE 802.16 standard) fulfill all ITU-R requirements for IMT-Advanced systems.

Taking into account the evaluation reports, ITU-R WP 5D decided in October 2010 that both submitted IMT-Advanced system proposals, LTE-A and WirelessMAN-Advanced, have successfully met all of the criteria established by ITU-R for the first release of IMT-Advanced, qualifying them as the first "true 4G systems" (ITU-R 2010c).

These two technologies will now enter the final stage of the IMT-Advanced process (step 8 of the list given at the beginning of this chapter), that is, the development of an ITU-R Recommendation specifying the in-depth technical standards for these radio technologies. This final step is expected to be completed in early 2012.

References

3GPP 2009a Further Advancements for E-UTRA (LTE-Advanced). Technical Specification Group Radio Access Network 36.912 v9.3.0, 3rd Generation Partnership Project (3GPP), http://www.3gpp.org/ftp/Specs/html-info/.

3GPP 2009b Requirements for Evolved UTRA (E-UTRA) and Evolved UTRAN (E-UTRAN). Technical Specification Group Radio Access Network 36.913 v9.0.0, 3rd Generation Partnership Project (3GPP), http://www.3gpp.org/ftp/Specs/html-info/.

3GPP 2009c Spacial channel model for Multiple Input Multiple Output (MIMO) Simulations. Technical Specification Group Radio Access Network 25.996 v9.0.0, 3rd Generation Partnership Project (3GPP), http://www.3gpp.org/ftp/Specs/html-info/.

3GPP 2010a Evolved Universal Terrestrial Radio Access (E-UTRA) and Evolved Universal Terrestrial Radio Access Network (E-UTRAN); Overall description; Stage 2. Technical Specification Group Radio Access Network 36.300 v9.5.0, 3rd Generation Partnership Project (3GPP), http://www.3gpp.org/ftp/Specs/html-info/.

3GPP 2010b LTE; Evolved Universal Terrestrial Radio Access (E-UTRA); Further Advancements for E-UTRA Physical Layer Aspects. Technical Specification Group Radio Access Network 36.814 v9.0.0, 3rd Generation Partnership Project (3GPP), http://www.3gpp.org/ftp/Specs/html-info/.

Döttling M, Mohr W and Osseiran A 2009 Radio Technologies and Concepts for IMT-Advanced. John Wiley & Sons, Ltd, Chichester.

ITU-R 2003 Framework and overall objectives of the future development of IMT-2000 and systems beyond IMT-2000. Recommendation ITU-R M.1645-E, International Telecommunications Union Radio (ITU-R), http://www.itu.int/rec/R-REC-M.1645/en.

ITU-R 2008a Background on IMT-Advanced. Document IMT-ADV/1-E, International Telecommunications Union Radiocommunication (ITU-R) Study Groups Working Party 5D, http://www.itu.int/md/R07-IMT.ADV-C-0001/en.

ITU-R 2008b Invitation for submission of proposals for candidate radio interface technologies for the terrestrial components of the radio interface(s) for IMT-Advanced and invitation to participate in their subsequent evaluation. Circular Letter 5/LCCE/2, International Telecommunications Union Radiocommunication (ITU-R) Study Groups Working Party 5D, http://www.itu.int/md/R00-SG05-CIR-0002/en.

ITU-R 2008c Requirements related to technical performance for IMT-Advanced radio interface(s). Report ITU-R M.2134, International Telecommunications Union Radio (ITU-R), http://www.itu.int/publ/R-REP/en.

ITU-R 2009a Acknowledgement of candidate submission from 3GPP proponent (3GPP organization partners of ARIB, ATIS, CCSA, ETSI, TTA AND TTC) under Step 3 of the IMT-Advanced process (3GPP technology). Document IMT-ADV/8-E, International Telecommunications Union Radiocommunication (ITU-R) Study Groups Working Party 5D, http://www.itu.int/md/R07-IMT.ADV-C-0008/en.

ITU-R 2009b Correction of typographical errors and provision of missing texts of IMT-Advanced channel models in Report ITU-R M.2135. Document IMT-ADV/3-E, International Telecommunications Union Radiocommunication (ITU-R) Study Groups Working Party 5D, http://www.itu.int/md/R07-IMT.ADV-C-0003/en.

ITU-R 2009c Guidelines for Evaluation of Radio Interface Technologies for IMT-Advanced. Report ITU-R M.2135-1, International Telecommunications Union Radio (ITU-R), http://www.itu.int/publ/R-REP/en.

ITU-R 2010a Evaluation of IMT-Advanced Candidate Technology Submissions in Documents IMT-ADV/4 and IMT-ADV/8 by TCOE India. Document IMT-ADV/16-E, International Telecommunications Union Radiocommunication (ITU-R) Study Groups Working Party 5D, http://www.itu.int/md/R07-IMT.ADV-C-0016/en.

ITU-R 2010b Evaluation of IMT-Advanced Candidate Technology Submissions in Documents IMT-ADV/6, IMT-ADV/8 and IMT-ADV/9 by WINNER+ Evaluation Group. Document IMT-ADV/22-E, International Telecommunications Union Radiocommunication (ITU-R) Study Groups Working Party 5D, http://www.itu.int/md/R07-IMT.ADV-C-0022/en.

ITU-R 2010c ITU Paves Way for Next-generation 4G Mobile Technologies – ITU-R IMT-Advanced 4G Standards to Usher New Era of Mobile Broadband Communications. Press release, International Telecommunications Union Radio (ITU-R), http://www.itu.int/net/pressoffice/press_releases/2010/40.aspx.

Network GTSGRA 2008a IMT-Advanced Workshop, Shenzhen, http://www.3gpp.org/ftp/workshop.

Network GTSGRA 2008b LTE-Advanced Workshop, Prague. Available online at http://www.3gpp.org/ftp.

WINNER+ 2010 Celtic Project CP5-026 Final conclusions on end-to-end performance and sensitivity analysis (ed. Werner M). Public Deliverable D4.2, Wireless World Initiative New Radio – WINNER+, June 2010, http://projects.celtic-initiative.org/winner+/index.html (accessed June 2011).

WINNER-II 2008 IST-4-027756 WINNER II Channel Models. Public Deliverable D1.1.2, Wireless World Initiative New Radio – WINNER, April 2008, http://projects.celtic-initiative.org/winner+/index.html (accessed June 2011).

11

Future Directions

Afif Osseiran, Jose F. Monserrat and Werner Mohr

It has been forecast that the increase in users' demand for mobile Internet data applications will overload Third Generation (3G) networks capacity in the coming years and will force International Mobile Telecommunications Advanced (IMT-Advanced) deployment to be optimized. In addition to the overall increase in throughput demand, there is a need for improving the ubiquity of the Quality of Service (QoS) indicators – allowing the mobile users to experience high QoS values in any geographical position, while minimizing consumption of radio resources and energy. Future developments of IMT-Advanced systems include more efficient physical mechanism to deal with the instable radio propagation and crowded radio frequency spectrum. Nevertheless, while physical layer improvements are already close to their physical limits and only the higher order Multiple-Input Multiple-Output (MIMO) mechanism seems to be able to improve system performance in terms of spectral efficiency, there is still huge potential to maximize efficiency at the radio resource and optimize interference management. There have been further developments in mobile communication systems to manage multicell and multiuser MIMO, together with new paradigms of network deployment including advanced relaying and Heterogeneous Networks (HetNets). In HetNets, cooperative Radio Resource Management (RRM) will consist first of coordinated scheduling/beamforming and next of joint transmission techniques. Despite these improvements, it is obvious that IMT-Advanced and beyond IMT-Advanced wireless high bit rate demands will require a high density of base stations. Since the conventional cellular system design is too expensive to deploy such a density, it is foreseen that relaying techniques (with radio backhaul links) will play an important role.

Previous chapters, presented the state of the art of techniques for IMT-Advanced and beyond. Several techniques within the areas of Radio Resource Management, Multiple-Input Multiple-Output, Coordinated Multipoint transmission or reception, Relaying, Network Coding and Device-to-Device communications have been surveyed, showing their potential. Many of those technological areas are evolving to meet the continuing increase in the data rate demands. In this short chapter, the main challenges and limits of those techniques are analyzed. This chapter also addresses green and energy efficient communications, a new research area with high potential.

Mobile and Wireless Communications for IMT-Advanced and Beyond, First Edition.
Edited by Afif Osseiran, Jose F. Monserrat and Werner Mohr.
© 2011 Afif Osseiran, Jose F. Monserrat and Werner Mohr. Published 2011 by John Wiley & Sons, Ltd.

11.1   Radio Resource Allocation

Future wireless systems will need to provide high data-rate services with high flexibility and reliability to guarantee QoS to the large number of end users. The principal techniques to be included in
next-generation wireless systems include several aspects of Advanced RRM that have a direct impact
on each user's performance, as well as the overall performance of the network. Within Advanced
RRM concept main topics are interference coordination, support for contiguous or noncontiguous
Carrier Aggregation (CA) and multidimensional scheduling.

Concerning interference coordination, decentralized interference avoidance using Busy Bursts
(BuBs) was proposed in Chapter 2. Although the scheme yields an impressive sevenfold increase
in cell-edge user throughput, it assumes a Time Division Duplex (TDD) system. In addition, it comes
at the expenses of introducing a low latency feedback channel and intercellular synchronization.
Hence, it would be interesting to extend the interference avoidance using BuBs to a Frequency Division Duplex (FDD) system and to conceive methods allowing robust intercellular synchronization.

With respect to CA, the new research lines focus on fractional soft handover in CA (Chang et al.
2009), flexible carrier aggregation for home base station (Li et al. 2009) and neighbor carrier signal
strength estimation (Yuan et al. 2010). It is worth noting that the use of large fragmented bandwidth
puts stringent *hardware limitations* on the system. Some of these limitations are: the antenna size,
the antenna radiation efficiency for wide bandwidth to be processed, the intermodulation distortion,
the size of the Fast Fourier Transform (FFT), and so forth. Therefore, new solutions should be found,
such as reconfigurable and more efficient antennas.

Finally, the multidimensional scheduling – frequency-time-space-transmitter selection – is quite a
difficult optimization problem that so far has been only very partially tackled in WINNER+ (2010).
Besides, the underlying objective of fulfilling QoS must be extended to upper layers to guarantee
an appropriate end-to-end experience. In this sense, there are a lot of procedures to be designed
concerning IP Multimedia Subsystem (IMS) interworking and real-time flow management. Other
topics like resource management in relaying and peer-to-peer scenarios are areas where little activity
has occurred but they need further research in the coming years before they can be completely adopted
in technological standards.

11.2   Heterogeneous Networks

The term Heterogeneous Network (HetNet) has been introduced in the Third Generation Partnership Project (3GPP) to denote advanced cell structures where additional nodes (pico-/femto- cells,
relays) are deployed within the local area range, mainly in hot spots. HetNets aim to improve overall
capacity as well as providing cost-effective coverage, a capacity extension, and a green radio solution. Note that this definition is different to that of classical heterogeneous networks understood as
a composition of multiple Radio Interface Technologys (RITs), for this reason the 3GPP choose the
nomenclature HetNet.

HetNets have been gaining much momentum in recent years and there is an urgent need to better understand their particularities in radio planning issues and cooperation strategies among nodes.
Indeed 3GPP LTE-Advanced (LTE-A) has recently started a new study item to investigate their
deployment (3GPP 2009). One of the main problems to solve is interference mitigation between
high- and low-power nodes, which is much more significant than in homogeneous networks. In order
to allow a feasible coexistence, specialized Inter-Cell Interference Coordination (ICIC) mechanisms
are needed (3GPP 2010c). It is agreed that, at their current state, classical ICIC fails to prevent severe (DL) interference at control channels and it is not fully effective with data channels. Related
research work on femtocells reveals that advanced MIMO techniques as well as Coordinated Multipoint transmission or reception (CoMP) for femtocells can minimize interference problem. This is

a challenging task because the control signaling between the two tiers is limited or does not exist. A possible solution to the femto-to-femto interference problem would use spectrum sensing to identify unused Physical Resource Blocks (PRBs). Macro-to-femto interference could also use spectrum sensing; the Home eNB (HeNB) would choose only those resources where there is no interference. The concept of spectrum sensing is not new but the characteristics of the femtocells likely require a different approach to sensing and utilizing unused spectrum blocks.

Another group of investigations deals with load balancing among the macro- and lower cell layers. These mainly focus on those parameters that affect the cell (re)selection procedure and indeed cell (re)selection methods have been identified as another key ingredient for HetNets (3GPP 2010c). Work dealing with modifications to the cell range extension (3GPP 2010a) is important in this field.

HetNets are expected to increase drastically the number of nodes to supervise. Thus, with the aim of reducing or at least not increasing OPerational EXpenditures (OPEX), research must be directed towards autonomous processes. This is why techniques like HetNets are likely to be used in the context of Self-Organized Networks (SONs).

11.3   MIMO and CoMP

A major challenge for wireless communication systems is how to allocate resources among users across space (including different cells), frequency and time dimensions and jointly design all the transceivers with different system optimization objectives. This remains unresolved for a large variety of optimization criteria, especially when combined with practical modulation and coding schemes. The problem is a difficult nonconvex combinatorial problem with integer constraints and finding jointly optimal solutions is most likely not possible. Therefore, efficient suboptimal solutions are required (Gesbert et al. 2010).

In MIMO, a major challenge is also to keep the Channel State Information (CSI) signaling overheads over the air at a minimum, which calls for decentralized schemes that are primarily based on locally available information and are robust to heavily quantized channel information. Thus, the design of transmission methods that are robust to CSI uncertainty, and sounding strategies that provide a good tradeoff between performance and overhead are needed.

In cellular systems, the evolution of multiantenna processing is heading towards CoMP techniques, where multiple Base Stations (BSs) cooperate in order to provide spatial access to the users. It is therefore a requirement for the practical future MIMO methods to support the CoMP paradigm, both from the signaling and the precoding point of view.

The CoMP systems represent a topic of wide interest in the research community. As for the MIMO systems, the challenge of CoMP is to design a system robust to CSI estimation impairments, and to be able to design powerful CSI estimation techniques for different mobility scenarios. An example is the high mobility scenario in conjunction with Joint Processing (JP) CoMP.

It is expected that future research in wireless systems will mainly focus on several aspects of CoMP: architectural, theoretical and practical.

From an architectural point of view, there are simplifications to be sought: decentralized approaches, distributing complexity among nodes and clustering. In particular, dynamic clustering is considered a viable solution for future CoMP methods. The introduction of HetNets, then, could be an option to exploit coordination also among different cells in a multilayer environment.

From a theoretical point of view, the most important question to be solved is that of the achievable tradeoff between the amount of information exchanged among the nodes and the expected performance gain (Gesbert et al. 2010). Another related important tradeoff, with achievable performance, is between the amount of feedback and performance gain.

Finally, synchronization of the cooperative entities in CoMP schemes is another research topic that deserves attention in the future, especially for the schemes requiring higher coherence.

11.4 Relaying and Network Coding

Relaying is still an active and promising research area. In particular, the scientific community is giving increasing attention to the idea of extending relaying to the multihop case – where there are more than two hops – and using mobile devices as relays instead of fixed relays.

The daunting task of multihop relaying has already been considered by the IEEE and 3GPP standardization bodies for implementation in the near future. The need to implement multihopping resides in the fact that in some scenarios the coverage extension provided by a single-hop relay does not suffice. One problem in multihop relaying is the difficulty of synchronization due to the unavailability of wired backhaul by construction or economic reasons.

The use of relaying in a mobile framework such as high-speed trains is very attractive from the economic point of view. Mobile relaying implies that the RN is moving. An example of mobile RN could be a device mounted on top of a car, a bus or a train, where a large number of users are located and would like to experience a wireless high data-rate network access. In this case, the backhaul link must cope with extreme conditions such as frequent handovers. The mobile Relay Node (RN) will complicate cell planning and long-term interference mitigation in networks. Hence there is an opportunity to study and propose ideas that can tackle these difficulties.

The research community continues to have high interest in the application of Network Coding (NC) to wireless communications. The plethora of publications on that topic did not solve all the related challenges (Li 2009). Hence, there are many opportunities that are still unexplored. More specifically, some areas with high potential for future research are:

- to jointly optimize source coding, channel coding and Network Coding;
- to optimally use Network Coding for retransmission under a unicast scenario;
- to apply Network Coding to a multi-cell context for instance in a CoMP scenario;
- to investigate whether the Network Coding operation shall be applied at the signal or bit or symbol level;
- to design resource allocation mechanisms for Network Coding-based wireless networks.

Finally NC can be applied to emerging areas such as cognitive radio and power/energy efficient networks.

11.5 Device-to-Device Communications

There are still some open research questions that need to be solved to enable Device-to-Device (D2D) underlay communications and eNodeB (eNB)-controlled D2D communications in general. The solution presented in Chapter 9 was limited to a single eNB that controls the D2D connections. However, two users might be located in different cells. Hence, resource allocation and mode selection needs to be coordinated across cells.

The cellular network can help to make the D2D session transparent to the UE. The approaches related to session setup assume that UEs are served either by the same PDN gateway or by the same Mobile Management Entity MME, that is, UEs served by the same operator. It is atypical for the whole circle of friends or customers to be served by the same operator and the same technology. Therefore, a solution that works across different operators would greatly enhance the user acceptance of D2D communications.

Moreover, future research should identify options to further increase the data rates of direct communication. Promising directions could be related to CA, which is currently standardized for Release 10 (Rel-10) of Long Term Evolution (LTE) in 3GPP (3GPP 2010b). Further, D2D communication could utilize additional bands that are not in use by cellular systems or license-exempt bands.

Finally, D2D communication could be an energy efficient way of implementing a light-weight femto eNB or HeNB. The light-weight HeNB will be idle when the members of its closed subscriber group are not nearby. The wide-area eNB will activate both the light-weight HeNB and a User Equipment (UE) in the vicinity and facilitate the D2D connection setup.

11.6   Green and Energy Efficiency

Data-communication traffic is currently increasing annually by 50% to 100%. It is expected that this growth will continue in the coming years. Energy consumption in communication networks is expected to increase, related to the growth in traffic. Studies indicate that Information and Communication Technologies (ICT) systems and devices contribute about 2% of global CO_2 emissions (EU Commission 2009). Energy consumption also results in a significant economic cost factor for the operation of communications networks. These issues are increasingly reaching the political agenda with the Report of the Brundtland Commission already in 1983 to the United Nations (Ban Ki-moon 2007; Barroso 2008; Brundtland 1983; Reding 2008). Reports from international organizations, such as Laitner and Ehrhardt-Martinez (2007), identify opportunities and challenges. Therefore, future ICT networks and systems have to be designed to be more energy efficient in order to reduce the CO_2 footprint and to reduce cost of network operation.

Today's mobile communication networks are designed under the paradigm of high spectral efficiency and peak throughput with respect to the limited available frequency spectrum for mobile and wireless communications and for economic deployment concepts. Energy consumption was not the major concern in the past. However, to meet the challenges of increasing energy efficiency in communication networks and systems it is necessary to investigate potential paradigm shifts and new technologies. In mobile communication networks the main source of energy consumption are the BSs. Figure 11.1 shows a rough estimate of the contributions of different network elements with

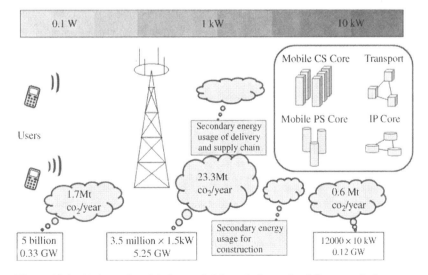

Figure 11.1   Estimate for global annual CO_2 emissions of mobile network elements

respect to their typical energy consumption and number in deployed networks. This estimation is based on the current number of world mobile subscribers.

Different aspects of energy savings in communication systems like energy efficient wireless transmission techniques, network architecture and protocols, backbone networks and cognitive opportunistic spectrum sharing schemes are under investigation – see IEEE 2010 for examples. Moreover, a research initiative called Green Touch was launched in 2010 to investigate communication networks from a holistic perspective (Green Touch n.d.).

With respect to Figure 11.1, there are two major research areas with high potential for energy savings in cellular mobile communication systems:

- network deployment concepts and
- BS implementation and transmission techniques.

The use of smaller cells provides some potential to reduce the overall power consumption in mobile networks. Depending on the propagation conditions and the relation between the necessary primary power for the RF-related and the non-RF-related part of the nodes, an optimum cell size can be expected. Multihop transmission offers an even bigger potential for power savings. Research on energy efficiency by means of network deployment concepts is related to HetNets (section 11.2) and Relaying (section 11.4).

Technology advanced in BS implementation, power amplifier design, cooling and power management has already resulted in significant savings.

A major potential for energy savings can be expected from the tradeoff between spectral efficiency and carrier bandwidth. Today's systems are designed for high spectral efficiency in order to achieve high peak throughput rates with respect to scarce frequency spectrum, which implies high necessary Signal-to-Noise ratios. According to Shannon's law (Shannon 1984) signal power and bandwidth can be traded against each other. A paradigm shift to lower order modulation schemes can reduce the necessary power consumption in a network significantly however at the expense of more necessary frequency spectrum. Research challenges are in the design of tradeoffs between these parameters and the associated sharing techniques.

References

3GPP 2009 Scenarios for Heterogeneous Network for LTE-A. Technical Specification Group Radio Access Network R1-093787, 3rd Generation Partnership Project (3GPP), http://www.3gpp.org/ftp/.

3GPP 2010a Cell Association Analysis in Outdoor Hotzone of Heterogeneous Networks. Technical Specification Group Radio Access Network R1-101083, 3rd Generation Partnership Project (3GPP), http://www.3gpp.org/ftp/.

3GPP 2010b LTE; Evolved Universal Terrestrial Radio Access (E-UTRA); Further Advancements for E-UTRA Physical Layer Aspects. Technical Specification Group Radio Access Network 36.814 v9.0.0, 3rd Generation Partnership Project (3GPP), http://www.3gpp.org/ftp/Specs/html-info/.

3GPP 2010c Techniques to Cope with High Interference in Heterogeneous Networks. Technical Specification Group Radio Access Network R1-100702, 3rd Generation Partnership Project (3GPP), http://ftp.3gpp.org/ftp/.

Ban Ki-moon 2007 Address to the High-Level Segment of the UN Climate Change Conference in Bali Indonesia, December 12, http://www.un.org/.

Barroso J 2008 ICT Industry has a Major Role to Play in the European Economy of the 21st Century. CeBIT Trade Fair in Hannover, March 3, http://europa.eu/.

Brundtland Commission 1983 Report Our Common Future. United Nations 1983 http://www.un-documents.net/ocf-02.htm.

Chang J, Li Y, Feng S, Wang H, Sun C and Zhang P 2009 A Fractional Soft Handover Scheme for 3GPP LTE-Advanced System *Proc. ICC 2009 – IEEE Int. Conf. Commun.*, pp. 1–5.

EU Commission 2009 *Communication from the Commission to the European Parliament, the Council, the European Economic and Social Committee and the Committee of the Regions on mobilising Information and Communication Technologies to facilitate the transition to an energy-efficient, low-carbon economy.* Technical Report COM(2009) 111 final, Commission of the European Communities, http://ec.europa.eu/.

Gesbert D, Hanly S, Huang H, Shamai Shitz S, Simeone O and Yu W 2010 Multi-Cell MIMO Cooperative Networks: A New Look at Interference. *IEEE J. Select. Areas Commun.* **28**(9), 1380–1408.

Green Touch n.d. *Green Touch Initiative*, http://www.greentouch.org/.

IEEE 2010 Energy Efficiency in Communications (eds Zhang H., Gladisch A., Pickavet M., Tao Z. and Mohr W.) *IEEE Commun. Mag.* **48**, 48–79.

Laitner J and Ehrhardt-Martinez K 2007 Advanced Electronics and Information Technologies: The Innovation-Led Climate Change Solution. Technical report, American Electronics Association Europe (AeA Europe), http://www.itaa.org.

Li J, Liu Y, Duan J and Liang X 2009 Flexible carrier aggregation for home base station in IMT-Advanced System *Proc. 5th International Conference on Wireless Communications, Networking and Mobile Computing*, pp. 1–4.

Li Y 2009 Distributed coding for cooperative wireless networks: An overview and recent advances. *IEEE Commun. Mag.* **47**(8), 71–77.

Reding V 2008 *The potential of ICT to contribute to energy efficiency and a European low carbon economy* Speech 08/183 during launch of the EICTA report: "High Tech – Low Carbon", Brussels, April 8, http://europa.eu.

Shannon C 1984 Communication in the presence of noise. *Proc. IEEE* **72**(9), 1192 – 1201.

WINNER+ 2010 Celtic Project CP5-026 *Final Innovation Report* (eds Svensson T., Zinovieff E.). Public Deliverable D1.9, Wireless World Initiative New Radio – WINNER+, April 2010. http://projects.celtic-initiative.org/winner+/index.html (accessed June 2011).

Yuan P, Xiao D, Han J and Jing X 2010 Neighbor carrier signal strength estimation for carrier aggregation in LTE-A *Proc. WASE Intern Conference onInformation Engineering*, vol. 1, pp. 284–287.

Appendices

Appendix A

Resource Allocation

A.1 Dynamic Resource Allocation

A.1.1 Utility Predictive Scheduler

The main novelty introduced with the Utility Predictive Scheduler (UPS) is the formulation of its rate-based utility function, shown in Figure A.1. The scheduling priority of each user depends on the rate increase due to the allocation of the considered Physical Resource Block (PRB) to user, weighted by average user's throughput, similarly to Proportional Fair (PF) scheduler, with the users with the highest weighted rate being selected. Although the utility function is rate dependent, the design of the function is based on three parameters: α, β, and γ, as shown in Figure A.1. These parameters can be used to change the slope of selected sections of employed utility function to prioritize User Equipment (UE)s affected by higher latency, thus providing the support for different Quality of Service (QoS) classes. Moreover, even when considering only Best Effort (BE) traffic, this design of the scheduler and its utility function leads to a slight gain in spectral efficiency, as shown in WINNER+ (2009a,d).

A.1.2 Resource Allocation with Relays

A very interesting approach to resource allocation with QoS support in relay-enhanced network has been proposed in WINNER+ (2009c). The HurrY-Guided-Irrelevant-Eminent-NEeds (HYGIENE) scheduling algorithm brings urgency on top of relaying, which means that it gives the priority to urgent UEs and then to relayed UEs. A rushing entity classifier is introduced, which determines whether the UEs are of urgent class or not. Then, in the second processing step, the Base Station (BS) identifies UEs that require relaying. Based on the above mentioned classification scheduling, priority values are assigned to UEs according to Table A.1.

The resource allocation is jointly performed for a group of two consecutive time slots. First, relayed urgent UEs are scheduled, with the Real-Time (RT) packet being prioritized according to their remaining time-to-live, and the PRBs allocated in order to maximize the spectral efficiency. For allocation of PRBs to Non-Real-Time (NRT) packets a PF scheduler is used. In the second step, the urgent non-relayed UEs are scheduled according to the same allocation policy as above. Finally, the non-urgent UEs are scheduled in the third step according to the Max C/I policy WINNER+ (2009c).

Mobile and Wireless Communications for IMT-Advanced and Beyond, First Edition.
Edited by Afif Osseiran, Jose F. Monserrat and Werner Mohr.
© 2011 Afif Osseiran, Jose F. Monserrat and Werner Mohr. Published 2011 by John Wiley & Sons, Ltd.

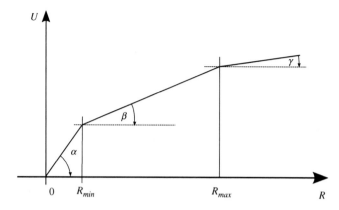

Figure A.1   Utility function of UPS and its parameterization - mapping data rate to utility value

Table A.1   Priority rules of HYGIENE scheduler

Type of user	Scheduling priority
Urgent relayed user	3
Urgent nonrelayed user	2
Nonurgent user	1

Figure A.2 presents the performance evaluation of the HYGIENE algorithm in comparison with several well-known schedulers: Maximum Carrier to Interference (MCI), PF, Modified-Largest Weighted Delay First (M-LWDF) and Exponential Delay Fairness (EDF). One can notice the performance improvement for the delay-sensitive RT services, such as the Voice over IP (VoIP).

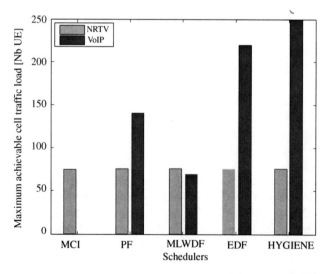

Figure A.2   Maximum achievable cell capacity with PF, MCI, M-LWDF, EDF and HYGIENE schedulers – mixed real-time traffic scenario

A.2 Multiuser Resource Allocation

A.2.1 PHY/MAC Layer Model

Resources are divided into Resource Blocks (RBs) occupying a given bandwidth and time, which can be allocated flexibly to the K UEs. A scenario where UEs travel with potentially high velocities is assumed. The high dynamics of the time varying channel prohibit the utilization of instantaneous Channel State Information at the Transmitter (CSIT). However, long-term CSIT that includes distance-dependent path loss and log-normal shadowing is assumed to be available.

Given long-term CSIT the data rate served to UE k is given by:

$$R_k = \alpha_k R_{\text{max},k} \tag{A.1a}$$

where $R_{\text{max},k}$ accounts for the rate if UE k is assigned all available slots exclusively. Additionally, the constraints

$$0 \le \alpha_k \le 1 \quad \forall k \in \mathcal{K} \quad \text{and} \quad \sum_{k \in \mathcal{K}} \alpha_k = 1 \tag{A.1b}$$

need to be fulfilled with $\mathcal{K} = \{1, \ldots, K\}$ being the set of all UEs.

A.2.2 APP Layer Model

The generic application characteristic resembles a bounded logarithmic relation between perceived quality and data rate, as illustrated in Figure A.3, described by the MOS as a function of the data rate R_k of UE $k \in \mathcal{K}$.

$$\text{MOS}_k(R_k) = \begin{cases} 1 & : R_k \le R_{1.0,k} \\ \text{MOS}_{0,k} \log \frac{R_k}{R_{0,k}} & : R_{1.0,k} < R_k < R_{4.5,k} \\ 4.5 & : R_k \ge R_{4.5,k} \end{cases} \tag{A.2a}$$

with

$$\text{MOS}_{0,k} = \frac{3.5}{\log\left(R_{4.5,k}/R_{1.0,k}\right)} \tag{A.2b}$$

$$R_{0,k} = R_{1.0,k} \left(\frac{R_{1.0,k}}{R_{4.5,k}}\right)^{\frac{1}{3.5}} \tag{A.2c}$$

and

$$0 \le R_{1.0,k} < R_{4.5,k} \quad \forall k \in \mathcal{K} \tag{A.2d}$$

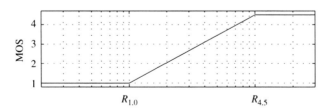

Figure A.3 Generic application characteristic for one example application class

The semilogarithmic plot of Figure A.3 visualizes the related parameters: the parameter $MOS_{0,k}$ determines the slope of $MOS_k(R_k)$ while $R_{0,k}$ shifts the curve along the x-axis.

The application characteristic of UE k, denoted by $\{U\}_k = U_k$ in Figure 2.5, can be parameterized by only two parameters, $\{R_{1.0,k}, R_{4.5,k}\}$, or alternatively $\{MOS_{0,k}, R_{0,k}\}$.

A.2.3 Optimization Problem

The objective of application driven multiuser resource allocation is to assign the available resources over the shared wireless medium described by the Physical (PHY)/Medium Access Control (MAC) layer model (A.1) running different applications modeled by (A.2). The share of resources for each UE, α_k, and the associated application data rates, R_k, are determined by solving an optimization problem between link and APP layer, as illustrated in Figure 2.5.

One commonly used utility function is to maximize the sum of the Mean Opinion Score (MOS) of all UEs, thereby maximizing the average perceived service quality. The corresponding optimization problem is formulated as in A.3 (Saul and Auer 2009):

$$\arg\max_{\boldsymbol{\alpha}} \sum_{k \in \mathcal{K}} MOS_k(\alpha_k) \qquad \text{s. t.} \qquad 0 \leq \alpha_k \quad \forall k \in \mathcal{K} \quad \text{and} \quad \sum_{k \in \mathcal{K}} \alpha_k = 1 \qquad (A.3)$$

where $\boldsymbol{\alpha} = [\alpha_1, \ldots, \alpha_K]^T$.

Alternatively, the max-min approach distributes resources such that all UEs experience the same utility degradation, which can be cast in the following optimization problem (Radunovic and Le Boudec 2007):

$$\arg\max_{\boldsymbol{\alpha}} \min_{k \in \mathcal{K}} MOS_k(\alpha_k) \qquad \text{s. t.} \qquad 0 \leq \alpha_k \quad \forall k \in \mathcal{K} \quad \text{and} \quad \sum_{k \in \mathcal{K}} \alpha_k = 1 \qquad (A.4)$$

Provision of max-min fairness in terms of UE perceived quality implies that one single UE that cannot achieve a good perceived quality, for example, due to a poor wireless channel and/or a demanding application, forces all other UEs to share this poor experience. Furthermore, from an operator's point of view it might be desirable to provide premium services with a higher quality only to some UEs, which contradicts the idea of "equal loss".

To mitigate these shortcomings, a resource allocation scheme is presented in (Saul 2008), which is described by the following optimization problem:

$$\arg\max_{\boldsymbol{\alpha}} \min_{k \in \mathcal{K}_{act}} f_{P,k}\left(MOS_k(\alpha_k)\right) \qquad \text{s. t.} \qquad 0 \leq \alpha_k \quad \forall k \in \mathcal{K} \quad \text{and} \quad \sum_{k \in \mathcal{K}} \alpha_k = 1 \qquad (A.5)$$

where $\mathcal{K}_{act} \subset \mathcal{K}$ is the subset of UEs that satisfy the specified MOS constrains, and $f_{P,k}$ is a strictly monotonic function that allows the operator to differentiate UEs, for example, to ensure that the MOS of a premium service is superior to the ordinary service quality. For instance, the condition that the MOS of UEs k_1 exceeds the MOS of UE k_2 by ΔMOS, is obtained by setting the inverse priority function to $f_{P,k_1}^{-1} = f_{P,k_2}^{-1} + \Delta MOS$.

In order to provide at least minimum service quality an access control is established, in the way that $\mathcal{K}_{act}$ out of $\mathcal{K}$ UEs are served, while the remaining UEs are denied access. By incorporating admission control to the resource allocation problem, service guarantees of $MOS_{min,k}$ for admitted UEs $k \in \mathcal{K}_{act}$ are established, according to the desired UE priorities, that is $f_{P,k_1}\left(MOS_{min,k_1}\right) = f_{P,k_2}\left(MOS_{min,k_2}\right)$, $\forall k_1, k_2 \in \mathcal{K}_{act}$.

The optimization problem (A.5) can be solved by finding the roots of an equation system. The following algorithm determines the admitted UEs $\mathcal{K}_{act}$ and the share of resources α:

Step 0:   Initialize, $\mathcal{K}_{\mathrm{act}} = \mathcal{K}$
Step 1:   Drop UEs until equation system is solvable (see step 1a–c)
Step 1a: Solve $\mathrm{MOS}_k(\alpha_k) = f_{\mathrm{P},k}^{-1}, k \in \mathcal{K}_{\mathrm{act}}$
Step 1b: If $\sum_{k \in \mathcal{K}_{\mathrm{act}}} \alpha_k \leq 1$, continue with step 2
Step 1c: Drop UE $k_{\mathrm{drop}} = \arg\max_k \alpha_k$, that is, $\mathcal{K}_{\mathrm{act}} \to \mathcal{K}_{\mathrm{act}} \setminus k_{\mathrm{drop}}$,
            and continue with step 1a
Step 2:   Solve equation system $\mathrm{MOS}_k(\alpha_k) = f_{\mathrm{P},k}^{-1}$,
            $k \in \mathcal{K}_{\mathrm{act}}$, and $\sum_{k \in \mathcal{K}_{\mathrm{act}}} \alpha_k = 1$

It may happen that the available resources are insufficient to serve all UEs with the desired quality $\mathrm{MOS}_{\min,k}$. In such cases some UEs cannot be served, which is realized in the algorithm step 1c.

Implications for the System Architecture

For Cross-Layer Optimization (CLO) between link and Application (APP) layer, different timescales are involved. As timescales on the link layer are several orders of magnitude smaller than those of the APP layer, in order to limit the signaling overhead between optimizer, link and APP layer, the optimizer should be placed close to the BS. For tuning of the APP layer data rates on a shorter timescale, it is desirable that transcoding is supported at the BS, and/or scalable video codecs (ITU-T 2007; Schwarz and Marpe 2007) are deployed.

A.2.4   Simulation Results

In order to evaluate the performance of the considered resource allocation schemes that aim to maximize the perceived quality as described in Section A.2.3, computer simulations are conducted. We consider a Long Term Evolution (LTE) downlink in the 10 MHz bandwidth Frequency Division Duplex (FDD) mode. By employing Orthogonal Frequency Division Multiple Access (OFDMA), UEs can be assigned to RBs in time and frequency. The Wireless World Initiative New Radio (WINNER) typical urban macrocell channel model (model C2, WINNERII 2007) is used, comprising path loss, shadowing and time-variant frequency-selective fading. The average Signal-to-Interference-plus-Noise Ratio (SINR) is constrained such that transmission at least with the lowest supported Modulation and Coding Scheme (MCS) is feasible. Taking into account velocities of UEs of 50 km/h, only long-term Channel State Information (CSI) at the transmitter is assumed to be available. While the same modulation and coding scheme is chosen for all RBs assigned to one UE, PRB's assigned to different UEs will typically use a different MCS. Furthermore, taking into account fast Automatic Repeat-reQuest (ARQ) at the BS, it is assumed that packets are always received error free. The applications are modelled by a bounded logarithmic utility (A.2), with a minimum required data rate of $R_{1.0} = 300$ kbps and a desired data rate of $R_{4.5} = 3$ Mbps.

Figure A.4 show the CDF of the MOS and the data rate, respectively, for the CLO techniques. There are $K = 16$ UEs with a guaranteed service quality of $\mathrm{MOS}_{\min} = 3.0$, which corresponds to a data rate of 1.1 Mbps. While the conventional max-min MOS technique (A.4) can serve only 81% of the UEs with $\mathrm{MOS}_{\min} = 3.0$, a significantly higher number of UEs achieves at least $\mathrm{MOS}_{\min} = 3.0$ for the max-min MOS utility (A.5) that incorporates admission control. This gain is achieved at the expense of not serving 1.3% of the UEs, who suffer a bad channel, for example due to strong shadowing at the cell edge. The max-sum MOS technique (A.3) serves more than 50% of the users with the best possible quality of MOS = 4.5. These are the users with good channels, for which high data rates can be achieved with comparably modest resource utilization. On the other hand, more than 12% of the users are not served with at least $\mathrm{MOS}_{\min} = 3.0$, and 6.5% of the users are not served at all.

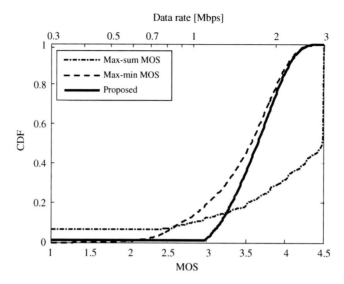

Figure A.4 CDF of UE perceived quality (Saul 2008). Reproduced by permission of © 2008 IEEE

A.3 Busy Burst Extended to MIMO

For performance evaluation the downlink of a hexagonal cell deployment is considered. The BS transmitter selects one beam from a set of fixed beamforming vectors given by a Discrete Fourier Transform (DFT) codebook. Channels models are taken from WINNER scenario C1 (Döttling et al. 2009, Chapter 2). A full buffer traffic model is considered. Perfect time and frequency synchronisation of the network is assumed. The simulation parameters shown in Table A.2, are taken from

Table A.2 System level simulation parameters

Parameter	Value
Carrier centre frequency	3.95 GHz
System bandwidth B	89.84 MHz
Number of subcarriers N_c	1840
Frame duration	0.6912 ms
OFDM symbols/frame	30
Total OFDM symbol length	22.48 μs
RB size	15 (time) × 8 (frequency) = 120
Number of RB/frame	2 (time) × 230 (frequency)
Number of sectors/cell	3
Number of antenna elements/sector	4
Average number of UEs/cell U	10
Transmit power per RB P	16.4 dBm
Elevation antenna gain A_e	14 dBi
Azimuth antenna element gain	$-\min\left[12\left(\frac{\theta}{\theta_{3dB}}\right)^2, A_m\right]$ [dB] where, $A_m = 20$ and $\theta_{3dB} = 70°$
Noise level N	-117.8 dBm/RB
Number of snapshots	50
Simulation duration per snapshot	50 ms

the WINNER Time Division Duplex (TDD) mode (WINNERII 2006). Link adaptation is assumed, where the modulation scheme is adaptively controlled based on the achieved SINR.

A.4   Efficient MBMS Transmission

A.4.1   Service Operation

Concerning streaming services, the system must ensure maximum coverage for its transmission. Therefore, more robust modulation and coding schemes are selected at the expenses of a reduced bit rate capability. The file delivery case is much more challenging and thus we focus on this service. For the rest of the section, file delivery case is the transmission of a 2 MB file.

The file download service in the Multimedia Broadcast Multicast Service (MBMS) consists of three phases:

User Service Discovery/Announcement phase   The User Service Discovery/Announcement phase provides information on available MBMS services. The responsible entity establishes a connection with all users in the multicast/broadcast group using the paging procedure to reach them. Through this end-to-end connection, the server knows a priori the exact number of users together with their current channel states to be served and informs them about the characteristics of the transmission.

Initial MBMS file transmission phase   The BS must select point-to-point (p-t-p) transmission or point-to-multi-point (p-t-m) transmission in order to maximize the spectral efficiency of the MBMS service. The switching criteria between p-t-p and p-t-m mode is based on the number of users eligible for p-t-m transmission. In order to identify the optimum transmission mechanism for MBMS service at every given time, the BS needs to estimate the number of users interested in the MBMS service, which is accomplished in the discovery phase. The switching criteria can be either dynamic — based on the channel estimation reports submitted by UEs, or static — based on off-line analysis, in which case the selection threshold between p-t-p and p-t-m is expressed in terms of a predefined number of users Θ (see Figure A.5).

In case of selecting the p-t-m mode, the transmission duration has to be configured, so to ensure that the file is successfully received by, in this case, 95% of the users. In order to achieve this, there are two possible options:

- Transmit the file using only MBMS until the 95% acquisition probability is reached (referred to as Conventional File Delivery).
- Transmit data with MBMS during the time required to achieve a certain acquisition probability and then use a repair phase that will serve the remaining users (referred to as Hybrid Delivery).

The decision must be made with the aim of optimizing the acquisition probability of the initial MBMS transmission phase while minimizing the delivery time.

Post-delivery Repair Phase   This phase only applies when using hybrid delivery. The purpose of this phase is to repair erroneous received files after the initial MBMS transmission. UEs not able to recover the file notify the minimum set of data packets required to repair the file or simply the total number of correctly received packets (3GPP 2009b). This information is useful to determine if p-t-p or p-t-m retransmission is preferred and is sent to the broadcast server using the end-to-end connection established with each user.

To avoid congestion, error reporting messages from UEs can be distributed over time within a backoff window and across multiple repair servers. The backoff window should be large enough to prevent congestion, but should not increase the duration of the repair phase.

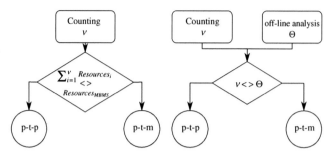

Figure A.5 Dynamic switching criteria (left) and static switching criteria (right)

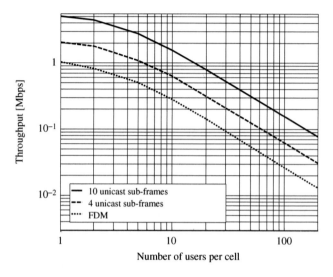

Figure A.6 Average user throughput versus the number of users per cell for unicast users with FDM

As mentioned before, UEs start the repair phase using dedicated p-t-p connections, but if the number of active users in this phase is high enough it is possible to employ a p-t-m connection with MBMS. However, during the initial file transmission there is no communication between the UEs and the server. Therefore, once the initial MBMS transmission is finished, the server does not have any information about the number of users that have not received the file and the amount of repair data needed by each of them. This information can be estimated in the beginning of the postdelivery repair session using application layer feedback.

With the aim of serving users with bad channel conditions, the new MBMS transmission should use a more robust MCS. The duration of this transmission would depend on the amount of repair data requested by the users at the time of the decision.

A.4.2 Frequency Division Multiplexing (FDM) Performance

For the FDM case, from the 10 MHz available, half will be dedicated to unicast, that is, 5 MHz, and the same for multicast. In terms of MBMS performance, this FDM distribution is equivalent to allocating five subframes from the ten available to MBMS. As shown in Figure A.6, FDM is never a good option for file delivery due to lower frequency diversity.

Appendix B

Spectrum Awareness

B.1 Spectrum Sensing

Sensing based on *Energy Detection* has been one of the most common approaches due to the reduced computational and implementation complexity this method exhibits. Moreover, it does not require any knowledge of the nature of the primary signal, thus being suitable for general-purpose systems. The signal is detected by comparing the output of the energy detector with a threshold, which depends on the noise floor (Urkowitz 1967). The election of an appropriate threshold is not trivial, see for example (Gelabert et al. 2009), and largely impacts the detection errors that come in the form of misdetection (i.e. not detecting a present primary signal) and false alarm (i.e. detecting a nonpresent primary signal). The former error may potentially induce interference between the primary and secondary systems whereas the latter error may cause spectrum underutilization. Despite the versatile operation of the energy detector and its reduced complexity, it reveals a poor performance under low Signal-to-Noise Ratio (SNR) values, see Tang (2005), and for detecting spread spectrum signals (Cabric et al. 2004; Yucek and Arslan 2006). In Sahin et al. (2009) energy detection is considered for the uplink of an Orthogonal Frequency Division Multiple Access (OFDMA) system and compared to ESPRIT (estimation of signal parameters by rotational invariance techniques) algorithms. Results indicate a considerably better performance of the energy detector compared to ESPRIT, especially when the subcarrier assignment changes frequently. In Pawelczak et al. (2007), energy detection is used for Opportunistic Carrier Aggregation (OCA) in a OFDMA system, where collision probability and interference are evaluated.

On the other hand, *Wave-form Based Sensing* (sometimes called *Coherent Sensing*) exploits known patterns, such as preambles, pilot patterns, spreading sequences, etc. by correlating the received signal with a copy of itself, see, for example, Sahai et al. (2006) and Tang (2005). Although this method is only applicable to systems exhibiting known signal patterns, it reveals improved performance with respect to the energy detector – see Tang (2005), and such improvement increases with the length of the known signal pattern.

Additionally, *Cyclostationarity Detection* is a method for detecting primary user signals by exploiting the cyclostationarity due to mean and autocorrelation periodicity of the received signals (Cabric et al. 2004; Fehske et al. 2005; Kim et al. 2007). These features are detected by analyzing the spectral correlation function, which is able to differentiate the modulated signal energy from the noise energy (since noise is a wide-sense stationary signal with no correlation).

Mobile and Wireless Communications for IMT-Advanced and Beyond, First Edition.
Edited by Afif Osseiran, Jose F. Monserrat and Werner Mohr.
© 2011 Afif Osseiran, Jose F. Monserrat and Werner Mohr. Published 2011 by John Wiley & Sons, Ltd.

Several additional spectrum sensing methods include *Matched-Filtering* (Tandra and Sahai 2005), *Multitaper Spectral Estimation* (Haykin 2005) and *Wavelet Transform Based estimation* (Tian and Giannakis 2006).

B.2   Geo-Location Databases

Location methods can be based on self-localization, where the device automatically collects this information using a Global Positioning System (GPS) or broadcast beacons from one or more base stations. The position can be also obtained from an external source that accurately knows the device's location. In addition, network-based approaches can be used to determine the position of a device by communicating with other unlicensed devices that may have a known position. These methods can be used individually or jointly. The utilization of location information in cognitive radio networks is addressed in Celebi and Arslan (2007).

On the other hand, the considered database will require a database provider, an access method and a model for updating the database information. The database provider will ensure the reliability of the database, its maintenance and operation. The database can be accessed over the Internet, which provides pervasive, flexible and low-cost access, assuming that secondary devices are Internet capable. Other access means such as dedicated channels provided by the secondary network infrastructure are also possible.

Database queries should be simple for the unlicensed devices and provide flexibility for the database to respond. For example if the unlicensed device entered its position, position error, and Equivalent Isotropically Radiated Power (EIRP) transmit power, then the database could compute which channels could be used without interference based on the location of licensed receiver service areas (Brown 2005). Database queries by the secondary device should be aligned with the database update periodicity. In this sense, there will be a tradeoff between limited reactivity in long-term queries and excessive overhead due to short-term queries.

The Institute of Electrical & Electronics Engineers (IEEE) 802.22 standard for cognitive radio access considers the use of geolocation and databases for the operation in geographically unused spectrum allocated to the television broadcast service (Stevenson et al. 2009).

B.3   Beacon Signaling

According to Brown (2005), beacons may be categorized into four types: *per transmitter*, *area transmitter*, *unlicensed signaling*, and *receiver detection* beacons. Each alternative presents its implementation and performance advantages and drawbacks. The *per transmitter* approach implies that each licensed user transmits a beacon alerting unlicensed users of its presence. This beacon may include information on the location, transmit power, etc. Alternatively, the *area transmitter* approach advocates for devoting a small amount of bandwidth dedicated to carry transmission information announcing utilized frequency bands, locations, and coverage areas of the licensed transmitters. On the other hand the *unlicensed signaling* approach assigns some frequency band such that unlicensed devices exchange information about licensed transmitters. This may be used in conjunction with other methods, such as spectrum sensing, in order to provide enhanced spectrum awareness. Finally, *receiver detection* beacons are intended to identify potential interference at the receivers rather than identifying licensed transmitting sources. In this sense, Brown (2005) proposed several methods for detecting receivers.

Appendix C

Coordinated MultiPoint (CoMP)

C.1 Joint Processing Methods

C.1.1 Partial Joint Processing

In Figure C.1, the average sum-rate per cell obtained by the different transmission schemes is plotted when moving away from one of the Base Stations BSs of the cluster.

C.1.2 Dynamic Base Station Clustering

In this section, a technique is highlighted in which the Base Station (BS) clusters are created in a dynamic way, that is, in each time slot t the sets of coordinated BSs are generated in order to maximize a given objective function (Boccardi et al. 2008). Such an approach can be seen as an extension of Papadogiannis et al. (2008), with the difference that User Equipments (UEs) and clustering are jointly selected in order to maximize the weighted sum rate.

In order to describe the scheme, let us define $C_n(t), n = 1, \ldots, N$ as the set of BS indexes belonging to the nth cluster at the time slot t and $U_m(t), m = 1, \ldots, M$ as the set of UE indexes scheduled for transmission in a given cluster at the time slot t. Define $\mathcal{C}(t) = \{C_1(t), \ldots, C_N(t)\}$ and $\mathcal{U}(t) = \{U_1(t), \ldots, U_M(t)\}$ respectively as the BS clustering and UE allocation at the tth time slot. For sake of clarity we drop the dependence on the time slot t. The physical layer technique is then described by the antenna weights $\mathcal{W}$ and the power allocation $\mathcal{P}$, and with these definitions we denote the throughput achievable in the nth cluster with $R_n(\mathcal{C}, \mathcal{U}, \mathcal{W}, \mathcal{P})$. Scheduling, BS clustering, calculation of the beamforming coefficients and power allocation are realized in a Central Unit (CU), which could be a new physical node or one BS acting as the CU.

The proposed technique can be summarized as follows:

- **Phase I**. Each BS sends the channel estimates to the CU.
- **Phase II**. Based on the Channel State Information (CSI) and on the scheduling requirements, the CU jointly creates the clusters of collaborating BSs, schedules the UEs in these clusters and calculates the beamforming coefficients and the power allocation.
- **Phase III**. The CU sends to the BSs beamforming coefficients, power allocation, indexes of the coordinated cells and indexes of the selected UEs. At this point, the BSs belonging to the same cluster need to share the data of the selected UEs between them.

Mobile and Wireless Communications for IMT-Advanced and Beyond, First Edition.
Edited by Afif Osseiran, Jose F. Monserrat and Werner Mohr.
© 2011 Afif Osseiran, Jose F. Monserrat and Werner Mohr. Published 2011 by John Wiley & Sons, Ltd.

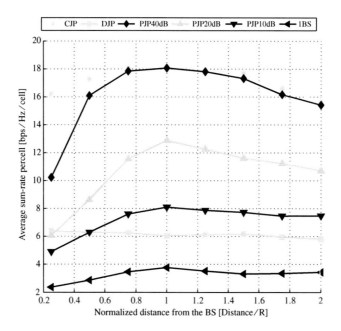

Figure C.1 Average sum-rate per cell versus normalized distance for $M = 6$ UEs and an edge-of-cell SNR of 15 dB (Thiele et al. 2010)

With respect to full network coordination (Centralized Joint Processing (CJP)), the proposed technique allows the reduction of signaling due to data sharing, while requiring the same amount of signaling due to channel estimates exchanges.

The problem of joint clustering and UE selection can be formalized as follows

$$
\begin{cases}
\max_{\mathcal{C}, \mathcal{U}} \sum_{n=1}^{N} R_n \left(\mathcal{C}, \mathcal{U}, \mathcal{W}, \mathcal{P} \right) \\
C_i \cap C_j = \phi, \forall C_i, C_j \in \mathcal{C}, i \neq j \\
U_k \cap U_l = \phi, \forall U_k, U_l \in \mathcal{U}, k \neq l
\end{cases}
\tag{C.1}
$$

The objective function $\sum_{n=1}^{N} R_n (\mathcal{C}, \mathcal{U}, \mathcal{W}, \mathcal{P})$ is a function of both the BS clusters and of the UEs scheduled in each cluster. For example, under a Zero Forcing (ZF) precoding assumption the optimum must at the same time minimize the intercluster interference and select a quasiorthogonal set of UEs to be scheduled in each cluster. The two constraints are related respectively to the assumption of nonoverlapping clusters and to the assumption that each UE cannot be served at the same time by BSs belonging to different clusters.

The optimal solution of (C.1) requires a brute force search over the sets of UEs and possible BS clusters. In the following we propose a suboptimal approach based on the idea of restricting the search space. This approach consists of two different stages, an offline stage and an online stage.

Offline phase. The candidate clusterings are chosen off-line taking into account path loss and shadowing, and more generally also average UEs distribution and average channel estimates.

Online phase. At each time frame t the central node estimates the weighted sum rate achievable for each cluster. This sum-rate estimation involves the choice of a candidate set of UEs to be scheduled, with a brute force UE selection or with a greedy UE selection technique, and the calculation of the power allocation. Finally, the clustering that maximizes the weighted sum-rate and the associated set of UEs, beamforming coefficients and power allocation are used for transmission in the tth time slot.

To evaluate the scheme, a system simulator has been used with 19 single antenna BSs and wraparound, 30 UEs per BS, and edge-of-cell Signal to Noise Ratio (SNR) of 15 dB. Each single-antenna UE is dropped with uniform probability inside each cell. Fairness is guaranteed by a Proportional Fair (PF) scheduler. The reference SNR is defined as the SNR at the cell vertex. The channel has been modeled considering Rayleigh small-scale fading and path loss, with path loss exponent equal to 4.5. Cf. WINNER+ (2009b) for further simulation details. The results are shown in Table 6.3.

C.2   Coordinated Beamforming and Scheduling

C.2.1   Decentralized Coordinated Beamforming

Let us assume that each UE is equipped with a single receive antenna. Hence the channel matrix between the UE m and BS l, $\mathbf{H}_{l,m}$ is reduced to a vector and will be denoted $\mathbf{h}_{l,m}$. Going from Equation 6.3, the receiver precombining SINR of UE m can be shown to be given by

$$\Gamma_m = \frac{\left| \sum\limits_{l \in \mathcal{B}_m} \mathbf{h}_{l,m} \mathbf{w}_{l,m} \right|^2}{N_0 + \sum\limits_{i=1, i \neq m}^{M} \left| \sum\limits_{l \in \mathcal{B}_i} \mathbf{h}_{l,m} \mathbf{w}_{l,i} \right|^2} \tag{C.2}$$

Note that full carrier phase synchronization is assumed between all BSs $l \in \mathcal{B}_m$ that serve a given UE m. Note that, for fixed $\gamma_m \ \forall\ m$, the constraints of (6.10) can be expressed as a generalized inequality with respect to the second-order cone (Bengtsson and Ottersten 2001; Boyd and Vandenberghe 2004; Wiesel et al. 2006) and (6.10) can be solved as a Second Order Cone Programming (SOCP). The generalized formulation shown in (6.10) can accommodate multiple scenarios ranging from coherent multicell beamforming with $|\mathcal{B}_m| > 1$ (in the extreme case $\mathcal{B}_m = \mathcal{B} \ \forall\ m$) to the coordinated single-cell beamforming case where $|\mathcal{B}_m| = 1 \ \forall\ m$. Note that the formulation above needs a complete channel knowledge of $\mathbf{h}_{l,m}$ between all pairs of m and l, and hence, it requires a centralized solution.

In the following, the decentralized method proposed in Tölli et al. (2011) is briefly summarized, where the beamformer vectors are obtained locally relying on coupled real-valued intercell interference parameters exchanged between adjacent BSs. For simplicity, attention is paid to the coordinated single-cell beamforming case, where $|\mathcal{B}_m| = 1 \ \forall\ m$.

The intercell interference term from BS l to UE m is denoted as $\zeta_{l,m}$, and the term with inequality is relaxed as

$$\zeta_{l,m}^2 \geq \sum_{i \in \mathcal{U}_l} \left| \mathbf{h}_{l,m} \mathbf{w}_{l,i} \right|^2 \tag{C.3}$$

The SINR formula in (C.2) is modified as

$$\Gamma_m = \frac{\left| \mathbf{h}_{l_m,m} \mathbf{w}_{l_m,m} \right|^2}{N_0 + \sum\limits_{l \in \bar{\mathcal{B}}_m} \zeta_{l,m}^2 + \sum\limits_{i \in \mathcal{U}_{l_m} \setminus m} \left| \mathbf{h}_{l_m,m} \mathbf{w}_{l_m,i} \right|^2} \tag{C.4}$$

where the index of serving BS for UE m is denoted as l_m, and $\bar{\mathcal{B}}_m = \{\mathcal{B} \setminus l_m\}$. Now, (6.10) can be reformulated for the special case $|\mathcal{B}_m| = 1 \ \forall \ m$ as:

$$\min. \quad \sum_{l=1}^{L} \sum_{m \in \mathcal{U}_l} \|\mathbf{w}_{l,m}\|_2^2$$

$$\text{s. t.} \quad \Gamma_m \geq \gamma_m, \forall \ m \tag{C.5}$$

$$\sum_{i \in \mathcal{U}_l} |\mathbf{h}_{l,m} \mathbf{w}_{l,i}|^2 \leq \zeta_{l,m}^2, \ \forall \ m \notin \mathcal{U}_l, \forall \ l$$

where the optimization variables are $\mathbf{w}_{l,m} \in \mathbb{C}^{N_T}, m = 1, \ldots, M, l = 1, \ldots, L$ and $\zeta_{l,m} \in \mathbb{R}_+, \forall \ m \notin \mathcal{U}_l, \forall \ l$. The second constraint in (C.5) guarantees that the intercell interference generated from a given BS l cannot exceed the UE specific thresholds $\zeta_{l,m} \ \forall \ m \notin \mathcal{U}_l$.

Observe that the BSs are coupled in the SINR constraints in (C.5) by the interference terms $\zeta_{l,m}$. If the interference terms were fixed, (C.5) would be decoupled and the transmitted power could be separately minimized at each BS. Next, the coupled SINR constraints are addressed by introducing local auxiliary variables and additional equality constraints. Thus, the coupling in the SINR constraints is transferred to coupling in the equality constraints, which can then be decoupled by dual decomposition (Boyd et al. 2007; Palomar and Chiang 2006). The decoupling is further simplified by the fact that each intercell interference term $\zeta_{l,m}$ couples exactly two (adjacent) BSs, that is, the serving BS l_m and the interfering BS l. Thus, it is enough to enforce the two local copies to be equal $\zeta_{l,m}^{(l)} = \zeta_{l,m}^{(l_m)} \ \forall \ m, l \in \bar{\mathcal{B}}_m$, where the superscript $(\cdot)$ indicates the local version of the term $\zeta_{l,m}$.

In order to obtain a distributed algorithm, a standard dual decomposition approach (Boyd et al. 2007; Palomar and Chiang 2006) is taken where the consistency constraints are relaxed by forming the partial Lagrangian. Finally, the independent BS specific subproblems can be solved as

$$\min. \quad \sum_{m \in \mathcal{U}_l} \|\mathbf{w}_{l,m}\|_2^2 + \boldsymbol{\xi}_l^T \boldsymbol{\zeta}^{(l)}$$

$$\text{s. t.} \quad \Gamma_m^{(l)} \geq \gamma_m, \forall \ m \in \mathcal{U}_l \tag{C.6}$$

$$\sum_{i \in \mathcal{U}_l} |\mathbf{h}_{l,m} \mathbf{w}_{l,i}|^2 \leq \zeta_{l,m}^{(l) \, 2}, \ \forall \ m \notin \mathcal{U}_l$$

where the optimization variables are $\mathbf{w}_{l,m} \in \mathbb{C}^{N_T} \ \forall \ m \in \mathcal{U}_l$ and $\boldsymbol{\zeta}^{(l)}$. Vectors $\boldsymbol{\zeta}^{(l)}$ and $\boldsymbol{\xi}_l$ include the local versions of interference terms and the corresponding Lagrange multipliers, respectively. The subproblems in (C.6) can be locally solved as SOCPs in each BS l with knowledge of the consistency price vectors $\boldsymbol{\xi}_l$. More details can be found in Tölli et al. (2009, 2011).

The master problem for the dual decomposition can be solved using the subgradient method (Boyd et al. 2007; Palomar and Chiang 2006). Since the subgradient of g at $\xi_{l,m}$ is simply the consistency constraint residual $\zeta_{l,m}^{(l)} - \zeta_{l,m}^{(l_m)}$, the master problem can be solved iteratively and independently at each BS with the following updates (Boyd et al. 2007):

$$\xi_{l,m}(t+1) = \xi_{l,m}(t) + \mu \left(\zeta_{l,m}^{(l)}(t) - \zeta_{l,m}^{(l_m)}(t) \right), \forall \ b, m \tag{C.7}$$

where t is the iteration index and μ is a positive step size. This requires exchanging local versions of $\zeta_{l,m}^{(l)}$ between the serving BS l_m and the interfering BS l.

Since the original problem (C.5) is convex, for each fixed channel realization, the algorithm is guaranteed to converge exactly to the optimal solution where $\zeta_{l,m}^{(l)} = \zeta_{l,m}^{(l_m)}, \ \forall \ m, l \in \bar{\mathcal{B}}_m$ as long as the step size μ is diminishing as a function of time, for example, $\mu(t) = c/\sqrt{t}$, where $c > 0$ (Boyd et al. 2007). In scenarios with time-correlated fading, a fixed step size is obviously more practical. In such a case, the algorithm is guaranteed to converge within some range of the optimal value, where the range decreases with the step-size parameter μ (Boyd et al. 2007).

Note that the intermediate iterates $\zeta^{(l)}_{l,m}(t)$ in the dual decomposition do not necessarily result in feasible solutions, $\Gamma_m \geq \gamma_m$, as $\zeta^{(lm)}_{l,m}(t) \neq \zeta^{(l)}_{l,m}(t)$. Thus, $\Gamma_m < \gamma_m$ can represent a solution of (C.6) for some m. At the cost of solving one additional subproblem per BS, however, *a feasible set of beamformers* $\mathbf{w}_{l,m}$ *can be always achieved* by using the average values $\zeta_{l,m}(t)$ in (C.6) for each BS. As a result, a feasible set of UE specific SINR targets $\gamma_m \; \forall \; m$ can be guaranteed even if the backhaul exchange rate of $\zeta^{(l)}_{l,m}(t)$ between BSs is relatively low, as demonstrated in (Tölli et al. 2009, 2011). Finally, the distributed algorithm is summarized in Algorithm C.1.

Algorithm C.1 Decentralized coordinated multicell beamforming

---

0. Initialize $\boldsymbol{\xi}_l(0)$ with some values, for example, $\boldsymbol{\xi}_l(0) = \mathbf{0}$, and set $t = 0$.
1. Solve (C.6) and transmit the resulting $\zeta^{(l)}_{l,m}$ (via backhaul signaling) to the coupled neighboring BSs, $\tilde{B}_m \; \forall \; m \in \mathcal{U}_l$ and $l_m \; \forall \; m \notin \mathcal{U}_l$.
2. Update the local consistency prices $\xi_{l,m}(t)$ as in (C.7).
3. Calculate the average intercell interference terms as in $\zeta_{l,m}(t) = \frac{1}{2}(\zeta^{(lm)}_{l,m}(t) + \zeta^{(l)}_{l,m}(t))$ and use the fixed average values $\zeta_{l,m}(t)$ in (C.6) to get a feasible set of beamformers $\mathbf{w}_{l,m} \; \forall \; m \in \mathcal{U}_l$.
4. Set $t = t + 1$ and go to Step 1 (until desired level of convergence).

An example of the time evolution of the sum power in a scenario depicted in Figure 6.8 is plotted in Figure C.2, where the behavior of the distributed, centralized (ideal) and ZF beamforming cases is compared in a time-correlated fading scenario with 10 dB SINR constraints. The example demonstrates that the distributed algorithm performs nearly as well as the centralized solution even at relatively high velocities, especially with low SINR targets. In this particular example, two BSs exchange at each reporting instant four real-valued interference terms. The normalized velocity is set

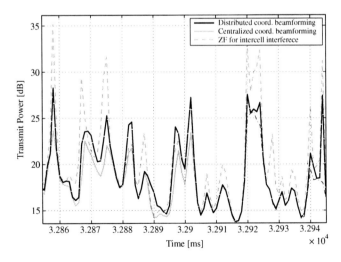

Figure C.2 Time evolution of the distributed algorithm with 10 dB SINR target, $T_s f_d = 0.05$ (Tölli et al. 2011). Reproduced by permission of © 2011 IEEE

Table C.1 Simulation details for worst companion method

Simulator details	
Channel Model	3GPP SCME 3D
Intersite distance	500 m
TX Antennas	4, vertical polarized with $\lambda/2$ antenna distance
TX Codebook	DFT-based with tapering
RX Antennas	2
Number of cells	57
Number of UEs per cell	15
Scheduler	α-PF
α values	0.5, 0.6, 0.8, 1.0, 1.3, 2.0, 3.0
Transmission rank	1
Cooperation set dimension	0–6

at $T_S f_d = 10^{-1}$, where f_d is the Doppler shift and T_S is the signaling period for the exchange of $\zeta_{l,m}^{(l)}$ between BSs. For example, $T_S = 2$ ms with 2 GHz center frequency corresponds to the velocity of 27 km/h.

C.2.2 Coordinated Scheduling via Worst Companion Reporting

The performance of the worst companion scheme with cyclically prioritized scheduling has been evaluated with a 3GPP-compliant simulator and the system parameters are summarized in Table C.1.

In Figure C.3, the performance of the worst companion reporting algorithm is studied, with different number of UEs. We note that both the cell edge and the cell average gain grow with the UE number. For example for 10 (see C.3(a)) and 20 (see C.3(b)) UEs the gain at the cell edge is respectively 29% and 41%. This is due to the combined effect of multiuser diversity and the fact that with a higher number of UEs the performance at the edge gets lower, so worst companion reporting has a higher impact.

C.3 Test-Bed: Distributed Realtime Implementation

In modern cellular networks, there is a general tendency to use distributed signal processing. The adaptation to the wireless channel can be much faster if it is performed directly in the serving BS. For Downlink (DL) Coordinated Multipoint transmission or reception (CoMP), the overall delay is reduced in the closed transmitter adaptation loop if the waveforms are generated at the serving BS. Ideas for a distributed implementation of DL CoMP in a limited cluster of coordinated BSs are developed in Jungnickel et al. (2008a), Ng et al. (2008), Papadogiannis et al. (2009), Thiele et al. (2009) and Zirwas et al. (2006, 2009). Base stations are synchronized using the Global Positioning System (GPS). User Equipment estimates the multicell CSI in the Downlink using the Release 8 (Rel-8) Cognitive Radio System (CRS) but estimation is now limited to the BSs in the cluster. User Equipment deliver CSI feedback to its serving BS. Next, BSs in the cluster exchange the CSI as well as scheduled data over a low-latency signaling network. Weights for the joint beamforming are computed redundantly at each BS. The relevant set of weights is applied to the data signals and, in this way, the transmitted waveforms can be computed locally; the desired signals sum up constructively while the mutual interference inside the cluster is canceled.

For local computation of the waveforms, scheduled data and the CSI are needed. Data instead of IQ samples transfer is a light burden for the backhaul. For distributed DL CoMP, latency is more related

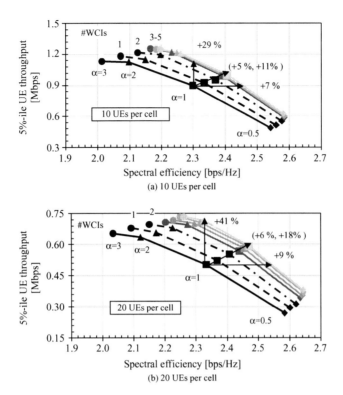

Figure C.3 Cell edge throughput versus spectral efficiency for the worst companion coordinated scheduling method

to the ongoing aging process of the CSI while it is exchanged over the backhaul. A few milliseconds may be tolerated for slowly moving UEs.

Distributed Synchronization

A first requirement of DL CoMP is that the physical layer and radio front ends are tightly synchronized. A precise reference clock is available everywhere where the Line of Sight (LoS) to the sky is free when using an advanced GPS receiver. All necessary clock signals locally at each BS are generated so that the entire radio network is fully synchronized (Jungnickel et al. 2008b). Clock signals comprise frame and sampling clock, the Common Public Radio Interface (CPRI) clock used between base band and Radio Frequency (RF) as well as IF and low phase-noise RF carrier frequencies.

Cell-Specific Pilots

Once in each 10 ms radio frame, the BSs provide cell-specific pilots denoted as CRS from which the UEs can estimate the multicell DL channel. A comb structure in the frequency domain is used where the comb of active subcarriers is shifted to identify the cell. Specific shifts are reused in the network similar to the well-known frequency reuse, so that the overhead due to the introduction of CRS remains limited.

CSI Feedback

Safe transmission using conservative modulation and coding schemes is mandatory for the CSI feedback. Quadrature Phase Shift Keying (QPSK) modulation and code rate 1/2 in 10 MHz bandwidth are used thus splitting the 20 MHz uplink band between the cells. For each pilot in the comb, the CSI for the entire 2×4 channel matrix is fed back. While using the full 20 MHz bandwidth for CoMP, one out of four CSI over Ethernet packets contains the feedback for 5 MHz bandwidth. The Ethernet packets are delivered as regular uplink traffic to the BS. Feedback of the uncompressed complex-valued CSI for the 2×4 MIMO DL channel to both BSs requires (4 Tx) $\times$ (2 Rx) $\times$ (144 pilots) $\times$ $(2 \times 16$ bit) / 10 ms = 3.68 Mbps, which results in 4.6 Mbps including overhead for higher layers. Time stamps derived from the GPS are introduced in order to ensure that the multicell channel matrix is built from consistent CSI packets at the precoder in each BS.

Synchronous Data Exchange

A technical requirement for distributing the coherent precoding operation in the network is that the data flows are strictly synchronized at the inputs of the local precoders in Figure 6.11 for all BS cooperating in the cluster. The scheduled data after the Medium Access Control (MAC) layer processing have been copied and forwarded the copy over the X2 interface to the other BSs in the CoMP cluster. The maximum load of 300 Mbps has been realized on X2 by exchange of six encoded bits per symbol, while still ignoring the actual MCS used. Load-aware compression of the scheduled data would be implemented in a real network to lower this rate. In this way, the load on X2 becomes highly dynamic and it depends on the traffic in the other cell.

Backhaul Network

Note that the distributed CoMP architecture depicted in Figure 6.11 where both data and CSI are exchanged, turned out as an essential enabler for CoMP trials in the field where distributed UE and BS locations are considered. Only the feedback of the served UE is decoded at the serving BS and then the BSs exchange the CSI. In this way, the CSI feedback is more robust since it is transmitted over the uplink for which power control and timing advance are optimized.

However, complex data flows in the network behind CoMP need to be organized: these are the application-related data for each UE, the CSI feedback and the exchange of scheduled data between the BSs. Therefore all flows needed for distributed DL CoMP are consequently organized using standard Ethernet protocols. This enables the use of widely available low-cost equipment based on the IEEE 802.3 standard. As a general approach, an existing standard extension, namely Virtual Local Area Networks (VLANs) described in IEEE 802.1q (IEEE 2006) is taken into consideration to organize the flows. VLAN-enabled switches are used to multiplex and demultiplex the different types of traffic. Virtual Local Area Networks are based on a label, denoted as tag, based on which a given packet is switched selectively from one input port to one or more output ports in the Ethernet switch. The use of VLANs has a negligible impact on the overall precoder delay. The delays introduced by the network are related to the maximum packet length of 12 μs for 1 Gbps Ethernet, and the propagation times over for example, 5 km optical fiber which takes 25 μs. These delays are negligible compared to the other delays present in the joint transmission loop due to the transmission of the feedback over a bandwidth-limited channel and the computation of the precoding matrices.

The use of VLANs can be extended to support the hybrid clustering approach mentioned in sub-section 6.1.2. Potential clusters can be preconfigured by the mobility management using predefined packet routing tables in each VLAN switch. Once a given cluster is formed, UE is informed about the tag to be used over the PDCCH. It broadcasts the CSI over the uplink into the preconfigured subnet identified by the tag. The serving eNodeB (eNB) uses a second tag for broadcasting the data to the other eNBs in the cluster. In this way, data and CSI packets find their own way through the network based on the tags and as much overhead as needed only is created in the backhaul network.

Local Precoder

The local precoder enables manifold multicell transmission schemes. It is realized in the transmitter signal processing pipeline by using a linear matrix-vector multiplication unit originally described in Jungnickel et al. (2005). Selectively for each subcarrier, the incoming IQ signal constellations of data streams from all BSs in a CoMP transmission are multiplied with a weight matrix before the signals are passed into the IFFT at each antenna.

There is an algorithmic part executed in a companion DSP for setting the weights. For DL CoMP, this algorithm assembles at first the CSI received over the feedback link from the UE in the own cell, or over X2 from other UEs in other cells. The resulting channel matrix includes all BSs and all UEs handled in the same cluster. Next the channel is interpolated in the frequency domain. Finally, the Zero Forcing precoding matrix is determined on each subcarrier.

Precoder Algorithms

Three different precoding algorithms have been used, one for interference-limited transmission where the combined precoder matrix in both cells is diagonal. Second, the inverse of the multicell multiuser channel was considered. Third, the other cells have been switched off in order to measure the isolated cell capacity as a reference.

Precoded Pilots

The precoding matrix **P** depends on the CSI feedback in other cells. It cannot be foreseen by a UE in a given cell how such external feedback in other cells influences the interference situation after CoMP has been applied from the network side. Consequently, a second set of pilots is needed, denoted as DM-Reference Signal (RS) in the Long Term Evolution (LTE)-Advanced standardization. The DM-RS enable the UE to estimate the effective, that is, the precoded channel **HP**. The DM-RS identify the streams provided to the UEs in each cell.

For the DM-RS the pilots used for single-cell transmission are fed through the precoder. The two antennas in the first cell and the two antennas in the second cell are handled as a larger four-antenna array. Note that the single-cell pilot concept used in one cell can be easily extended to cover many cells by introducing a so-called virtual pilot sequences (Thiele et al. 2008). In this way, the additional overhead required for the DM-RS in case of multicell operation can be limited, at the cost of reduced mobility.

Appendix D

Network Coding

D.1 Nonbinary NC based on UE Cooperation

Proof. (Theorem 8.3.1)

One possibility to measure the performance in the range of medium-to-high Signal to Noise Ratio (SNR), outage probabilities are considered, represented by the diversity order d. Recall from Chapter 5, d can be expressed as:

$$d \triangleq \lim_{\text{SNR} \to \infty} \frac{-\log P}{\log \text{SNR}} \tag{D.1}$$

where P is the relevant error probability under consideration (e.g. frame error probability or outage probability). Clearly, a received codeword is obtained as $Y_{i,j,k} = a_{i,j,k} X_{i,j,k} + n_{i,j,k}$, where $X_{i,j,k}$ and $Y_{i,j,k}$ are transmitted and received channel codewords, respectively; $n_{i,j,k}$ is additive white Gaussian noise with zero-mean and unit variance; $a_{i,j,k}$ denotes the fading channel gain. The index $i = 1, 2$ denotes the transmitting user U_1 or U_2, $j = 0, 1, 2$ denotes the receiving Base Station (BS), U_1 and U_2, respectively, and k denotes the time slot. The factors $a_{i,j,k}$s are i.i.d. random variables for different i, j or k with a Rayleigh distribution and unit variance. For reciprocal inter-user channels, $a_{i,j,k} = a_{j,i,k}$. With i.i.d. Gaussian codewords ($X_{i,j,k}$s), the Mutual Information (MI) between $Y_{i,j,k}$ and $X_{i,j,k}$ is $\text{MI}_{i,j,k} = \frac{1}{2} \log(1 + |a_{i,j,k}|^2 \text{SNR})$. Let R denote the rate of all channel codes. $X_{i,j,k}$ cannot be decoded correctly if $|a_{i,j,k}|^2 < g$, where $g = \frac{2^{2R}-1}{\text{SNR}}$. For Rayleigh fading, the corresponding outage probability of the link corresponding to $a_{i,j,k}$ is obtained as (Tse and Viswanath 2005):

$$P_e = Pr\{|a_{i,j,k}|^2 < g\} = 1 - e^{-g} \approx \frac{2^{2R}-1}{\text{SNR}} \tag{D.2}$$

The approximation holds for high SNRs. Without loss of generality, let us analyze the overall outage probability for U_1. If there is no outage in the inter-user channel, there are 4 different network coding blocks. An outage occurs only when the direct systematic codeword s_1 cannot be decoded, and two (or three) out of other three codewords (for s_2, $\alpha_1 s_1 + \beta_1 s_2$ or $\alpha_2 s_1 + \beta_2 s_2$) cannot be decoded at the BS. Since the $a_{i,j,k}$s are i.i.d, the overall outage probability is hence obtained as

$$P_0 = P_e \left(\binom{3}{2} P_e^2 (1 - P_e) + P_e^3 \right) \approx 3 P_e^3 \tag{D.3}$$

Mobile and Wireless Communications for IMT-Advanced and Beyond, First Edition.
Edited by Afif Osseiran, Jose F. Monserrat and Werner Mohr.
© 2011 Afif Osseiran, Jose F. Monserrat and Werner Mohr. Published 2011 by John Wiley & Sons, Ltd.

With probability P_e, the Relay Node (RN) cannot decode the partner's codeword. In this case, the BS performs Maximum Ratio Combining (MRC) and decodes. Thus, overall outage occurs when

$$\text{MI}_{MRC} = \frac{1}{2}\log(1 + (|a_{i,j,1}|^2 + |a_{i,j,2}|^2)\text{SNR}) < R \tag{D.4}$$

Then, the outage probability is $P_{MRC} = 0.5g^2 \approx \frac{(2^{2R}-1)^2}{2\text{SNR}^2}$. Combining above results, the total outage probability is

$$P_{o,1} = P_e P_{MRC} + (1 - P_e)P_0 \approx 3.5P_e^3 \tag{D.5}$$

Consequently, the diversity order is $d = 3$. If the inter-user channels are not reciprocal, we only need to separately consider the outage probability of two inter-user channels. By a similar analysis, the outage probability for U_1 is $P_{o,2} \approx 4P_e^3$. Hence, the diversity order is still 3. □

D.2 Multiuser and Multirelay Scenario

Proof. (Theorem 8.2)
For quasi-static fading channels a received codeword (base-band) is given as

$$Y_{i,j} = a_{i,j}X_{i,j} + n_{i,j} \tag{D.6}$$

where $X_{i,j}$ and $Y_{i,j}$ are the transmitted and received channel codewords, respectively. Let us assume that $X_{i,j}$ has power P_s, where $n_{i,j}$ is an additive white Gaussian noise sample with double-sided power spectral density $N_0/2$, and $a_{i,j}$ denotes the channel gain due to path-loss, shadowing and frequency nonselective fading. Here, the indices $i = 1, 2, 3, 4$ denote the transmitting nodes: U_1, U_2, RN1, and RN2, respectively, and $j = 0, 1, 2$ denote the receiving nodes: the BS , RN1 and RN2, respectively. The factors $a_{i,j}$s have zero-mean and are i.i.d. complex random variables. Without loss of generality, the factor $|a_{i,j}|$ is considered to have Rayleigh distribution and unit variance.

The outage probability for U_1 is given as follows. Identical considerations hold for U_2 due to symmetry. For U_1, an outage event occurs if the direct codeword s_1 cannot be decoded *and*, in addition, one or two blocks from s_2 and $s_1 \oplus s_2$ (with an MRC receiver) cannot be decoded either. Hence, for perfect Source Relay (SR) channels (denoted as the event B) the outage probability for U_1 is given as

$$P_o(U_1|B) = P_{sr,0}P_e(P_e + (1 - P_e)P_{MRC}) \approx P_e^2 \tag{D.7}$$

In a similar way all SR channel outage patterns can be evaluated and the outage probabilities are obtained. Here, an outage pattern denotes the collection of states of all SR channels that are either in outage or not. In summary, the outage probability for U_1 can be calculated as

$$\begin{aligned}
P_o(U_1) &= P_{sr,0}P_e(P_e + (1 - P_e)P_{MRC}) \\
&\quad + 2P_{sr,1}(P_e^2 + 3P_eP_{MRC}) \\
&\quad + P_{sr,2}(4P_e^2 + P_e + (1 + P_e)P_{MRC}) \\
&\quad + 2P_{sr,3}(P_e + P_{MRC}) + P_e^5 \\
&\approx P_e^2 \tag{D.8}
\end{aligned}$$

where $P_{sr,i} = P_e^i(1 - P_e)^{4-i}$, and $i \in \{0, 1, 2, 3\}$ is the probability that i SR channels are in outage. It is easy to see that the diversity order is $d = 2$. This is not an optimal result since the information of each user is transmitted through three paths. □

Proof. (Theorem 8.3.3)

Let us assume $(\alpha_1, \beta_1) = (1, 2)$ and $(\alpha_2, \beta_2) = (1, 1)$. In addition, assuming perfect source to RN channels, there are four different network codewords: $s_1, s_2, s_1 + s_2$ and $s_1 + 2s_2$. For U_1, an outage event occurs only when the codeword on the direct channel and two out of three codewords corresponding to $s_2, s_1 + s_2$ or $s_1 + 2s_2$ cannot be decoded at the BS. Since all channels are assumed to be independent the outage probability for perfect SR channels (denoted by the event B) is given as $P_{o,NNC}(U_1|B) = P_{sr,P} P_e(\binom{3}{2} P_e^2(1 - P_e) + P_e^3) \approx 3P_e^3$. Similarly, all possible SR channel patterns can be analyzed and then the corresponding outage probability is obtained. In summary, the overall outage probability for U_1 can be obtained as

$$
\begin{aligned}
P_{o,NNC}(U_1) = {} & P_{sr,P} P_e \left(\binom{3}{2} P_e^2(1 - P_e) + P_e^3 \right) \\
& + 2P_{sr,1} P_e (P_{MRC} + P_e) + 4P_{sr,1} P_e P_{MRC} \\
& + 4P_e^4(1 - P_e)^2 + P_e^3(1 - P_e)^2 + 2P_e^2(1 - P_e)^2 P_{MRC} + P_e^3(1 - P_e)^2 P_{MRC} \\
& + 2P_e^3(1 - P_e) P_{MRC} + P_e^4(1 - P_e) + P_e^5 \\
& \approx 6P_e^3
\end{aligned}
\tag{D.9}
$$

As the information of each user is transmitted through three independent paths, a maximum diversity order of $d = 3$ is obtained. Clearly, it is higher than the two-user, two-relay scheme using binary-Network Coding (NC). □

Appendix E

LTE-Advanced Analytical Performance and Peak Spectral Efficiency

E.1 Analytical and Inspection Performance Assessment by WINNER+

The Wireless World Initiative New Radio + (WINNER+) evaluation group addressed all evaluation characteristics for the 3GPP Long Term Evolution (LTE) Release 10 (Rel-10) and Beyond (LTE-Advanced) proposal assessment. This includes characteristics to be evaluated by means of analytical methods and by inspection. The results of these evaluations are summarized below.

E.1.1 Analytical Evaluation

Peak Spectral Efficiency

The Peak Spectral Efficiency (PSE) is defined in ITU-R (2008). It is basically the highest theoretical data rate normalized by bandwidth assignable to a single mobile station assuming error-free conditions. The WINNER+ Independent Evaluation Group (IEG) evaluated PSE for LTE-Advanced (LTE-A) Frequency Division Duplex (FDD) mode and Time Division Duplex (TDD) mode in Uplink (UL) and Downlink (DL). The evaluation configuration parameters provided in ITU-R (2009) with up to 4 Rx and 4 Tx antennas at the base station and up to 4 Rx and 2 Tx antennas at the mobile station were used. Configurations with up to eight antennas were also investigated for information purposes. The exact calculation is detailed out in section E.2. The FDD Radio Interface Technology (RIT) with a DL/UL PSE of 16.3/8.4 bps/Hz as well as the TDD RIT with a DL/UL PSE of 15.8/7.9 bps/Hz beat the requirement of 15/6.75 bps/Hz. Therefore both RITs clearly fulfill the PSE requirement.

Control Plane Latency

The idle-to-connected state transition can take less than 50 ms, and the dormant-to-active transition can take as little as 9.5 ms. Hence WINNER+ concluded from the analysis in (3GPP 2009a, Annex B) that both the FDD and the TDD RIT fulfill the control plane latency requirement of at most 100 ms.

Mobile and Wireless Communications for IMT-Advanced and Beyond, First Edition.
Edited by Afif Osseiran, Jose F. Monserrat and Werner Mohr.
© 2011 Afif Osseiran, Jose F. Monserrat and Werner Mohr. Published 2011 by John Wiley & Sons, Ltd.

User Plane Latency

Based on the assumption of 10% Hybrid Automatic Repeat-reQuest (HARQ) BLock Error Rate (BLER) a user plane latency of 4.8 ms is calculated for the FDD RIT. This result is well below the requirement given in Table 10.3. For the TDD RIT the HARQ round trip time depends on which of the seven possible UL/DL configurations is chosen. The resulting sum of UL and DL delays is in the range of between 5.2 ms and 6.2 ms for the total DL delay and between 6.1 ms and 9.1 ms for the total UL delay. Therefore the TDD RIT also fulfills the user plane latency requirement.

Intrafrequency and Interfrequency Handover Interruption Time

For the FDD RIT a total interruption time of 10.5 ms is calculated. This includes an average delay due to the RACH scheduling period of 0.5 ms assuming uniform start of waiting time between 0 ms and 1 ms.

The TDD RIT total interruption time depends on the TDD configuration as this has an influence on the RACH waiting time and the resulting waiting time for a DL slot. Minimizing the RACH waiting time to 1.1 ms by choosing configuration 0 with most UL slots yields a total interruption time of 15.5 ms while the total interruption time itself can be minimized to 12.5 ms with a higher RACH waiting time of 2.5 ms which is overcompensated by a smaller average waiting time for a DL subframe. With this result both RITs fulfill the interruption time requirement.

E.1.2   Inspection

The following characteristics to be evaluated by inspection are defined in ITU-R (2009) and evaluated by the WINNER+ IEG:

Bandwidth and Bandwidth Scalability

Both the FDD RIT and the TDD RIT fulfill the requirement to support a scalable bandwidth up to and including 40 MHz. This can be achieved, for example, by aggregating two 20 MHz component carriers. With aggregated multiple components bandwidth up to 100 MHz can be supported. Both the FDD RIT and the TDD RIT fulfill the requirement to support of at least three band-width values as 1.4, 3, 5, 10, 15 and 20 MHz component carrier bandwidths are supported.

Inter-system Handover

For both the FDD RIT and the TDD RIT WINNER+ concluded that inter system handover between the proposal FDD and TDD RITs and another system is supported, fulfilling the corresponding requirement.

Deployment Possible in at Least one of the Identified International Mobile Telecommunications (IMT) Bands

Based on the list of supported spectrum bands that is overlapping with the identified IMT bands it is clear that the FDD RIT and the TDD RIT support usage of at least one IMT spectrum band and thus, the requirement is fulfilled.

Support of a Wide Range of Dervices

By inspecting the FDD RIT and TDD RIT proposals and analyzing the required technical properties and comparing with the technology potential of the submission, it is concluded that the FDD RIT and TDD RIT support the required basic conversational service class, rich conversational service class and conversational low delay service class, and thus also support a wide range of services. Hence, WINNER+ concluded that the service requirements are fulfilled for the TDD RIT and the FDD RIT.

E.2   Peak Spectral Efficiency Calculation

The PSE is defined in ITU-R (2008). It is basically the highest theoretical data rate normalized by bandwidth assignable to a single mobile station assuming error-free conditions. The WINNER+ IEG evaluated PSE for LTE-A FDD mode and TDD mode in UL and DL. In addition to evaluation configuration parameters provided in (ITU-R 2009) with up to 4 Rx and 4 Tx antennas at the base station and up to 4 Rx and 2 Tx antennas at the mobile station, configurations with up to eight antennas were investigated for information purposes. From a mathematical point of view the PSE calculation is not demanding. It is simply the number of data bits that can be transmitted divided by the bandwidth and the time needed for that transmission. But LTE-A, like any other mobile radio system, needs overheads that do not contribute to the data rate. Reference and synchronization signals as well as broadcast channels and control signaling with channels carrying different indicators and control information from such overheads. Depending on the mode and the direction of transmission, different overhead types have to be taken into account. In TDD mode the guard period (GP), which separates DL and UL transmission in time domain adds additional overheads. For the PSE calculation one may additionally distinguish between different overhead types that add to the data rate or not. This topic was raised during a workshop organized by Third Generation Partnership Project (3GPP) for all IEGs end of 2009 and finally clarified by International Telecommunication Union – Radiocommunication Sector (ITU-R) in a liaison statement in 2010. A further topic was the handling of the GP duration in TDD mode and its influence on the time normalization for PSE calculation. In the following a detailed derivation of the WINNER+ IEG results for PSE with up to 4 Rx and 4 Tx antennas at the base station and up to 4 Rx and 2 Tx antennas at the mobile station is provided. Those results are then compared to the self-evaluation results provided by 3GPP.

E.2.1   FDD Mode Downlink Direction

In FDD mode, frame structure type 1 is used. UL and DL are separated in the frequency domain. Each frame is 10 ms long and consists of 10 subframes with a 1 ms duration. Each subframe consists of two slots with a 0.5 ms duration. With a normal cyclic prefix one slot equals the duration of seven resource elements in the time domain. Each resource element contains one Orthogonal Frequency Division Multiplexing (OFDM) symbol. One resource block has a length of one slot in time domain and 12 subcarriers in frequency domain. The subcarrier spacing is 15 kHz. One resource block pair has a length of two resource blocks in time domain and one resource block in frequency domain.

For a bandwidth of 20 MHz 100 resource blocks are used in the frequency domain. Each resource block has a frequency width of 12 resource elements. Each resource element uses a bandwidth of 15 kHz. This results in a bandwidth of $100 \cdot 12 \cdot 15\,\text{kHz} = 18\,\text{MHz}$ which can actually be used for transmission. The remaining bandwidth of $20\,\text{MHz} - 18\,\text{MHz} = 2\,\text{MHz}$ is used as guard band.

Each resource block pair spans 12 resource elements in frequency domain and 14 resource elements in time domain with normal cyclic prefix. Therefore each resource block pair consists of $12 \cdot 14 = 168$ resource elements. For a bandwidth of 20 MHz 100 resource blocks are used in the frequency domain. In the time domain, one frame consists of 10 resource block pairs. Therefore one frame consists of $100 \cdot 12 \cdot 14 \cdot 10 = \mathbf{168000}$ resource elements.

For calculating the peak spectral efficiency the overhead which does *not* contribute to the data rate has to be taken into account. This means that the number of resource elements not carrying data has to be subtracted from the number of resource elements per frame.

The resource elements not carrying data are used for the DL-Reference Signal (RS), the Physical Broadcast Channel (PBCH), for synchronization (Synchronization Channel (SCH)), and for L1/L2 control signaling (including Physical Control Format Indicator Channel (PCFICH) carrying Control Format Indicator (CFI), Physical Hybrid Automatic Repeat Request Indicator Channel (PHICH) carrying HARQ indicator and Physical Downlink Control CHannel (PDCCH) carrying Downlink Control Indicator (DCI)). A detailed calculation for these signals and channels is as follows:

- DL-RS
  A 4 × 4 antenna configuration is assumed. Cell-specific reference signals corresponding to four cell-specific antenna ports are used. For this configuration 24 reference symbols are used per resource block pair resulting in $100 \cdot 24 \cdot 10 = \mathbf{24000}$ resource elements per frame.
- PBCH
  The PBCH is transmitted every 10 ms in the first four OFDM symbols in the second slot of the first subframe and over the middle six resource blocks excluding eight resource elements per resource block which are reserved for the reference signals. Its Transmission Time Interval (TTI) is 40 ms, but the same information is repeated in all four frames of the TTI in four bursts to enable soft combining and enhance the demodulation performance. Every 10 ms burst is self-decodable. Overall, there are $6 \cdot (4 \cdot 12 - 8) = \mathbf{240}$ resource elements per frame, which are reserved for PBCH.
- SCH
  The SCH carries the synchronization signals PSS and SSS for frequency and timing acquisition and for physical layer ID determination, respectively. The PSS is transmitted in the fifth OFDM symbol in the first slot of each half-frame, that is, twice a frame over the middle six resource blocks. The SSS is transmitted in the sixth OFDM symbol in the first slot of each half-frame, that is, twice a frame over the middle six resource blocks. There is no overlapping with reference signals, that is, in all six resource blocks the whole bandwidth is used for PSS and SSS without interruption. Overall, there are $2 \cdot 6 \cdot 12 = 144$ resource elements per frame, which are reserved for PSS and the same amount for SSS, which results in $2 \cdot 144 = \mathbf{288}$ resource elements, which are reserved for SCH.
- L1/L2 control signaling
  One symbol L1/L2 control is assumed. In a realistic system where many users are to be supported, more symbols would be needed. For determination of peak spectral efficiency only one user is assumed. In this case the one symbol assumption is feasible. In each subframe the first OFDM symbol of all resource blocks is used spanning the entire system band excluding four resource elements per resource block, which are reserved for the reference signals. Overall, there are $100 \cdot (12 - 4) \cdot 10 = \mathbf{8000}$ resource elements per frame, which are used for L1/L2 control signaling.

The highest supported modulation format is 64-Quadrature Amplitude Modulation (QAM), where six data bits are mapped to one OFDM symbol, that is, each resource element carries six data bits. Taking into account that four layers are used with a 4 × 4 antenna configuration and the frame duration of 10 ms, the peak spectral efficiency is

$$PSE_{FDD,DL} = 4 \cdot 6 \cdot 100 \cdot (168000 - 24000 - 240 - 288 - 8000) \, \frac{bit}{s \cdot 20\,MHz} \qquad (E.1)$$

$$= 16.25664 \, \frac{bit}{s \cdot Hz} \approx 16.3 \, \frac{bit}{s \cdot Hz} \qquad (E.2)$$

E.2.2 FDD Mode Uplink Direction

As in the DL, for FDD UL one frame consists of **168 000** resource elements. For calculating the peak spectral efficiency, the overhead that does not contribute to the data rate has to be taken into account. This means that the number of resource elements not carrying data has to be subtracted from the number of resource elements per frame. The resource elements not carrying data are used for the reference signals (Demodulation (DM)-RS), the Physical Uplink Control CHannel (PUCCH), and the Physical Random Access Channel (PRACH).

- DM-RS
  The demodulation reference signal (DM-RS) is multiplexed in time with other channels such as Physical Uplink Shared CHannel (PUSCH) and PUCCH. Its purpose is to enable channel estimation for UL coherent demodulation/detection of the UL control and data channels. The PUSCH is used for UL data, but the PUSCH may also carry Acknowledge (ACK)/Negative Acknowledge (NACK) for DL data and Channel Quality Indicator (CQI)/Precoding Matrix Indicator (PMI)/Rank Indicator (RI). In each time slot one symbol duration is dedicated to DM-RS. Due to the single carrier waveform used in UL the whole bandwidth used for PUSCH cannot carry data during that DM-RS time period. With normal cyclic prefix this results in one out of seven symbols. Therefore the data rate of PUSCH is reduced by a factor of **6/7**.
- PUCCH
  The PUCCH carries Uplink Control Information (UCI). According to the 3GPP self-evaluation report two resource block pairs per subframe are assumed for PUCCH. This results in $2 \cdot 2 \cdot 12 \cdot 7 \cdot 10 = \mathbf{3360}$ resource elements per frame.
- PRACH
  The PRACH is used when the User Equipment (UE) and the eNodeB (eNB) are not synchronized. Synchronization on the UL is important to maintain the orthogonality of users, which minimizes UL intra-cell interference. According to the 3GPP self-evaluation report six resource-block pairs per frame are assumed for PRACH. This results in $6 \cdot 2 \cdot 12 \cdot 7 = \mathbf{1008}$ resource elements per frame.

As in the DL the highest supported modulation format is 64-QAM, where six data bits are mapped to one OFDM symbol, that is, each resource element carries six data bits. Two instead of four layers are considered for spatial-multiplexing. The resulting peak spectral efficiency is

$$\text{PSE}_{\text{FDD,UL}} = 2 \cdot 6 \cdot 100 \cdot (168000 - 3360 - 1008) \cdot \frac{6}{7} \frac{\text{bit}}{\text{s} \cdot 20\,\text{MHz}} \tag{E.3}$$

$$= 8.41536 \frac{\text{bit}}{\text{s} \cdot \text{Hz}} \approx 8.4 \frac{\text{bit}}{\text{s} \cdot \text{Hz}} \tag{E.4}$$

E.2.3   TDD Mode Downlink Direction

In TDD mode frame structure type 2 is used. UL and DL are separated in the time domain. Frame, subframe and slot duration is the same than in FDD mode. Resource blocks and resource block pairs or of the same size than in FDD mode.

For UL-DL separation in the time domain, some subframes are reserved for DL transmission and some are reserved for UL transmission. In addition to that, special subframes are used, which contain the three fields Downlink Pilot Timeslot (DwPTS), Guard Period (GP) and Uplink Pilot Timeslot (UpPTS). DwPTS is used for DL; GP is a guard period, and UpPTS is used for UL. How many DL subframes, UL subframes, and special subframes are used is defined by the UL-DL configuration. There exist seven UL-DL configurations with different numbers of UL and DL subframes per frame and with DL-to-UL switch-point periodicities of 5 ms (half-frame periodicity) and 10 ms (frame periodicity). In the self-evaluation report UL-DL configuration 1 is assumed. This configuration has a half-frame periodicity with four DL subframes, four UL subframes and two special subframes per frame. The lengths of the three special subframe fields are defined by the special subframe configuration. For normal cyclic prefix there exist nine special subframe configurations with different DwPTS, GP and UpPTS lengths. In the self-evaluation report special subframe configuration four is assumed. In this configuration 12 symbol durations are used for DwPTS, one symbol duration is used for UpPTS and the remaining symbol duration serves as GP. This is the smallest possible GP duration. The specifications allow for longer GP to support larger cell sizes.

As in FDD mode one frame consists of **168 000** resource elements. For calculating the peak spectral efficiency the overhead that does not contribute to the data rate has to be taken into account. This means that the number of resource elements not carrying data has to be subtracted from the number of resource elements per frame.

As opposed to FDD mode one has to distinguish between different subframe types. Both DL subframes and special subframes carry data but the overhead calculation is different:

- DL Subframes
  Four subframes per frame are DL subframes. This results in $100 \cdot 12 \cdot 14 \cdot 4 = \mathbf{67\,200}$ resource elements per frame. The resource elements to be subtracted for peak spectral efficiency calculation are listed below:
  – DL-RS
    As in FDD mode 24 symbols per resource block pair are used as reference signal. This results in $100 \cdot 24 \cdot 4 = \mathbf{9600}$ resource elements per frame.
  – PBCH
    This is exactly the same than in FDD mode. Note that the first subframe is always a DL subframe, independent of the UL-DL configuration. So the value of $6 \cdot (4 \cdot 12 - 8) = \mathbf{240}$ resource elements also holds independent of this type of configuration.
  – SCH
    There is no PSS located in the TDD DL subframes. For SSS the symbol positions are different but the amount of $2 \cdot 6 \cdot 12 = \mathbf{144}$ stays the same than in FDD mode.
  – L1/L2 control signaling
    Following the same argumentation than for FDD one symbol duration per subframe is assumed for L1/L2 control signaling. This results in $100 \cdot (12 - 4) \cdot 4 = \mathbf{3200}$ resource elements per frame.
- Special Subframes
  Two subframes per frame are special subframes. The DwPTS field is used for DL transmission and contains $100 \cdot 12 \cdot 12 = 14\,400$ resource elements. This results in $100 \cdot 12 \cdot 12 \cdot 2 = \mathbf{28\,800}$ DwPTS field resource elements per frame. The resource elements to be subtracted for peak spectral efficiency calculation are listed below. The remaining resource elements are used for data transmission.
  – DL-RS
    Although the DwPTS duration does not span the whole subframe, all reference signal positions are within that duration. The last two symbols of a subframe are not used for DL-RS. Therefore the number of resource elements reserved for reference signals is $100 \cdot 24 \cdot 2 = \mathbf{4800}$ per frame.
  – SCH
    For PSS the symbol positions are different but the amount of $2 \cdot 6 \cdot 12 = \mathbf{144}$ stays the same than in FDD mode. There is no SSS located in the TDD special subframes.
  – L1/L2 control signalling
    Following the same argumentation than for FDD one symbol duration per subframe is assumed for L1/L2 control signaling. This results in $100 \cdot (12 - 4) \cdot 2 = \mathbf{1600}$ resource elements per frame.

The highest supported modulation format is 64-QAM, where six data bits are mapped to one OFDM symbol, that is, each resource element carries six data bits. Taking into account that four layers are used with a 4×4 antenna configuration, the number of subframes used for DL transmission and the subframe duration of 1 ms, the peak spectral efficiency is

$$
\begin{aligned}
\mathrm{PSE_{TDD,DL}} = 4 \cdot {} & \frac{6 \cdot 1000 \cdot (67200 - 9600 - 240 - 144 - 3200)}{4 + 2 \cdot \frac{12 + f_{\mathrm{GP,DL}} \cdot 1}{14}} \frac{\mathrm{bit}}{\mathrm{s} \cdot 20\,\mathrm{MHz}} \\
+ 4 \cdot {} & \frac{6 \cdot 1000 \cdot (28800 - 4800 - 144 - 1600)}{4 + 2 \cdot \frac{12 + f_{\mathrm{GP,DL}} \cdot 1}{14}} \frac{\mathrm{bit}}{\mathrm{s} \cdot 20\,\mathrm{MHz}}
\end{aligned}
\tag{E.5}
$$

The factor $f_{\mathrm{GP,DL}}$ is the relative GP duration that is attributed to the DL. For the UL an equivalent factor can be defined, for which $f_{\mathrm{GP,UL}} = 1 - f_{\mathrm{GP,DL}}$ holds. The need for these factors is motivated in the following. In the calculation for $\mathrm{PSE_{TDD,DL}}$ without this factor the DL transmission duration of $4 + 2 \cdot \frac{12}{14}$ subframes would be taken as reference duration for peak spectral efficiency. By doing this the GP field duration would be ignored. Assuming that the same would be done for UL the sum of both reference durations would not sum up to the frame duration and part of the resources in time would be completely ignored when determining the peak spectral efficiency. To account for this the GP duration has to be attributed to either to the DL duration ($f_{\mathrm{GP,DL}} = 1$, $f_{\mathrm{GP,UL}} = 0$) or the UL duration ($f_{\mathrm{GP,DL}} = 0$, $f_{\mathrm{GP,UL}} = 1$) or to both durations. In the latter case it has to be clarified how the splitting ratio between the GP part that is attributed to DL and the GP part that is attributed to UL is to be defined.

For this splitting ratio, a fair approach is to take the ratio of DL transmission time to UL transmission time without GP. In the case at hand this leads to

$$
f_{\mathrm{GP,DL}} = \frac{4 + 2 \cdot \frac{12}{14}}{4 + 2 \cdot \frac{1}{14} + 4 + 2 \cdot \frac{12}{14}} = 0.5797101
\tag{E.6}
$$

times the GP durations within one frame that are attributed to the DL duration ultimately leading to

$$
\mathrm{PSE_{TDD,DL}} = 15.788304 \, \frac{\mathrm{bit}}{\mathrm{s} \cdot \mathrm{Hz}} \approx 15.8 \, \frac{\mathrm{bit}}{\mathrm{s} \cdot \mathrm{Hz}}
\tag{E.7}
$$

E.2.4 TDD Mode Uplink Direction

The overall number of resource elements in UL subframes is $100 \cdot 12 \cdot 14 \cdot 4 = \mathbf{67\,200}$. As the overhead calculation is very similar to the FDD mode UL, no detailed description is given here. For PUCCH two resource block pairs per subframe are spent resulting in $2 \cdot 2 \cdot 12 \cdot 7 \cdot 4 = \mathbf{1344}$ resource elements per frame. For PRACH six resource block pairs per frame are spent resulting in $6 \cdot 2 \cdot 12 \cdot 7 = \mathbf{1008}$ resource elements per frame. Due to DM-RS the data rate is reduced by a factor of $\mathbf{6/7}$. Additional overhead is due to SRS in UpPTS with a duration of one symbol in the special subframe. The resulting peak spectral efficiency is

$$
\mathrm{PSE_{TDD,UL}} = 2 \cdot \frac{6 \cdot 1000 \cdot (67200 - 1344 - 1008) \cdot \frac{6}{7}}{4 + 2 \cdot \frac{1 + f_{\mathrm{GP,UL}} \cdot 1}{14}} \frac{\mathrm{bit}}{\mathrm{s} \cdot 20\,\mathrm{MHz}}
\tag{E.8}
$$

$$
= 7.9350952 \, \frac{\mathrm{bit}}{\mathrm{s} \cdot \mathrm{Hz}} \approx 7.9 \, \frac{\mathrm{bit}}{\mathrm{s} \cdot \mathrm{Hz}}
\tag{E.9}
$$

Table E.1 DL peak spectral efficiencies

PSE in bps/Hz	FDD RIT	TDD RIT
ITU requirement	15	15
3GPP result for four-layer spatial multiplexing	16.3	16.0
WINNER+ result for four-layer spatial multiplexing	16.3	15.8
3GPP result for eight-layer spatial multiplexing	30.6	30.0
WINNER+ result for eight-layer spatial multiplexing	30.6	30.5

Table E.2 UL peak spectral efficiencies

PSE in bps/Hz	FDD RIT	TDD RIT
ITU requirement	6.75	6.75
3GPP result for two-layer spatial multiplexing	8.4	8.1
WINNER+ result for two-layer spatial multiplexing	8.4	7.9
3GPP result for four-layer spatial multiplexing	16.8	16.1
WINNER+ result for four-layer spatial multiplexing	16.8	15.8

E.2.5 Comparison with Self-Evaluation

Tables E.1 and E.2 summarize the results from equations (E.2,E.4,E.7,E.9) for DL and UL, respectively. Results for eight-layer spatial multiplexing in the DL and for four-layer spatial multiplexing in the UL are also provided. Those results can be obtained in a similar way as described above. A comparison with the International Telecommunication Union (ITU) requirements unveils that in all cases the requirements are clearly fulfilled.

References

3GPP 2009a Further Advancements for E-UTRA (LTE-Advanced). Technical Specification Group Radio Access Network 36.912 v9.3.0, 3rd Generation Partnership Project (3GPP), http://www.3gpp.org/ftp/Specs/html-info/.

3GPP 2009b Multimedia Broadcast/Multicast Service (MBMS): Protocols and Codecs. Technical Specification Group Services and System Aspects 23.402 v9.4.0, 3rd Generation Partnership Project (3GPP), http://www.3gpp.org/ftp/Specs/html-info/.

Bengtsson M and Ottersten B 2001 Optimal and suboptimal transmit beamforming. In L. C. Godara (Ed.) *Handbook of Antennas in Wireless Communications*. CRC Press, Boca Raton, FL.

Boccardi F, H.Huang and Alexiou A 2008 Network mimo with reduced backhaul requirements by MAC coordination *42nd Asilomar Conference*, pp. 1125–1129.

Boyd S and Vandenberghe L 2004 *Convex Optimization*. Cambridge University Press, Cambridge, UK.

Boyd S, Xiao L, Mutapcic A and Mattingley J 2007 Notes on decomposition methods. Course reader for convex optimization II, Stanford, http://www.stanford.edu/class/ee364b/.

Brown T 2005 An analysis of unlicensed device operation in licensed broadcast service bands *Proc. First IEEE International Symposium on New Frontiers in Dynamic Spectrum Access Networks*, pp. 11–29.

Cabric D, Mishra S and Brodersen R 2004 Implementation issues in spectrum sensing for cognitive radios *Proc. IEEE Thirty-Eighth Asilomar Conference on Signals, Systems and Computers*, vol. 1, pp. 772–776, Pacific Grove, CA.

Celebi H and Arslan H 2007 Utilization of location information in cognitive wireless networks. *IEEE Wireless Communications* **14**(4), 6–13.

Döttling M, Mohr W and Osseiran A 2009 *Radio Technologies and Concepts for IMT-Advanced*. John Wiley & Sons, Ltd, Chichester.

Fehske A, Gaeddert J and Reed J 2005 A new approach to signal classification using spectral correlation and neural networks *Proc. First IEEE International Symposium on New Frontiers in Dynamic Spectrum Access Networks*, pp. 144–150, Baltimore, MD, USA.

Gelabert X, Akyildiz IF, Sallent O and Agusti R 2009 Operating point selection for primary and secondary users in cognitive radio networks. *Computer Networks* **53**(8), 1158-1170.

Haykin S 2005 Cognitive radio: brain-empowered wireless communications. *IEEE Journal on Selected Areas in Communications* **23**(2), 201-220.

IEEE 2006 Virtual Bridged Local Area Networks. Standard IEEE 802.1Q-2005, http://standards.ieee.org/getieee802/download/802.1Q-2005.pdf.

ITU-R 2008 Requirements related to technical performance for IMT-Advanced Radio Interface(s). Report ITU-R M.2134, International Telecommunications Union Radio (ITU-R), http://www.itu.int/publ/R-REP/en.

ITU-R 2009 Guidelines for Evaluation of Radio interface Technologies for IMT-Advanced. Report ITU-R M.2135-1, International Telecommunications Union Radio (ITU-R), http://www.itu.int/publ/R-REP/en.

ITU-T 2007 Advanced video coding for generic audiovisual services. Recommendation ITU-T H.264, International Telecommunications Union Telecommunication Standardization Sector (ITU-T).

Jungnickel V, Forck A, Haustein T, Schiffermller S, von Helmolt C, Luhn F, Pollock M, Juchems C, Lampe M, Zirwas W, Eichinger J and Schulz E 2005 1Gbit/s MIMO-OFDM transmission experiments. *IEEE 62nd Vehicular Technology Conference VTC2005-Fall* pp. 861–866.

Jungnickel V, Thiele L, Schellmann M, Wirth T, Zirwas W, Haustein T and Schulz E 2008a Implementation concepts for distributed cooperative transmission. *42nd Asilomar Conference on Signals, Systems and Computers*.

Jungnickel V, Wirth T, Schellmann M, Haustein T and Zirwas W 2008b Synchronization of cooperative base stations. *IEEE International Symposium on Wireless Communication Systems 2008 (ISWCS2008)*.

Kim K, Akbar IA, Bae KK, Um JS, Spooner CM and Reed JH 2007 Cyclostationary approaches to signal detection and classification in cognitive Radio *Proc. IEEE International Symposium on New Frontiers in Dynamic Spectrum Access Networks*, pp. 212–215, Dublin, Ireland.

Ng BL, Evans J, Hanly S and Aktas D 2008 Distributed downlink beamforming with cooperative base stations. *Information Theory, IEEE Transactions on* **54**(12), 5491–5499.

Palomar D and Chiang M 2006 A tutorial on decomposition methods for network utility maximization. *IEEE J. Select. Areas Commun.* **24**(8), 1439–1451.

Papadogiannis A, Gesbert D and Hardouin E 2008 A dynamic clustering approach in wireless networks with multicell cooperative processing *Proc. IEEE ICC 2008*, pp. 4033–4037.

Papadogiannis A, Hardouin E and Gesbert D 2009 Decentralising multi-cell cooperative processing on the downlink: a novel robust framework. *EURASIP J. Wireless Commun. Networking Special Issue on Broadband Wireless Access*, Article ID 890685.

Pawelczak P, Prasad RV and Hekmat R 2007 Opportunistic Spectrum Multichannel OFDMA *Proc. ICC 2007 – IEEE Int. Conf. Commun.*, pp. 5439–5444, Glasgow, Scotland.

Radunovic B and Le Boudec JY 2007 A unified framework for max-min and min-max fairness with applications. *IEEE/ACM Trans. Networking* **15**(5), 1073–1083.

Sahai A, Tandra R, Mishra SM and Hoven N 2006 Fundamental design tradeoffs in cognitive radio systems *Proc. ACM First international workshop on Technology and policy for accessing spectrum*, p. 2, New York, NY, USA.

Sahin M, Guvenc I and Arslan H 2009 Opportunity detection for OFDMA-based cognitive radio systems with timing misalignment. *IEEE Transactions on Wireless Communications* **8**(10), 5300–5313.

Saul A 2008 Wireless resource allocation with perceived quality fairness *Asilomar Conference on Signals, Systems and Computers*, pp. 1557–1561.

Saul A and Auer G 2009 Multiuser resource allocation maximizing the perceived quality. *EURASIP J. Wireless Commun. and Networking*, Article ID 341689.

Schwarz H and Marpe D 2007 Overview of the Scalable Video Coding Extension of the H.264/AVC Standard. *Circuits and Systems for Video Technology, IEEE Transactions on* **17**(9), 1103–1120.

Stevenson C, Chouinard G, Lei Z, Hu W, Shellhammer S and Caldwell W 2009 IEEE 802.22: The first cognitive radio wireless regional area network standard. *IEEE Communications Magazine* **47**(1), 130–138.

Tandra R and Sahai A 2005 Fundamental limits on detection in low SNR under noise uncertainty *Proc. International Conference on Wireless Networks, Communications and Mobile Computing*, vol. 1, pp. 464–469, Maui, HI, USA.

Tang H 2005 Some physical layer issues of wide-band cognitive radio systems *Proc. First IEEE International Symposium on New Frontiers in Dynamic Spectrum Access Networks*, pp. 151–159, Baltimore, MD, USA.

Thiele L, Boccardi F, Botella C, Svensson T and Boldi M 2010 Scheduling-Assisted Joint Processing for CoMP in the Framework of the WINNER+ Project *Proc. Future Network & Mobile Summit 2010*, Florence, Italy.

Thiele L, Schellmann M, Schiffermüller S and Jungnickel V 2008 Multi-cell channel estimation using virtual pilots. *IEEE 67th Vehicular Technology Conference VTC2008-Spring*.

Thiele L, Wirth T, Haustein T, Jungnickel V, Schulz E and Zirwas W 2009 A unified feedback scheme for distributed interference management in cellular systems: Benefits and challenges for real-time implementation. *17th Euopean Signal Processing Conference (EUSIPCO2009)*. invited.

Tian Z and Giannakis GB 2006 A Wavelet Approach to Wideband Spectrum Sensing for Cognitive Radios *Proc. IEEE 1st International Conference on Cognitive Radio Oriented Wireless Networks and Communications*, pp. 1–5, Mykonos Island, Greece.

Tölli A, Pennanen H and Komulainen P 2009 Distributed coordinated multi-cell transmission based on dual decomposition *Proc. IEEE GLOBECOM 2009*, pp. 1–6, Honolulu, Hawai'i, USA.

Tölli A, Pennanen H and Komulainen P 2011 Decentralized minimum power multi-cell beamforming with limited backhaul signaling. *IEEE Trans. Wireless Commun.* **10**(2), 570–580.

Tse D and Viswanath P 2005 *Fundamentals of Wireless Communication*. Cambridge University Press, Cambridge.

Urkowitz H 1967 Energy detection of unknown deterministic signals. *Proceedings of the IEEE* **55**(4), 523–531.

Wiesel A, Eldar YC and Shamai S 2006 Linear precoding via conic optimization for fixed MIMO receivers. *IEEE Trans. Signal Processing* **54**(1), 161–176.

WINNER+ 2009a Celtic Project CP5-026 *First set of best innovations in advanced radio resource management* (ed. Saul A.). Public Deliverable D1.1, Wireless World Initiative New Radio - WINNER+, http://projects.celtic-initiative.org/winner+/index.html (accessed June 2011).

WINNER+ 2009b Celtic Project CP5-026 *Intermediate Report on CoMP (Coordinated Multi-Point) and Relaying in the Framework of CoMP* (ed. Mayrargue S.). Public Deliverable D1.8, Wireless World Initiative New Radio - WINNER+, November 2009, http://projects.celtic-initiative.org/winner+/index.html (accessed June 2011).

WINNER+ 2009c Celtic Project CP5-026 *Intermediate Report on System Aspect of Advanced RRM* (ed. Monserrat J. and Sroka P.). Public Deliverable D1.5, Wireless World Initiative New Radio - WINNER+, April 2009, http://projects.celtic-initiative.org/winner+/index.html (accessed June 2011).

WINNER+ 2009d Celtic Project CP5-026 *Results of Y1 proposed candidate proof-of-concept evaluation* (ed. Safjan K.). Public Deliverable D4.1, Wireless World Initiative New Radio - WINNER+, http://projects.celtic-initiative.org/winner+/index.html (accessed June 2011).

WINNERII 2006 *Test Scenarios and Calibration Cases Issue 2* (ed. Döttling M.). Public Deliverable D6.13.7, Wireless World Initiative New Radio – WINNER, http://www.ist-winner.org/ (accessed June 2011).

WINNERII 2007 *WINNER II Channel Models* (ed. Kyösti P, Meinilä J.). Public Deliverable D1.1.2, Wireless World Initiative New Radio – WINNER, http://www.ist-winner.org/ (accessed June 2011).

Yucek T and Arslan H 2006 Spectrum Characterization for Opportunistic Cognitive Radio Systems *Proc. IEEE Military Communications Conference*, pp. 1–6, Washington, DC.

Zirwas W, Mennerich W, Schubert M, Thiele L, Jungnickel V and Schulz E 2009 *Cooperative Transmission Schemes* CRC Press, Taylor & Francis Group.

Zirwas W, Schulz E, Kim JH, Jungnickel V and Schubert M 2006 Distributed organization of cooperative antenna systems. *European Wireless* Conference 2006, Athens, Greece.

Index

Mobile and Wireless Communications for IMT-Advanced and Beyond, First Edition.
Edited by Afif Osseiran, Jose F. Monserrat and Werner Mohr.
© 2011 Afif Osseiran, Jose F. Monserrat and Werner Mohr. Published 2011 by John Wiley & Sons, Ltd.

CPSIA information can be obtained at www.ICGtesting.com
Printed in the USA
LVOW121147280911

248157LV00004B/7/P